Progress in Colloid and Polymer Science

Recently Published and Forthcoming Volumes

Smart Colloidal Materials
Volume Editor: Richtering, W.
Vol. 133, 2006

Characterization of Polymer Surfaces and Thin Films
Volume Editors: Grundke, K., Stamm, M., Adler, H.-J.
Vol. 132, 2006

Analytical Ultracentrifugation VIII
Volume Editors: Wandrey, C., Cölfen, H.
Vol. 131, 2006

Scattering Methods and the Properties of Polymer Materials
Volume Editors: Stribeck, N., Smarsly, B.
Vol. 130, 2005

Mesophases, Polymers, and Particles
Volume Editors: Lagaly, G., Richtering, W.
Vol. 129, 2004

Surface and Colloid Science
Volume Editor: Galembeck, F.
Vol. 128, 2004

Analytical Ultracentrifugation VII
Volume Editors: Lechner, M. D., Börger, L.
Vol. 127, 2004

Trends in Colloid and Interface Science XVII
Volume Editors: Cabuil, V., Levitz, P., Treiner, C.
Vol. 126, 2004

From Colloids to Nanotechnology
Volume Editors: Zrinyi, M., Hórvölgyi, Z. D.
Vol. 125, 2004

Aqueous Polymer Dispersions
Volume Editor: Tauer, K.
Vol. 124, 2004

Trends in Colloid and Interface Science XVI
Volume Editors: Miguel, M., Burrows, H. D.
Vol. 123, 2004

Aqueous Polymer – Cosolute Systems
Volume Editor: Anghel, D. F.
Vol. 122, 2002

Molecular Organisation on Interfaces
Volume Editor: Lagaly, G.
Vol. 121, 2002

Lipid and Polymer-Lipid Systems
Volume Editor: Nylander, T.
Vol. 120, 2002

Progress in Colloid and Polymer Science

Editors: F. Kremer, Leipzig and W. Richtering, Aachen

Volume 132 · 2006

Characterization of Polymer Surfaces and Thin Films

Volume Editors:
Karina Grundke
Manfred Stamm
Hans-Jürgen Adler

Springer

The series Progress in Colloid and Polymer Science is also available electronically (ISSN 1437-8027)

- Access to tables of contents and abstracts is *free* for everybody.
- Scientists affiliated with departments/institutes subscribing to Progress in Colloid and Polymer Science as a whole also have full access to all papers in PDF form. Point your librarian to the Springerlink access registration form at http://www.springerlink.com

ISSN 0340-255X
ISBN-10 3-540-31241-2
ISBN-13 978-3-540-31241-3
DOI 10.1007/b104970
Springer Berlin, Heidelberg, New York

Springer is a part of Springer Science + Business Media

http://www.springer.com

Printed in Germany

Cover design: eStudio Calamar S. L., F. Steinen-Broo, Pau/Girona, Spain

Typesetting and production: LE-TEX Jelonek, Schmidt & Vöckler GbR, Leipzig

Printed on acid-free paper

Progr Colloid Polym Sci (2006) 132: V

PREFACE

Dresden Polymer Discussions have been organized by the Leibniz Institute of Polymer Research Dresden and the Institute of Macromolecular and Textile Chemistry of the Technische Universität Dresden now since 1986. Every two years, interested scientists, researchers and students from polymer research centres, universities and industry from all over the world discuss latest results, trends and new demands on research and development in different fields of polymer science.

Following this tradition the **10th Dresden Polymer Discussion** took place in the charming environment of the old town of Meißen near the Saxonian capital Dresden, from April 10 to 13, 2005. The meeting was dedicated to the topic "Characterization of Polymer Surfaces and Thin Films". Surface and thin film properties of polymer materials can be related to micro structure and molecular origin, which remains a central problem in physics, chemistry, biology and materials science. It forms the basis for technological control and improvement of existing as well as the design of new and innovative materials. To achieve this aim comprehensive characterization of surfaces and interfaces with a wide spectrum of analytical techniques is necessary.

The contributions of this volume were presented at the meeting and selected for publication in "Progress in Colloid and Polymer Science" covering a representative spectrum of surface sensitive techniques and their application to polymer surface and thin film characterization as well as recent examples of technologically relevant materials and process development.

February 2006

Karina Grundke
Manfred Stamm
Hans-Jürgen Adler

Progr Colloid Polym Sci (2006) 132: VI–VII

CONTENTS

Progr Colloid Polym Sci (2006) 132: 1–6
DOI 10.1007/2882_034

Published online: 10 March 2006

Sergei G. Kazarian
K. L. Andrew Chan

Sampling Approaches in Fourier Transform Infrared Imaging Applied to Polymers

Sergei G. Kazarian (✉) ·
K. L. Andrew Chan
Department of Chemical Engineering,
Imperial College London, SW7 2AZ, UK
e-mail: s.kazarian@imperial.ac.uk

Abstract FTIR imaging is a powerful tool that combines enhanced visualisation with chemical information. However, prior to applying this method to polymer systems one needs to appreciate the advantages and limitations of a particular imaging mode, i.e. its spatial resolution and field of view. Therefore, there is a need to discuss the differences between the two main FTIR imaging modes: transmission and Attenuated Total Reflection (ATR) and to review recent developments in macro FTIR imaging that include innovative introduction of a diamond ATR accessory for imaging applications.

Keywords Focal Plane Array · FTIR · Imaging · Infrared spectroscopy · Polymer blends

Introduction

Combining microscopy with chemical recognition represents a powerful approach to study heterogeneous polymeric materials. This combination has been enhanced further as a chemical imaging method using FTIR spectroscopy with an array rather than a single element detector [1–13]. Focal plane array detectors had been originally developed for applications in military technology (as a part of the smart missiles guidance systems) and astronomy (as a detection device for the imaging of galaxies). Fortunately, these array detectors (such as 64 × 64 or 128 × 128 elements arrays) are now available for materials research. These arrays allow one to simultaneously measure thousands of FTIR spectra from different locations in a sample. These FTIR spectra then provide information about the concentration of specific compound and their distribution in the sample within the measured field of view. Therefore, FTIR imaging can be thought of as a chemical photography [14]. FTIR imaging does not require the use of any dyes or labelling techniques and provides fast and reliable 2D imaging for enhanced visualisation of biological, pharmaceutical and polymeric materials. Nevertheless, the applicability of this method, when applied in transmission (which is still the most common approach in FTIR microscopy and imaging [15–17]) is restricted by the diffraction limit of light in air and by the necessity to microtome samples for transmission measurements. We have recently demonstrated that that micro-ATR imaging objective with germanium crystal allowed us to achieve spatial resolution the order of 3–4 μm which was below the diffraction limit of infrared light in air [18]. Importantly, this enhanced spatial resolution was achieved without the use of expensive synchrotron source which provided, for example, clear images of the medulla of the hair [19] and detected heterogeneity of the distribution of cholesterol esters in aortic plaques [20]. ATR imaging mode requires minimal sample preparation which is important for many biomedical samples and polymeric materials.

Our group was the first to report imaging with a diamond ATR accessory that provided a field of view of ca. 1 mm^2 and the spatial resolution of ca. 15 μm without the use of an infrared microscope [18]. The demonstration of the applicability of a diamond ATR accessory for FTIR imaging opened up a range of new opportunities in polymer research, from compaction of tablets [21–23] to studying phase separation in polymer blends subjected to supercritical fluids [24]. This imaging approach was successfully utilised for the study of dissolution of tablets in aqueous solutions [25]. We have also demonstrated macro

ATR imaging using an inverted pyramid ATR crystal made of ZnSe applied to drug release from polymer/drug formulations [14, 25]. The range of applications of the macro ATR imaging approach to polymeric materials is significant, and therefore, this paper introduces some technical aspects of the approach and compares this method to imaging in transmission. This should help polymer scientist to choose an approach that would be the most suitable for studying a particular system using FTIR imaging.

Experimental Sections

FTIR Measurements

Similar to conventional FTIR measurements, FTIR imaging with FPA detector can be operated in transmission or ATR mode.

Transmission Imaging. In transmission mode, the infrared light goes through the sample and the absorbance is measured by the detector. The transmission method usually yields spectra with a high signal to noise ratio (SNR) because the detector receives high infrared light intensity. However, sample thickness is important for this type of measurement. In order to avoid the infrared energy being totally absorbed, the sample needs to be sectioned to produce a thin film. A sample of uniform thickness is also necessary for accurate quantitative analysis. When imaging a heterogeneous sample by the transmission method, the domain size of the component of interest also needs to be considered when choosing the sample thickness. The thickness should be similar to or smaller than the domain size in order to produce meaningful results. Otherwise, the spatial resolution will be degraded by variation in the "Z-axis" (the axis that measures thickness of the sample). If the domain size of a sample is less than the sample thickness (see Fig. 1), any overlapping of domains in the

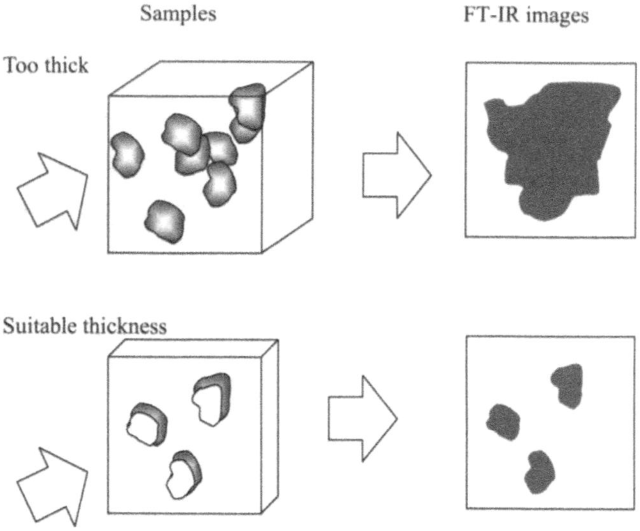

Fig. 1 A schematic diagram showing the possible resultant image of the same sample with different chosen thickness

Z-axis will produce a misleading image of the component's distribution in the sample. Therefore the thickness of the sample is not only important for preventing total absorption of infrared energy but also for the accuracy of the result.

To demonstrate this, a sample blend of polystyrene (PS) – low density polyethylene (LDPE) were microtomed cryogenically into films of 3 μm and 20 μm of uniform thickness. FTIR images of these films were measured in transmission under the infrared microscope (IRScope II, Bruker Optics) with an ×15 objective. The images of the distribution of LDPE are shown in Fig. 2. The images were generated by plotting the integrated values of the LDPE absorption band at 1466 cm^{-1} over the whole images area. The image of the sample with 3 μm thickness has shown clear and well defined domains of LDPE while the image

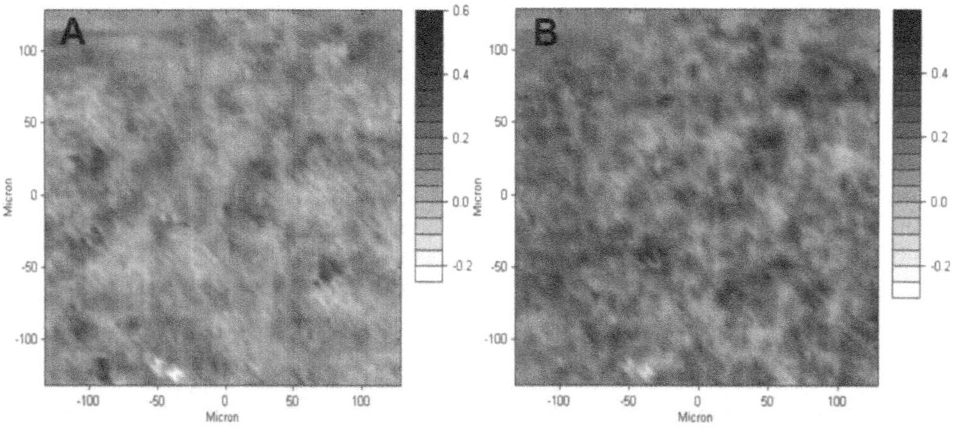

Fig. 2 Transmission FTIR images showing the distribution of LDPE in the films with the thickness of **A** 3 μm and **B** 20 μm

of the sample with 20 μm has shown a more crowded distribution due to superposition of LDPE domain in the Z-axis of the sample.

Spatial Resolution in FTIR Imaging

A microscope, specially designed for FT-IR measurements, is needed to measure infrared spectra of small samples or to capture infrared images with higher spatial resolution. High spatial resolution of imaging methods is very important to reveal the true distribution of different components in heterogeneous materials. The general perception is that the spatial resolution achieved with FT-IR imaging is poor (tens of micrometers) due to the relatively long wavelength of infrared radiation. Despite of this, many researches publications reported the spatial resolution of FTIR imaging of ca. 4–8 μm which is obtained by dividing the number of detector pixels by the width of the image. Consider a light ray focused on a small spot by a lens, diffraction can be observed when the spot size is smaller or equal to the wavelength of the light. A schematic diagram of a typical diffraction pattern is shown in Fig. 3. The area of the diffraction pattern spot is called the Airy Disc.

Most of the intensity of the signal from the spot falls in the region between $-r$ and r. Consider now there are two spots that are adjacent to each other. If the two spots are separated by a distance less than the radius r, the intensity of the light from the two spots is heavily overlapped and hence it is very difficult to distinguish from one point to the other. The two spots can not be resolved and appear as one bigger point. When the two spots are separated by a distance approximately equal to the radius r, the intensities of the light are partly overlapped and the two spots can just about be distinguished. When the two spots are separated by a distance more or equal to the diameter of the airy disc, the two spots are well separated and the two points are considered as resolved. The diagram shown in Fig. 4 illustrates these three situations.

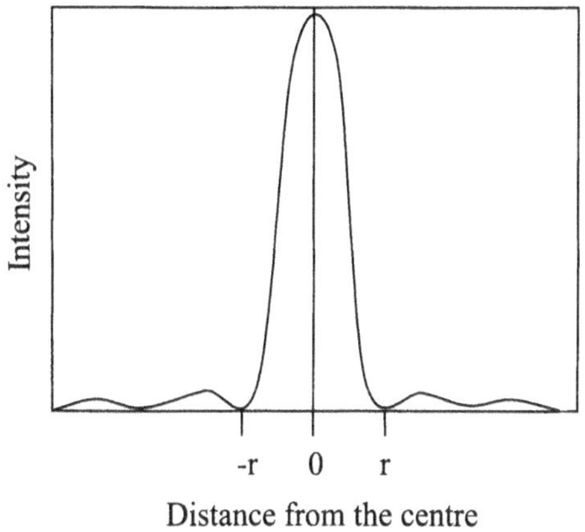

Fig. 3 Schematic diagram of the diffraction pattern of light

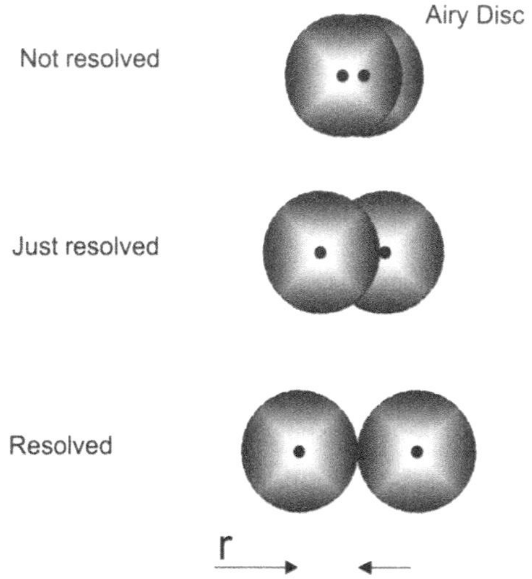

Fig. 4 Airy discs of two adjacent features with different separations. Fringes of the Airy Disc are not shown

The radius, r, of the Airy Disc is given by the Rayleigh criterion equation which is shown in Eq. 1.

$$r = 1.22\lambda(2\mathrm{NA})^{-1} \quad (1)$$

NA is defined as

$$\mathrm{NA} = n \sin\theta \quad (2)$$

where n is the refractive index of the imaging medium between the objective and the sample λ is the wavelength of light and θ is half the angular aperture.

It is clear that the spatial resolution (the separation between the two points required for them to be resolved, i.e. $2r$) is not merely dividing the number of detector pixels by the width of the image. The spatial resolution in transmission FTIR imaging should be in the order of 10–30 μm depending on the wavelength of the absorption band used.

ATR Imaging. ATR imaging with visible light can be observed most easily when a glass of water is held by one's hand, the skin or finger print that made contact with the glass could be observed through the water surface while the rest of the light is totally internally reflected. The area where the surface of glass has made contact with the skin would destroy the total reflection and formed an image of the finger print. An ATR image of a finger print was first demonstrated by Harrick [26]. He obtained the ATR image by illuminating a prism with a finger pressed on the large surface and then he focused the light that is internally reflected off the prism onto a photofilm. ATR-FT-IR spectroscopic imaging applied the same principle described by Harrick. Total internal reflection occurs when an electromagnetic wave (infrared is used in this application) travels from a higher refractive index material (the ATR crystal) to a lower refractive index material (the sample) at an angle

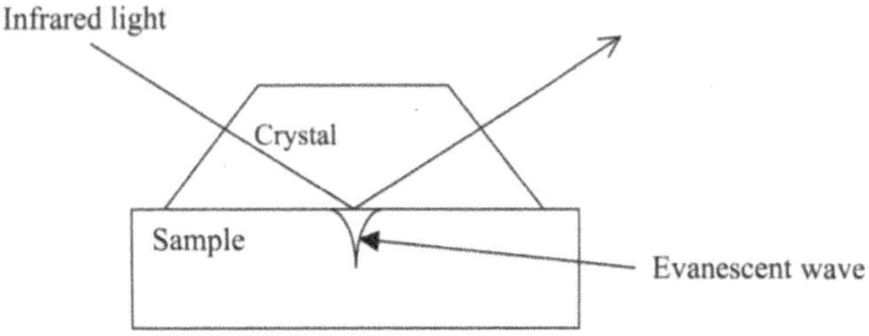

Fig. 5 Schematic diagram of an evanescent wave in the sample

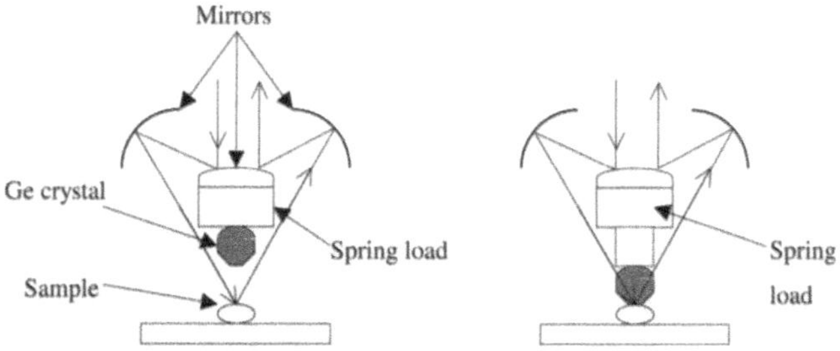

Fig. 6 Schematic diagrams showing the ATR objective in reflecting visible mode (*left*) and measuring mode (*right*). In the measuring mode, the Ge crystal is lowered to make contact with the sample. Different pressure can be applied onto the sample. (Drawings are not to scale)

greater than the critical angle. At the interface where the radiation is totally internally reflected, the radiation penetrates into the second medium of lower refractive index as an evanescent wave (Fig. 5). This wave decays exponentially with the depth of penetration, d_p, given by Eq. 3.

$$d_p = \lambda \left[2n_1\pi(\sin^2\theta - n_2^2 n_1^{-2})^{0.5}\right]^{-1} \qquad (3)$$

θ is the incident angle, n_1 and n_2 are the refractive indices of the ATR crystal and the sample respectively and λ is the wavelength of the incident light. If the second medium absorbs this radiation, the evanescent wave causes an attenuation effect on the incident wave. Hence by detecting the attenuated radiation, an infrared spectrum can be obtained.

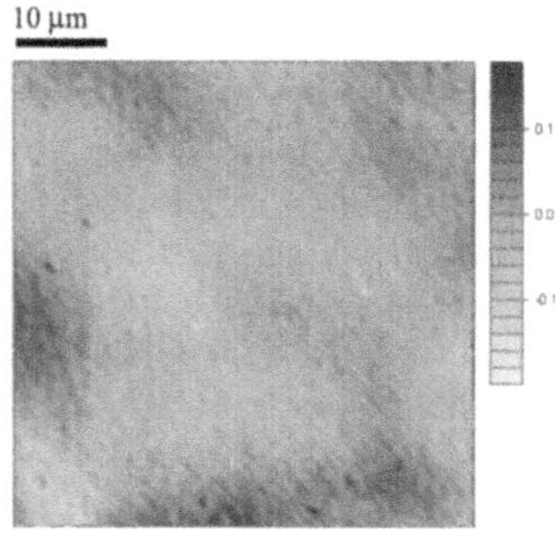

Fig. 7 ATR-FTIR image showing the distribution of LDPE in the PS/LDPE blend. The imaged area is 50 × 50 µm^2

Microscopic ATR Imaging

For micro-ATR measurements, the microscope operates in reflective mode. A specially designed objective is used to perform micro ATR spectroscopy. An ATR crystal (germanium is used in this system) is attached to the end of the objective which can be moved up or down. When the objective is in visible mode (Fig. 6, left), it operates as a visible microscope objective which allows the sampling area to be located with white light before an ATR measurement is taken. In measuring mode, the crystal is lowered to make contact with the sample as shown in Fig. 6 (right). The sampling surface of the ATR crystal is slightly convex to ensure good contact can be made between the sample and the crystal. There are different levels of contact pressure which can be chosen to optimise the contact between the sample and the crystal.

Since ATR image involved the sample being measured in contact with an ATR crystal of high refractive index, n, the spatial resolution is improved according to Eq. 1 and 2 [18]. When germanium is used as the ATR crystal, which has a refractive index of 4, the spatial resolution achieved can be improved up to four times when compared to transmission using the same optical objective. The field of view in micro ATR imaging is ca. 50 × 50 µm^2.

To demonstrate the difference between the micro ATR and transmission imaging approach, the same PS/LDPE blend shown in Fig. 2 was measured using the micro ATR system. The resultant image showing the distribution of LDPE is presented in Fig. 7. The image shows a much lower absorbance compared to the images in Fig. 2 because the pathlength using this approach is much shorter (ca. 0.8 µm compared to 3 or 20 µm in the corresponding transmission FTIR images). It is clear from the image in Fig. 7 that the micro ATR imaging approach offers a higher magnification for the smaller imaged area. The size of the LDPE domains was found to be between 8–30 µm.

Diamond ATR Imaging

When a larger field of view is desired, FTIR images can be acquired without the use of a microscope. A macro chamber is employed which allows different type of spectroscopic accessories to be used instead. FTIR imaging with a diamond ATR accessory (Golden Gate, Specac Ltd.) is a new innovative approach of FTIR imaging since the accessory itself was not designed for imaging purpose. The optical pathlength inside the accessory is shown in Fig. 8.

The first image obtained using the diamond ATR accessory was the surface of a finger because it was the most easily accessible sample with distinct features. The chemical image shown in Fig. 9 is generated using the amide I band at 1640 cm^{-1}. The fringes of the finger are shown clearly with high contrast. The quality of the spectra obtained by this method is shown by the spectrum extracted from an area indicated by the arrow (see

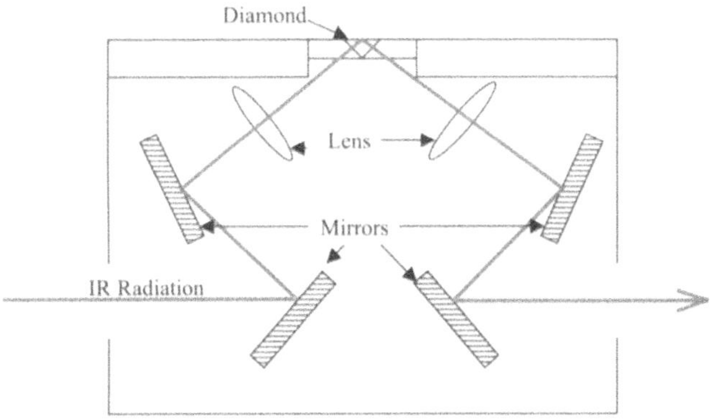

Fig. 8 A schematic diagram of the diamond ATR accessory

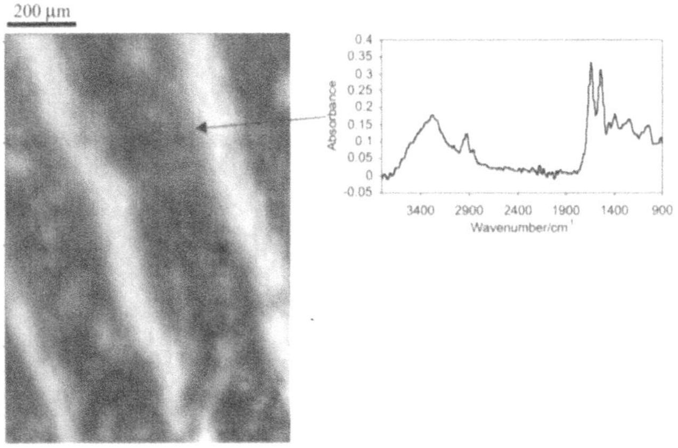

Fig. 9 FT-IR image of a finger surface acquired using the diamond ATR accessory. An extracted spectrum from the area indicated by the arrow is shown on the right

Fig. 9). The region of the spectrum between 1900 cm^{-1} and 2500 cm^{-1} where diamond absorbs too strongly has shown a higher noise level but no spectral artefact can be observed. A higher number of co-additions can be applied to improve the SNR in that region.

Conclusion

This paper summarises the advantages of FTIR imaging in ATR mode compared to FTIR imaging in transmission and also reviews two ATR imaging methods (micro and diamond ATR) with two different fields of view. ATR-FTIR imaging greatly broadens the range of materials amendable for study, and in particular, this approach has tremendous potential in applications to polymeric materials. In addition, FTIR imaging in ATR mode is uniquely suited for the imaging of aqueous systems or systems in contact with water due to the shallow penetration of the evanescent wave into the sample. The short pathlength (from a fraction to a few micrometers) results in obtaining images of thin surface layers of the samples without the need to microtome. It has been shown that imaging in transmission mode could provide spurious images if the thickness of the sample exceeds the size of the domains in the sample. Macro ATR-FTIR imaging opens opportunities for chemical imaging of patterned polymeric materials, colloid systems, emulsions and materials for high-throughput analysis. Opportunities exist to apply FTIR imaging in ATR and transmission modes under a controlled environment (relative humidity, vapour of organic solvent) which would assist the studies of polymer durability and performance under such conditions [16, 27, 28]. There is a great potential for applications of ATR-FTIR imaging to study dynamic polymer systems: dissolution, swelling, sorption, crystallisation, polymerisation and other processes. This method offers new opportunities to a polymer scientist and it is hoped that this article would facilitate novel applications of this chemical imaging technology in polymer and colloid science.

Acknowledgement We thank EPSRC and Pfizer for the provision of a CASE studentship for K.L.A. Chan. We thank Specac Ltd and Bruker Optics for advice.

References

1. Kidder LH, Haka AS, Lewis EN (2002) Instrumentation for FT-IR Imaging. In: Chalmers JM, Griffiths PR (eds) Handbook of Vibrational Spectroscopy, vol 2. Chichester: Wiley, p 1386–1404
2. Kazarian SG, Higgins JS (2002) Chemistry & Industry Issue 10:21–23
3. Bhargava R, Wang S-Q, Koenig JL (2003) Adv Polym Sci 163:137–191
4. Chalmers JM, Everall NJ, Schaeberle MD, Levin IW, Lewis EN, Kidder LH, Wilson J, Crocombe R (2002) Vib Spectrosc 30:43
5. Artyushkova K, Wall B, Koenig JL, Fulghum JE (2001) J Vac Sci Technol A 19:2791–2799
6. Koenig JL (1998) Microscopic Imaging of Polymers. ACS, Washington, DC
7. Koenig J (2002) Adv Mater 14:457–460
8. Miller-Chou BA, Koenig JL (2002) Macromolecules 35:440–444
9. Miller-Chou BA, Koenig JL (2003) Macromolecules 36:4851–4861
10. Salzer R, Steiner G, Mantsch HH, Mansfield J, Lewis EN. Fresenius (2000) J Anal Chem 366:712–726
11. Snively CM, Koenig JL (1999) J Polym Sci Polym Phys 37:2353
12. Snively CM, Koenig JL (1999) J Polym Sci B: Polym Phys 37:2261–2268
13. Sommer AJ, Tisinger LG, Marcott C, Story GM (2001) Appl Spectrosc 55:252–256
14. Kazarian SG, Chan KLA (2003) Macromolecules 36:9866–9872
15. Gupper A, Wilhelm P, Schmied M, Kazarian SG, Chan KLA, Reussner J (2002) Appl Spectrosc 56:1515–1523
16. Gupper A, Chan KLA, Kazarian SG (2004) Macromolecules 37:6498–6503

17. Gupper A, Kazarian SG (2005) Macromolecules 38:2327–2332
18. Chan KLA, Kazarian SG (2003) Appl Spectrosc 57:381–389
19. Chan KLA, Kazarian SG, Mavraki A, Williams DR (2005) Appl Spectrosc 59:149–155
20. Colley CS, Kazarian SG, Weinberg PD, Lever MJ (2004) Biopolymers 74:328–335
21. Chan KLA, Hammond SV, Kazarian SG (2003) Anal Chem 75:2140–2147
22. Van der Weerd J, Chan KLA, Kazarian SG (2004) VibSpectrosc 35:9–13
23. Van der Weerd J, Kazarian SG (2004) Appl Spectrosc 58:1413–1419
24. Kazarian SG, Chan KLA (2004) Macromolecules 37:579–584
25. Van der Weerd J, Kazarian SG (2004) J Contr Rel 98:295–305
26. Harrick NJ (1987) Internal Reflection Spectroscopy, Harrick Scientific Corporation, p 327
27. Chan KLA, Kazarian SG (2005) J Comb Chem 7:185–189
28. Chan KLA, Kazarian SG (2004) Vibrational Spectroscopy 35:45–49

Progr Colloid Polym Sci (2006) 132: 7–15
DOI 10.1007/2882_040

Published online: 8 March 2006

Reiner Salzer
Cordelia Zimmerer
Marco Kitsche
Gerald Steiner
Hans-Georg Braun

Molecular Imaging of Microstructured Polymer Surfaces

Reiner Salzer (✉) · Cordelia Zimmerer · Marco Kitsche · Gerald Steiner
Institute of Analytical Chemistry, Dresden University of Technology, 01062 Dresden, Germany
e-mail: reiner.salzer@chemie.tu-dresden.de

Hans-Georg Braun
Leibniz Institute of Polymer Research Dresden, Hohe Str. 6, 01069 Dresden, Germany

Abstract Within last decade both IR and Raman spectroscopy gained the potential of imaging the chemical heterogeneity of material surfaces. IR spectroscopic imaging offers the advantage of fast measurements with lateral resolution of a few micrometers. Due to the high acquisition rate of spectral data even dynamic systems can be studied at a time resolution of seconds. This report will demonstrate the unique opportunities by FTIR spectroscopic imaging to investigate material characterized by small thickness of the sample layer and non-crystalline state. SPR imaging is a new technique based on the classical SPR that combines high sensitivity with excellent lateral resolution. In the second part of this report we shall discuss the laterally resolved investigations of polymer material by a combination of FTIR imaging and SPR imaging. For the first time, pores in metal supported polymer membranes having a diameter of few micrometers could be generated by using information obtained via the new way of convoluting spectral image data. The approach can be transferred to applications having a domain size of few micrometers.

Keywords FTIR imaging · SPR imaging · Microstructured polymer · Data fusion

Introduction

Fourier-Transform Infrared (FTIR) spectroscopy as well as Raman spectroscopy are well established as methods for structural analysis of compounds in solution or when adsorbed to surfaces or in any other state. Analysis of the spectra provides information of qualitative as well as of quantitative nature. Very recent developments, FTIR imaging spectroscopy as well as Raman mapping spectroscopy, provide important information leading to the development of novel materials. If applied under optical near-field conditions, these new technologies combine lateral resolution down to the size of nanoparticles with the high chemical selectivity of a FTIR or Raman spectrum. These techniques now help us obtain information on molecular order and molecular orientation and conformation [1].

Here we report on the analytical methods applied for the optimization of the microstructuring process within polymer membranes on metal substrates. Two optical imaging techniques have been employed, FTIR imaging and surface plasmon resonance (SPR) imaging. FTIR imaging preferentially probes chemical molecular properties, SPR imaging mainly indicates physical bulk parameters at a much better lateral resolution. Finally we report our first results on an emerging digital technique called data fusion, which is used to combine the complementary information of two independent analytical techniques. Aim of the data fusion procedure described here is chemical information at improved lateral resolution.

In a previous study we demonstrated that natural ion channels can be integrated into thin polymer membranes and channel gating can be measured using the patch-clamp technique [2]. Ion channels were obtained from the electrical organ of the ray torpedo marmorata [3]. In those experiments [2], thin polyester membranes with small pores of 10 to 20 μm in diameter were used as matrix for the ion channels. Pores in the membrane were formed by ion beam bombardment and subsequent widening of the ion tracks by etching steps. A lipid bilayer of soy bean lipoid S 75 with cholesterol was stretched across the pores and, subsequently, ion channels were integrated by spreading an enriched lipid solution.

Patch-clamp measurements can well be performed under stable laboratory conditions but neither under field conditions nor in a highly parallelized manner like in a sensor array. These restrictions can be overcome by an optical detection scheme based on SPR. SPR permits label free optical detection, which is of great interest for protein biochips [4]. A recent development is SPR imaging, which enables rapid analysis of sensor arrays in combination with very high sensitivity [5]. The excitation of surface plasmons requires a prism for coupling light into an ultra-thin metal layer. The analyte to be measured is located adjacent to the metal layer, within the so-called evanescent field above the surface. The extent of coupling between photons and plasmons not only depends on the parameters of incident light (e.g., angle of incidence and wavelength), but also on the optical properties of the analyte layer. For a given optical arrangement and a given light source, the onset of resonance coupling causes a sharp minimum in the spectrum of the reflected light. This SPR minimum is subject to wavelength shifts upon slightest changes of optical parameters within the analyte layer.

The intensity of the evanescent field above the ultrathin metal layer decays by an exponential function. For this reason, reliable SPR imaging measurements of a sensor array require direct contact between the metal layer and the polymer scaffold containing the ion channels or any other sensing element. Bringing into contact the kind of porous polymer membranes described in the first paragraph and the metal layer does not provide the required direct contact due to the surface roughness of both items. The required sensitivity and durability of the metal – polymer compound can only be achieved by (i) formation of the polymer membrane directly on the metal surface and (ii) subsequent microstructuring of this membrane. The generation of completely empty micropores in polymer membranes on metal substrates has not yet been described in the literature.

Experimental

SF6 glass prisms were thoroughly cleaned, rinsed with organic solvents, treated in a plasma cleaner and, subsequently, heated in a vacuum chamber to a temperature of approx. 100 °C. A 50 nm gold film was deposited. During deposition the vacuum chamber was evacuated to 10^{-3} Pa. The thickness of the metal layer on the prism was controlled in-situ during the evaporation process using a oscillating quartz gauge.

Poly(methyl)methacrylate (PMMA) was chosen as polymer membrane. PMMA is a versatile polymer that is well suited for many applications in microelectronics. PMMA is most commonly used as a high resolution photo resist for e-beam lithography as well as for UV and X-ray microlithography [6]. Moreover, PMMA is known for its good biocompatibility, optical homogeneity and good adhesion on gold [7]. Thin membranes are obtained by spin coating of standard PMMA (molecular weight 950 000) dissolved in chlorobenzene. All PMMA membranes were baked at 180 °C for approx. 1 hour. Membrane thickness depends upon solid polymer concentration and spin speed [8]. The thickness was measured by an optical reflectometer (FTPadv, Sentech, Berlin Germany) equipped with a microscope (IX 70 Olympus). Thickness and refractive index were calculated by simulation of the spectrum according to the Fresnel formulas.

Microstructured patterns of PMMA on gold films were generated by electron beam lithography (LEO Gemini DSM 982, and electron beam lithographic system ELPHY, Zeiss Germany). The development process was carried out according to standard procedures. In a first step, the quality of microstructure was probed by SEM.

SPR images were captured using a home-built instrument [9]. A HeNe laser (Lasos GmbH, Jena, Germany) serves as monochromatic light source at a wavelength of 633 nm. A polarizer permits measurements both with s-polarized and p-polarized light, the latter is used as reference. A 60° SF6 prism (n_D = 1.8) is used as Kretschmann ATR coupler. The reflected light is focused by a glass lens and examined by means of a CCD camera (KP-M1EK/Hitachi GmbH, Rodgau, Germany). The CCD camera captures 756 × 581 pixels on an area of 10 × 10 mm.

Microstructured films were characterized by scanning electron microscopy (LEO GEMINI DSM 982/Carl Zeiss NTS GmbH, Oberkochen, Germany) at low voltages (0.9–1 keV) in order to avoid sample charging.

The topology of the microstructure was investigated by atomic force microscopy (DualScope/DME, Herlev, Denmark) in non-contacting mode. The scan speed of the cantilever was set to 50 μm/s at a constant force of 0.16 nN.

The Bruker Hyperion FTIR imaging spectrometer (Fig. 1) is designed to operate both in step-scan and continuous-scan modes. The microscope houses both a conventional single element MCT detector and a focal plane array (FPA) detector for imaging measurements. The sample area captured for one image is 270 × 270 μm^2 under the microscope or 4 × 4 mm in the macro sample compartment. The additional macro sample compartment is a particular accessory of this instrument, it turned out to be indispensable for the investigation of larger sample areas. Computer-controlled x;y-stages permit sequential recording of images, which can be pooled in the subse-

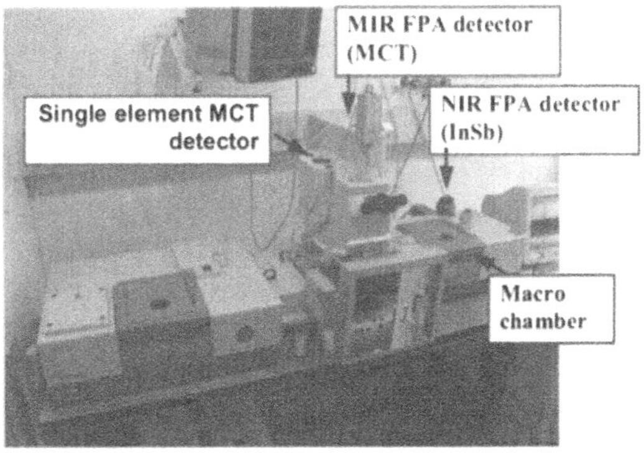

Fig. 1 FTIR Imaging Spectrometer Bruker IFS 66/S (*left*) equipped with a Bruker Hyperion Microscope (*mid*) and a particular macro chamber for investigation of larger objects (*right*)

quent evaluation procedure in order to render larger sample areas.

The currently used FPA has 64 × 64 detector elements, i.e. during one scan it probed the investigated sample area by 4096 complete FTIR spectra. Elaborated chemometric methods are needed to extract the molecular information buried in the wealth of data. The usual procedure starts with filtering out bad spectra (e.g., ruptures in the biological sample of Fig. 2a), followed by data reduction and classification routines. Non-supervised clustering methods provide sound results for the identification of slightest chemical differences across the investigated sample area. All spectra in the obtained data set are color-coded according to their cluster membership (Fig. 2b). Finally, all pixels of the sample image will be color-coded according to the cluster-membership of their FTIR spectra (Fig. 2c).

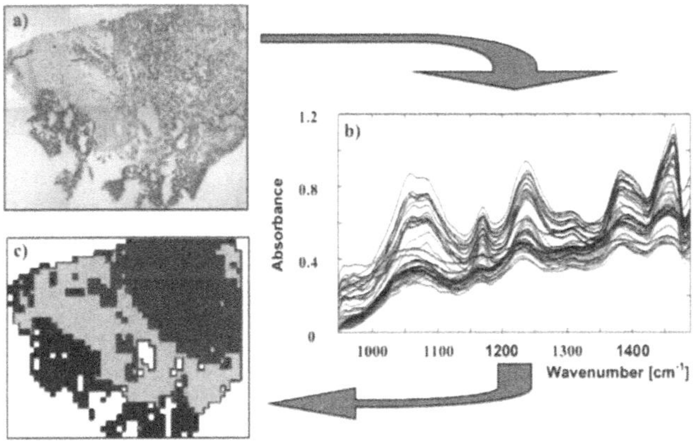

Fig. 2 Generation of spectroscopic FTIR images. **a** Native sample, here a thin section of biological tissue. **b** FTIR spectra are collected at every pixel of the native sample. "Bad" spectra, e.g. from areas outside the sample, have already been removed. **c** "Good" spectra have been subject to a chemometric cluster analysis. Pixels are color-coded according to the cluster membership (spectral staining). Slightest chemical differences across the sample are now clearly revealed

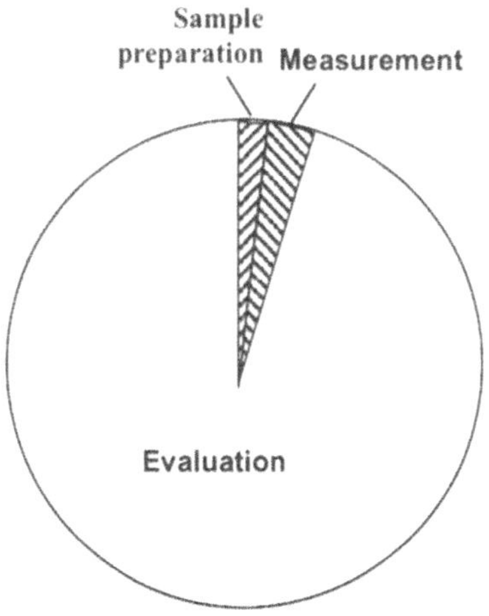

Fig. 3 The measurement of FTIR images does not require much sample preparation. Acquisition of spectra is very fast due to the huge amount of elements in the detector array. By far most time is spent for the evaluation of the molecular data. The figure gives a semi-quantitative impression of the size of the time slices

The basic difference between an common FTIR spectrometer and a FTIR imaging spectrometer consists in the layout of the detector. The FPA of the imaging spectrometer is made of the same material (MCT) as detectors of common FTIR spectrometers, it only houses 4096 individual detector elements arranged in a 64 × 64 array. This results in a 4096-fold increase in spectral data amount. Speed of data transfer and time needed for evaluation of thousands of spectra are current the limiting factors to capture and to evaluate FTIR images. A semi-quantitative display of time fractions during measurement of FTIR images is shown in Fig. 3.

Results and Discussion

Generation of Micropores in Metal Supported Polymer Membranes

Ion channels play an essential role in medical diagnostics and drug development. Such applications require the integration of ion channels together with a lipid bilayer into an artificial microstructured polymer membrane (Fig. 4). The polymer membrane is attached to a metal coated optical prism. The membrane contains micropores of approximately one micrometer in diameter. Lipid bilayers are stretched across the pores. The bilayers host the receptor molecules. After activation of an ion channel, thousands of ions stream into the cavity below the ion channel. The change of ion concentration can easily be detected by SPR measurements.

The properties of the polymer membrane must be adjusted to specific applications in order to: (i) constitute a matrix of polymer-free pores, (ii) form a reser-

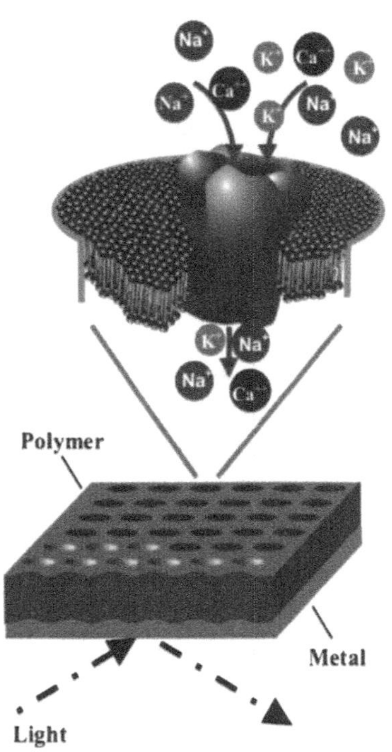

Fig. 4 Ion channel biosensor. The ion channels and their surrounding lipid bilayer are accommodated in a microstructured polymer membrane. Gating of the ion channel by an analyte results in an influx of ions into the pore. The concentration change is detected by surface plasmons, which are excited by light in the underlying metal layer

voir for the ions that flow through the ion channel and (iii) ensure a stable linkage between the polymer membrane and the metal layer. The quality of the obtained microstructures is usually assessed by scanning electron microscopy (SEM) or a similar microscopic technique. These microscopic techniques mainly provide images of the topology of the sample, they do not depict the distribution of chemical features like domains within the layer or like cross-linking of polymer chains. It is also not possible to assess, whether the micropores get completely down to the gold surface. Any small residue of PMMA remaining inside the pore will result in a distinct loss in detector sensitivity due to reduced contrast within the SPR image, which in turn results in a failure to monitor smaller changes in ion concentration within the pore.

Uniformity and thickness of the thin polymer membranes are crucial parameters for the performance of the sensor. Membranes of uniform thickness are expected to be produced conveniently by spin coating. Microscopic interferometry in reflection mode is a relative simple and non-destructive method to determine the exact thickness of thin polymer membranes. The thickness of all membranes was tested by a commercial reflectometer equipped with a standard microscope.

Micopores in gold-supported PMMA membranes were generated by electron beam lithography. The quality of the micropores was assessed by scanning electron microscopy (SEM). The SEM image in Fig. 5a gave rise to the impression of perfect pores, but the SPR image (Fig. 5b) shows only poor contrast. This PMMA layer was 0.8 µm thick. The poor contrast in the SPR image indicates imperfect micropores of rather ill-defined shape and/or volume. Imperfections with respect to the volume are caused in particular by polymer residues inside pores. Since the SPR signal is mainly influenced by matter in regions of high intensity of the evanescent field – i.e. in regions closest to the gold surface – polymer residues are expected to be located in the proximity of the gold layer. Only substance in this region is able to evocate such a low contrast within the SPR image. This presumption is supported by AFM measurements (Fig. 5c). The AFM image indicates perfectly shaped pores, as already seen in the SEM image. Both AFM and SEM images provide superior lateral resolution but do not provide any clue about the chemical nature of the investigated matter, in particular of the chemical nature or of the composition of the matter at the bottom of the pore. The AFM image does not provide means to decide, whether the bottom layer of the pore is made of remaining polymer or whether it shows the bare gold substrate.

SEM, SPR, and AFM measurements rely of totally different physical parameters of the sample. One particular difference between SEM and AFM measurement on one hand and SPR measurement of the other hand is the observation direction: SEM and AFM images are taken from top of the sample, whereas SPR images are measured from the back side of the polymer membrane. The penetration depth of the evanescent field (decay of the field intensity to $1/e$ of its size directly above the gold surface) is smaller than the thickness of the polymer membrane. In other words, SPR probes regions of the polymer membrane different from regions probed by SEM or AFM.

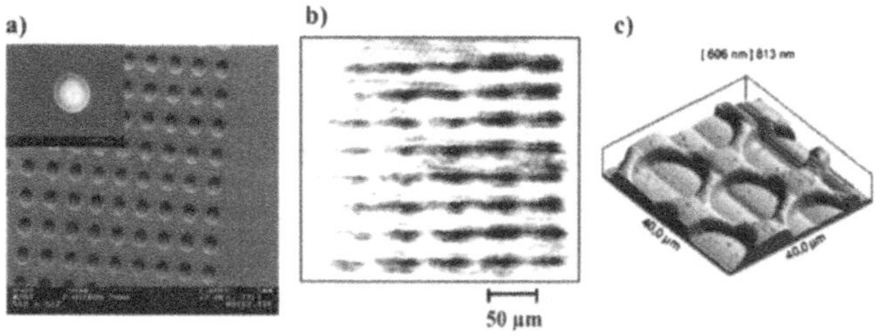

Fig. 5 Evaluation of a microstructured PMMA membrane generated by electron beam lithography. **a** The electron microscopic image indicates a perfect array of micropores. The *insert* shows a magnified view on a single pore. **b** The SPR image of the PMMA membrane reveals insufficient microstructuring. **c** The AFM image indicates a perfect topology of the micropores. No conclusion about the chemical nature of the bottom layer of the pore can be derived from the AFM data

The presumption for gold-supported polymer membranes of remaining matter inside pores generated by electron beam lithography can merely be verified by chemical arguments. However, SPR imaging provides only physical information. On the other hand, FTIR imaging spectroscopy is capable of probing membranes on a molecular level but it can not achieve the high lateral resolution in the high nanometer range as SPR imaging [10]. If samples permit, FTIR imaging measurements may be performed at ATR (ATR – Attenuated Total Reflectance) conditions. According to the higher refractive index of the ATR crystall, FTIR-ATR imaging enables improved lateral resolution [11]. The ATR technique was not employed in this study mainly due to the need to access the surface of the investigated sample and due to the profile of the evanscent field across the PMMA membrane. The latter causes overestimation of surface-near layers compared to lower regions of the membrane. All measurement in the current report have been made by transmission FTIR imaging. A combination of both FTIR and SPR imaging should provide a possibility to obtain as well as physical and chemical information of the sample at previously not achievable lateral resolution in the low micrometer range.

There are several reasons, why PMMA residues remain at the pore bottom. One important consideration are backscattered electrons, which may lead to enhanced cross linking among polymer chains. Another issue are surface reactions between gold and PMMA. Such reactions may be initiated by the electron beam.

The infrared stretching vibration of the carbonyl group, $\nu(C=O)$, appears as a very strong band. This mode is well suitable for FTIR spectroscopy to detect PMMA down to a thickness of 2 nm [12], hence a FTIR image should be able to reveal the presence of polymer residues remaining within pores. The spectroscopic FTIR image shown in Fig. 6a was taken from a microstructured PMMA membrane of similar size like shown in Fig. 5. The investigated area is 270 μm × 270 μm in size. The image consists of 64 × 64 pixels, each pixel holds a complete FTIR spectrum. All spectra were subject to a non-supervised cluster analysis. The cluster analysis resulted in 9 significant clusters. The membership of any of the individual FTIR spectra in Fig. 6a to one of the 9 clusters is indicated by shades of gray. Lighter shades represent clusters in regions of lowest PMMA abundance (pore regions). Darker shades represent clusters in regions of higher PMMA abundance (web regions). All spectra assigned to a particular cluster are very similar in number, location and size of their vibrational bands. For this reason, every cluster can well be represented by its centroid spectrum, a kind of mean spectrum. The carbonyl stretching bands of the centroid spectra of the 9 clusters of Fig. 6a are shown in Fig. 6b. The strongest bands in Fig. 6b belong to clusters of the PMMA web with its bulk properties. Medium sized bands belong to the sloped pore walls. These clusters indicate the transition from bulk material to the pores. Evidently, none of the clusters exhibits spectra without a carbonyl band. The lowest spectral trace reveals that the carbonyl group still absorbs inside the pores. This clearly indicates the presence of PMMA residues at the bottom of the pore.

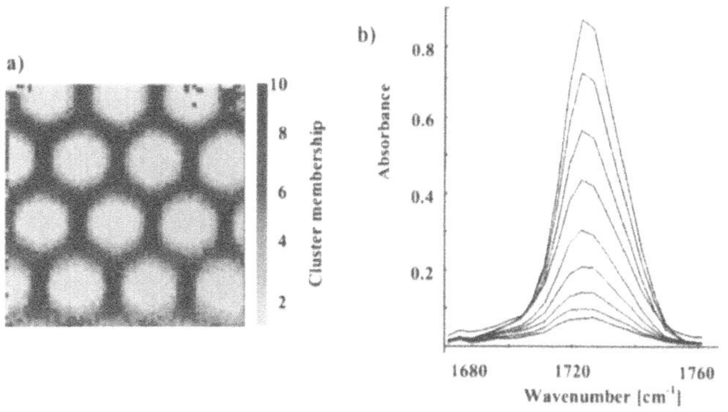

Fig. 6 FTIR evaluation of a microstructured PMMA membrane (sample size 270 × 270 μm^2) generated by UV lithography. **a** The FTIR image comprises 64 × 64 pixels, i.e. 4096 individual FTIR spectra. All FTIR spectra have been subject to a cluster analysis. Cluster membership is indicated by gray-scale coding. **b** Average (centroid) spectra for the 9 clusters in the FTIR image in the carbonyl spectral region. Spectra indicating higher absorbance belong to the clusters in the PMMA web. Less absorbing cluster are located within the pores. The absorbance never reaches zero. This clearly indicates remaining PMMA even in the middle of the pores

The carbonyl bands of the centroid spectra in Fig. 6b not only reveal the presence of PMMA residues inside the pores of the microstructured polymer membrane. Closer inspection of the band location indicates a slight shift of the band maxima. The maxima of the stronger bands of the bulk regions appear at slightly lower frequencies than the maxima of the weaker bands (pore regions). This frequency upshift of 5 cm^{-1} indicates differences in the chemical properties of carbonyl groups in bulk and in pore regions. Bands of clusters in pore regions appear at higher frequencies, indicating stronger carbonyl bonds for the residues within the pore. High-energetic radiation used for microstructuring obviously enabled polar moieties like $C=O$ groups to interact with the metal layer beneath the polymer. It should be noted that common microscopic techniques are neither able to reveal such small amounts PMMA at the bottom of the pore nor do they provide a chemical picture for the distribution of chemical non-equivalencies across the polymer membrane (Fig. 7). This knowledge is prerequisite for the optimization of the microstructuring process. In order to achieve high detection sensitivity for the sensor array, the remaining PMMA has to be removed as completely as possible.

Subsequent plasma treatment, modified development processes or lower beam energy during lithography are promising possibilities to produce pores totally free of polymer. Any progress can sensitively be indicated by FTIR spectroscopic imaging. The identification by FTIR imaging spectroscopy of the chemical reasons for the for-

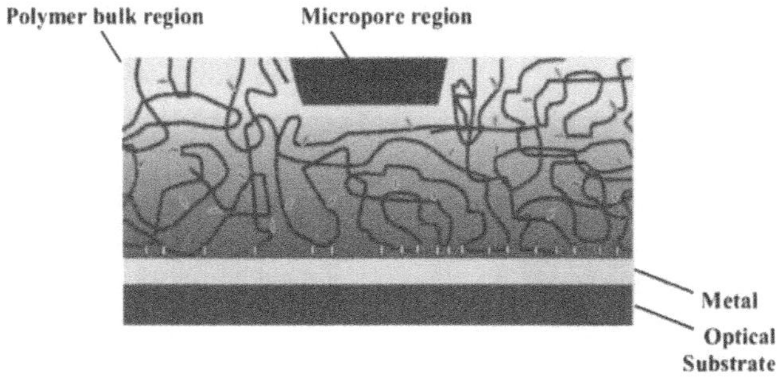

Fig. 7 Sketch of chemical heterogeneity after microstructuring the PMMA membrane derived from C = O stretching bands. C = O groups within the bulk are depicted *dark with a white hemline* whereas C = O groups interacting with the metal substrate are *white with a dark hemline*

mation of PMMA residues at the bottom of the pores in the polymer membrane provided the necessary basis for the optimization of the microstructuring process. The surplus of energy during the microstructuring process available for reactive groups to form strongly bound residues is to a large extend provided by the electrons used for irradiation. This basic conclusion was verified by replacing the gold film beneath the PMMA membrane by a silver film. Gold has a much stronger backscattering efficiency for electrons than silver. The quantitative evaluation by FTIR imaging spectroscopy of the amount of PMMA residue inside the pores revealed a distinct reduction for the silver substrate compared to the gold substrate (Fig. 8). But even in case of silver substrates there are too many electrons available, and they initiate the formation of too much residue inside the pores. For this reason we replaced electron beam lithography by UV lithography. The lower energy content of the UV photons results in the demanded decrease in PMMA residue inside the pores. The decrease is so large that even gold can be used again. Gold has the advantage of superior stability in sensor applications.

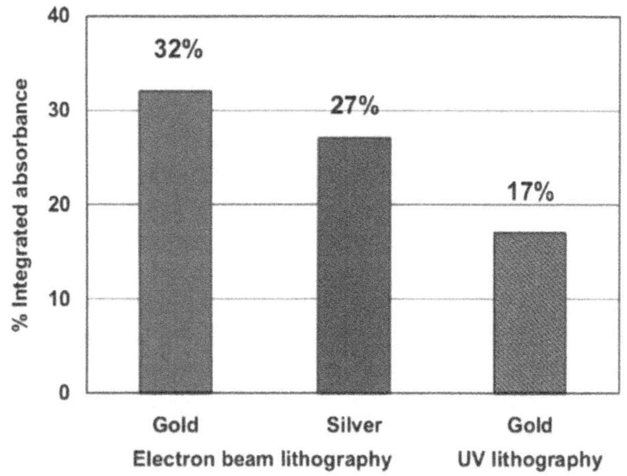

Fig. 8 Integrated absorbance of the C = O stretching band (1760–1680 cm^{-1}) in micropore regions compared to bulk regions of the PMMA membrane. The evaluation is based on the centroid spectra of the spectral clusters in the pore center and across the bulk region

The SPR image of an optimized PMMA membrane on gold substrate is shown in Fig. 9. It should be remembered that a SPR image reveals the backside of the polymer membrane. SEM and AFM images (top views) of this sample exhibit a similar perfect impression like those in Fig. 5, they need not to be shown again. The sample in Fig. 9 is superior to that of Fig. 5 both in terms of smaller pore size, of higher contrast between pore and bulk regions, and of pore shape. The latter is still rather ragged, further improvement of the pore shape may be desirable. Nevertheless, pore diameters are now in the range of merely three micrometers. This corresponds to roughly one tenth of the initial situation depicted in Fig. 5. Most of the pores in Fig. 9 exhibit a similar dark grade of contrast. Moreover, the majority of pores shows a constant contrast across its cross-section. Both facts point to the absence of larger amounts of residues inside the pores, either evenly distributed across the pore or fused together in grains at the bottom of the pore.

The comparison of the FTIR image in Fig. 6 and the SPR image in Fig. 9 clearly reveals strengths and weaknesses due to underlying physical principles of the corresponding methods. The FTIR image provides rich chemical detail, but the lateral resolution of the spectroscopic image controlled by the diffraction limit does not permit to identify the chemical background of the ragged shape of the pores. On the other hand, SPR lacks chemical specificity but provides physical information for a correspondingly large sample area at a correspondingly large resolution. The combination of the information content of the two analytical techniques may provide the information needed for the generation in supported polymers of high-quality pores of diameters even below 1 μm.

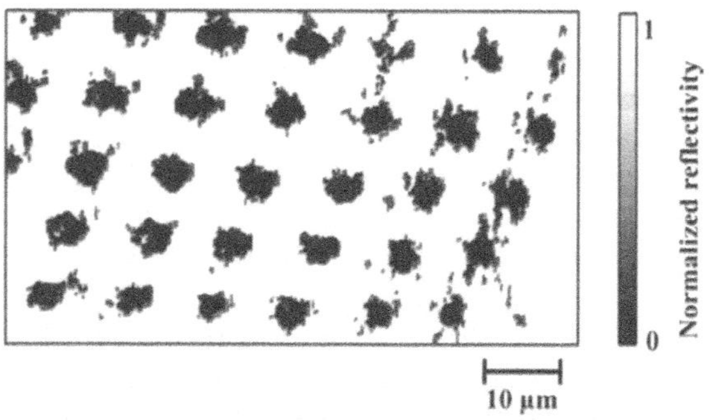

Fig. 9 SPR image of a PMMA membrane on gold after improvement of the microstructuring procedure

Data Fusion FTIR-SPR

The simultaneous use of multisource data can provide a more reliable view on the observed object. In order to exploit the information content, sophisticated color-related and numerical data fusion techniques have been developed [13]. These techniques aim at the incorporation of complementary information for the initially independent results into a new combined data set in such a way that an information surplus can be retrieved from the combined data set. This information surplus remains inaccessible if individual data sets are evaluated separately.

Data fusion of spectroscopic images is a rather new approach. Its terminology is not yet settled, which gives rise to some misunderstandings. A common misconception is related to the terms "image fusion" and "data fusion". The general definition of "image fusion" refers to the conventional combination of two or more different images to form a new image by using a certain algorithm. Other expressions are sensor fusion [14], image merging [15], multisensor data fusion [16] and image integration [17]. Image fusion is already well established in high-tech medical diagnosis like functional imaging [18]. Basically the term "image fusion" describes the mere overlay of at least two images, i.e. the combination of the contrast values of the initial images. On the other hand, "data fusion" refers to the convolution of information contained in the data sets. In case of spectroscopic images this information content consists of a complete spectrum for every pixel in the image. Data fusion implies the use of ancillary (i.e. non-imaging) information other than remotely sensed images. The term "data fusion" (or "information fusion") refers to a different, higher level of image processing, where the images have already been transformed into knowledge or information before being fused. The particular algorithms for data fusion have to be based on the physical background of the measured information. For this reason, data fusion of FTIR and SPR results requires the development of a approach dedicated to the particular information content of the particular techniques. In the illustrated case (Fig. 10), the C = O str absorbance in the FTIR image was chosen to judge about the pore quality. The C = O str absorbance sensitively indicates the amount of PMMA at the particular pixel position. The two extreme cases sketched in Fig. 10 are bulk and pore regions in the PMMA membrane. On the contrary, FTIR does not provide clues about the presence of electrolyte inside the pore, i.e. whether the pore is completely filled by the adjacent liquid phase or not. Information about the filling status of the pore can easily be obtained from SPR data, whereas SPR is not able to differentiate between the polymer bulk and polymer pores perfectly filled by electrolyte. This example compares very well to the underlying principle of the meanwhile well established 2D techniques in chromatography [19]. In case of 2D chromatography, Fig. 10 would represent a plate with the axes "FTIR Imaging" and "SPR Imaging" replaced by the description "eluent 1" and "eluent 2". The sketched three spots on the plate would represent chromatographic fractions. The two fractions at the bottom could not be separated by using merely eluent 1, the two fractions could not be separated by using merely eluent 2. The separation of all three fractions can only be obtained by 2D chromatographic techniques but not by two independent chromatographic runs even under identical conditions to those of the 2D chromatographic experiment. This example illustrates the potential of data fusion of two independent spectroscopic techniques. Data fusion may provide access to previously overlaid information.

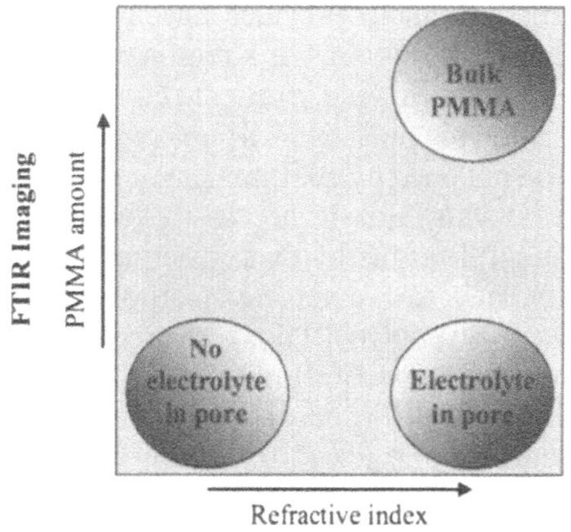

Fig. 10 Access to hidden information by data fusion: contrast in the SPR image depends on the refractive index; C = O str intensity in the FTIR image indicates the PMMA amount in the image voxel. FTIR differentiates very well between pore and bulk regions but not between pores w/o electrolyte. SPR differentiates very well between pores w/o electrolyte but not between filled pore and bulk regions

The sketch in Fig. 10 is oversimplified in so far as it highlights only three boundary situations, bulk polymer on one hand and filled or empty pores on the other hand. In terms of information content, every corner of the square in Fig. 10 represents different physical or chemical properties. An appropriate graphical representation of a real application requires color coding. Every corner of the square, i.e. every property of the polymer investigated, has to be assigned a characteristic color. The interior areas of a square like that in Fig. 10 exhibit color mixing with every tint representing a unique combination of the investigated properties of the microstructured polymer.

The subsequent investigations are focused on the filling state of the micropores in the PMMA membrane, whereas the preceding chapter referred to residues at the bottom of the micropores. Micropores have to be filled by electrolyte for their intended application as ion channel biosensor. The available imaging techniques revealed variations among the pore volumes after admission of electrolyte, but none of these methods provided unequivocal clues what caused the variations. For this reason we employed the new data fusion technique in order to combine the molecular information obtained by FTIR imaging and the physical information obtained by SPR imaging.

Figure 11 illustrates the necessary steps in order to obtain meaningful results by data fusion of FTIR and SPR images. At first, the complete microstructured PMMA sample was measured by FTIR spectroscopic imaging. The sample area of approx. 500 μm × 500 μm required 4 FTIR images to be taken (Fig. 11a). FTIR images were pooled and subsequently inspected for an initial quality test of the microstructuring process. After the quality check an SPR image was taken from the area to be investigated (Fig. 11b). Due to the different optical layouts of the instruments, the rough images have to be corrected for different observation angles and to remove possible overlay between the segments of the pooled FTIR images. The images in Figs. 11a and 11b already underwent the necessary correction steps, i.e. they are deskewed with respect to the exact location of the pixels. It is important to note that only the location of the pixels got deskewed, whereas the spectral data assigned to the pixels was not changed at all. The spectral data were then subject to data fusion. Depending upon the evaluation strategy and the amount of spectral data utilized, the result of the data fusion can be multidimensional like in every multivariate evaluation. In Fig. 11c, the information content behind every pixel had to be reduced to two dimensions in order to show it by gray-coded values. Details for some pores are given in Fig. 11d.

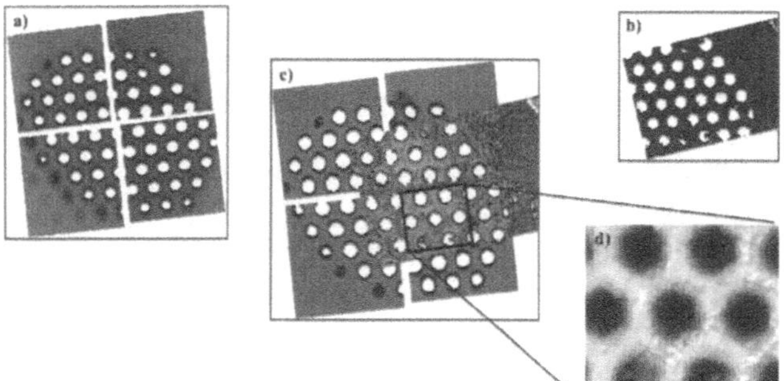

Fig. 11 Data fusion of FTIR and SPR images. **a** Pooled FTIR bright field images of a microstructures polymer membrane. **b** SPR image of a segment of the same polymer membrane. **c** Fused data of **a** and **b**. **d** Enlarged segment of **c** showing non-empty pores. The chemical nature of the pore content can be deduced from the color coding of the fused data

The gray-scale has been inverted in Fig. 11d for clarity, i.e. pores completely filled with electrolyte are now dark. The PMMA web clearly stands out in Fig. 11d. The pores in the upper row are well shaped, and they are completely filled by electrolyte. In contrast, the pores in the bottom row are well shaped as well but hold inclusions. The distinct gray-shades of the inclusions point to different chemical nature of the inclusions. In the original fully-colored image the inclusions exhibit totally different colors, which permit the assignment of the inclusions in the bottom left pore to an air lock, whereas the inclusion in the pore to the right is a polymer residue. It should be emphasized that the size of this polymer inclusion is too small to be laterally resolved in the FTIR image.

Conclusion

FTIR spectroscopic imaging provides new opportunities for the investigation of material having small layer thickness, non-crystalline state and domain size of few micrometers. Samples may even exhibit all three properties at once, as it is the case for microstructured polymer membranes. This is a particular advantage of the new measuring approach. Domain sizes may even be reduced to the upper nanometer range if Raman spectroscopic mapping is employed [20].

Microstructuring of the surface of a polymer membrane on solid support is commonly performed by various lithographic techniques. In previous applications is was not necessary or it was not tested whether the microstructured pores extend down to the neat support surface or whether the lower part of the pore remains filled with bulk polymer of with residues of the lithographic process. It could be shown by FTIR spectroscopic imaging for the first time, that residues remain inside the pores, which chemical origin the residues have and how the microstructuring process can be optimized towards completely empty pores, even if the diameter of the pores does not exceed a couple of micrometers.

SPR imaging is a new and versatile technique for sensor applications. Combination of the physical information provided by SPR imaging with the chemical information obtained by FTIR imaging by a process termed data fusion revealed information content so far hidden. Fused data provide insight into the chemical content of the mircrostructured pores after their contact with an adjacent liquid phase. Information based on fused FTIR/SPR data provides the necessary clue for further sensor applications of microstructured polymer membranes on metal substrates.

References

1. Salzer R, Steiner G, Kano A, Richter T, Bergmann R, Rodig H, Johannsen B, Kobelke J (2003) SPIE Proc 5047:362
2. Salzer R, Li J, Rautenberg C, Friedrich St (2001) Macromol Symp 164:239
3. Schrattenholz A, Godovac-Zimmerman J, Schäfer HJ, Albuquerque EX, Maelicke A (1993) Eur J Biochem 216:671

4. Steiner G, Zimmerer C, Friedrich St, Salzer R, Braun HG (2003) Screening 26:32
5. Steiner G (2004) Analyt Bioanalyt Chem 379:328
6. Cerrina F (1997) In: Choudhury PR (ed) Handbook of Microlithography, Micromachining, and Microfabrication. SPIE Press, p 253
7. Patnaik A, Li C (1998) J Appl Phys 83:3049
8. Zimmerer C, Roy J, Braun HG, Kitsche M, Steiner G, Salzer R (2003) SPIE Proc 5047:403
9. Steiner G, Sablinskas V, Hübner A, Kuhne Ch, Salzer R (1999) J Mol Struct 509:265
10. Salzer R, Steiner G, Mantsch HH, Mansfield J, Lewis NE (2000) Fresenius J Anal Chem 366:712
11. Chan KLA, Kazarian SG (2003) Appl Spectrosc 57:381
12. Balastre M, Berquier JM (1999) Langmuir 15:8691
13. Pohl C, van Genderen JL (1998) Int J Remote Sens 19:823
14. Phillips BE, Keys LD (1990) Image Proc Remote Sens 4:238
15. Carper WJ, Lillesand TM, Kiefer RW (1990) Photogramm Eng Remote Sens 56:459
16. Blodgett CF, Franklin SE (1993) Comput Geosci 19:577
17. Ehlers M, Welch R (1998) Int J Remote Sens 9:873
18. Felix R, Lemke AJ, Niehues SM, Amthauer H, Rohlfing T (2004) RöFo 176:1811
19. Link AJ, Eng J, Schieltz DM, Carmack E, Mize GJ, Morris DR, Garvik BM, Yates JR III (1999) Nat Biotechnol 17:676
20. Krafft C, Knetschke T, Siegner A, Funk RHW, Salzer R (2003) Vib Spectr 32:75

Progr Colloid Polym Sci (2005) 132: 16–22
DOI 10.1007/2882_024

Published online: 9 March 2006

S. Busche
M. Kasper
T. Mutschler
N. Leopold
G. Gauglitz

Interaction Behaviour of the Ultramicroporous Polymer Makrolon® by Spectroscopic Methods

S. Busche (✉) · M. Kasper · T. Mutschler · N. Leopold · G. Gauglitz
Eberhard-Karls-University, Institute of Physical and Theoretical Chemistry,
Auf der Morgenstelle 15, 72076 Tübingen, Germany
e-mail: stefan.busche@ipc.uni-tuebingen.de

Abstract Makrolon®, a commercially available polycarbonate with a glassy ultramicroporous structure (mean pore-volume 0.1 nm^3), was often employed as sensitive layer for optical sensors in recent years. Due to the definite pore volume-distribution, it can be used as a size-selective sensitive layer. The interaction behaviour of Makrolon of different layer-thicknesses under the influence of methanol, ethanol and 1-propanol is characterized by Spectral Ellipsometry (SE), Surface Plasmon Resonance Spectroscopy (SPR) and Reflectometric Interference Spectroscopy (RIfS). Sorption into the polymer layer follows a superposition of Henry- and Langmuir-type behaviour; this was discussed depending on the size of the respective analyte. An adequate layer-thickness for the sensor methods RIfS and SPR is proposed.

Keywords Reflectometric Interference Spectroscopy · Spectral Ellipsometry · Surface Plasmon Resonance Spectroscopy · Ultramicroporous Polymer

Introduction

A chemical sensor is a device that transforms chemical information into an analytically useful signal. Chemical sensors contain two basic functional units: a receptor part and a transducer part. The receptor part is usually a sensitive layer, therefore a well founded knowledge about the mechanism of interaction of the analytes of interest and the selected sensitive layer has to be achieved. Various optical methods have been exploited in chemical sensors to transform the spectral information into useful signals which can be interpreted as chemical information about the analytes [1]. These are either reflectometric or refractometric methods. Optical sensors based on reflectometry are reflectometric interference spectroscopy (RIfS) [2] and ellipsometry [3,4]. Evanescent field techniques, which are sensitive to changes in the refractive index, open a wide variety of optical detection principles [5] such as surface plasmon resonance spectroscopy (SPR) [6–8], Mach-Zehnder interferometer [9], Young interferometer [10], grating coupler [11] or resonant mirror [12] devices. All these optical methods use sensitive layers; in case of chemosensors, polymers are established as receptors.

In recent years, optical sensory applications of the ultramicroporous polymer Makrolon as sensitive layer for the discrimination of mixtures of different classes of compounds have been reported. The first implementation of Makrolon in literature was the quantification of the ozone-depleting refrigerant R22 in the presence of its most important substitute R134a using a RIfS-array [13]. The interaction characteristics of polymers taking advantage of different discrimination principles were determined. It was not necessary to reach the equilibrium between polymer and gaseous phase. Binary mixtures of the two refrigerants were measured, and the sensor signals were evaluated by artificial neural networks. Additionally, measurements with a low cost set-up were conducted. This enables a fast and economical monitoring at recycling stations. By evaluating different time-points of the sorption of the analytes into the polymer, the different diffusion kinetics of the analytes can be taken advantage of. Using artificial neural networks, binary mixtures of methanol and ethanol could be quantified with a single-sensor SPR set-up [14]. This concept was adapted for the quantification of binary mixtures of the above mentioned freones with a single-sensor SPR set-up. By using a thin sensitive layer (60 nm), the duration of sorption and regeneration could be minimized [15]. Numerous concentrations could be measured in a short time. Different approaches to the multivariate calibration of the two refrigerants are reported in literature [16, 17]. Single-sensor application of RIfS applying time-resolved measurements could be adopted. Ternary mixtures of R22, R134a and R32 and R143a respectively were quantified effectively [18]. Quantification of quaternary mixtures of alcohols was performed by SPR as well as RIfS. The different kinetics for layers of different thicknesses was utilized in a RIfS-array. Nevertheless, it was shown that quantification of quaternary mixtures can be performed by single-sensor set-ups successfully for both methods [19].

In this study, for an accurate understanding of the interaction behaviour of the ultramicroporous polymer Makrolon, as a receptor, under the influence of three alcohols is investigated by three transduction methods: Spectral ellipsometry (SE), RIfS and SPR.

These three methods are compared since each of them provides complemantary information. SE offers the possibility to determine absolute values of the refractive index n and physical layer thickness d by fitting a simulation to measured quantities for adequate layer systems. SPR is highly sensitive towards changes in the refractive index. RIfS presents itself as a straightforward method for the determination of changes in the optical thickness ($n \cdot d$).

These three optical methods have different requirements to the sensitive layer (optical transparency, homogeneity, interference pattern for reflectometry, resonance conditions for SPR etc.). Thus, the interaction behaviour of sensitive layers of different thicknesses and its effects on refractive index as well as on layer thickness under the influence of analytes must be known in order to build the most effective sensor device in respect to sensitivity, long term stability, speed, cost-performance ratio etc.

The permeation of a gas through a porous polymer is generally described by equations based on the kinetic theory of gases. The sorption isotherm described by Eq. 1 is concave to the pressure axis and is commonly observed for a penetrant gas in a glassy polymer. It is composed of Henry's law and Langmuir-terms [20]:

$$C = C_D + C_H = k_D p + \frac{C_H b p}{1 + b p} \qquad (1)$$

where C_D is the penetrant concentration dissolved by ordinary dissolution (Henry's law mode), k_D is the Henry's law parameter and C_H is the penetrant concentration described by the Langmuir equation. According to [21] the Langmuir adsorption occurs into the unrelaxed domains (ultramicropores) of the glassy polymer. The maximum penetrant concentration (Langmuir capacity constant) is given by C'_H and b is the affinity constant.

Makrolon has a mean free pore-volume of 0.1 nm^3 and the width at half-height of the corresponding distribution is 0.04 nm^3 [22]. The polymer layer reacts to the exposition of the analyte molecules by swelling and by changes of the refractive index. Due to the pore-volume distribution (see Fig. 1), the interaction kinetic depends on the molecule size [23]. The analytes used in this work are methanol with a size smaller, ethanol with a size almost equal to and 1-propanol with a size bigger than the mean free pore-volume.

At low analyte concentrations, the pores will be filled first, causing an increase of refractive index, since the analyte molecules with a refractive index $n_D = 1.329$ for methanol, $n_D = 1.361$ for ethanol and $n_D = 1.385$ for 1-propanol replace the air in the pores with a refractive index of 1 (all refractive indices are given for a wavelength of 589.3 nm corresponding to the D-line in the sodium spectrum and a temperature of 20 °C). An increase of layer thickness should also be observed, since the analyte molecules interact with the pores and cause a swelling. After almost all pores are filled with analyte molecules, additional molecules will begin to sorb into the polymer matrix and cause a swelling of the polymer backbone [20]. Thus, the swelling behaviour of the Makrolon corresponds to a superposition of Langmuir's law describing the filling of the pores and Henry's law describing the sorption into the polymer matrix. Different sized analytes will cause different swelling behaviours and thus different superpositions of both laws.

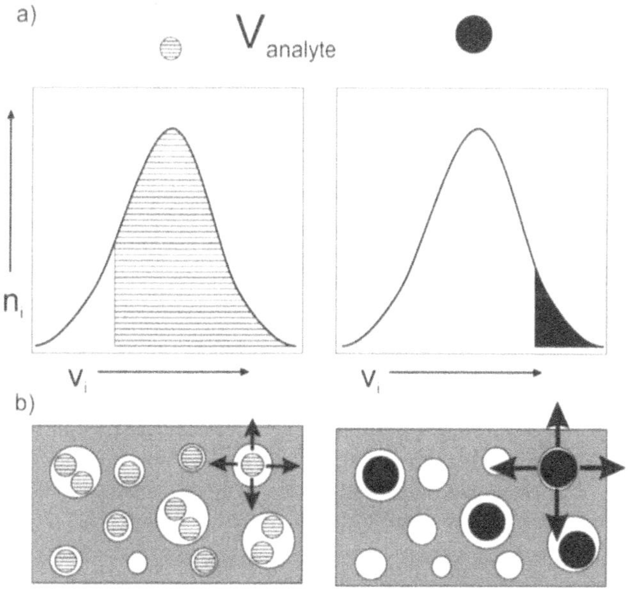

Fig. 1 Pore-volume distribution of the ultramicroporous polymer and the accessible volume for analytes in dependence of the molecule volume (**a**). Scheme of interaction between pores and different sized molecules: for smaller analytes more pores are accessible and the interaction between molecules and polymer backbone is weaker (**b**)

At higher concentrations, when all available pores are filled, a further swelling of the polymer layer should be observed. The refractive index of Makrolon is $n_D = 1.586$ and because of the sorption of the analytes with a lower refractive index into the polymer matrix, it should decrease. By the observation of changes of the refractive index on the one hand and the determination of changes of the layer thickness on the other hand, statements about the interaction behaviour will be made.

Another objective of this work is the interaction behaviour of layers of different thicknesses under the influence of the same analyte. Thus, Makrolon layers between 35 nm and 455 nm were prepared and characterized under the effect of the above named analytes with the considered detection methods.

Methods and Material

Spectral Ellipsometry

In Spectral Ellipsometry, changes in the state of polarisation of white light upon reflection at surfaces are monitored. This enables separation of the refractive index and physical thickness by modelling of a layer system. To handle correlations between layer thickness and refractive index, adequate layer thicknesses have to be guaranteed to avoid physically unreasonable solutions of the fitting due to local minima.

SE-measurements were done with a Spectral Ellipsometer ES4G from Sopra, with a Xenon high-pressure 1 2174-01 light source from Hamamatsu, a double monochromator with a slit width of 400 μm, a spectral range from 230–930 nm and a spectral resolution of 0.05 nm. The evaluation of the measurements was carried out by Film Wizard 32 with the Cauchy-model. The observation of the relative changes in the refractive index at 589.3 nm was sufficient to monitor the sorption of the analytes.

Surface Plasmon Resonance Spectroscopy

The SPR-technique exploits the fact that, at certain conditions, surface plasmons of metallic layers can be excited by radiation. The shift of the resonance wavelength or the shift of the angle of incidence is influenced by analytes interacting with the sensitive layer.

For SPR-measurements a self-made apparatus based on the Kretschmann configuration [14] was used. The metallic layer was prepared by evaporating silver onto the glass prism (half cylindrical, ground 20 × 40 mm) using a vacuum evaporation system (Pfeiffer Vacuum GmbH, Wetzlar, Germany) at high vacuum (10^{-6}–10^{-7} mbar) and at a rate of about 0.6–1 nm s^{-1}. The deposition of approximately 50 nm of silver was monitored by a crystal oscillator.

The conditions of the incident angle and the kind of glass were chosen to realize a resonance wavelength of about 700 nm. As the refractive index is highly temperature-dependent, the temperature is kept at a constant value by a thermoregulator.

Reflectometric Interference Spectroscopy (RIfS)

In the RIfS-method, one part of the radiation is reflected at the interface of a thin layer, whereas the other penetrates the layer and is there reflected at the other interface. These two partial reflected beams become superimposed and form an interference pattern resulting in constructive or destructive interference. This interference pattern depends also on the optical thickness of the layer, which is given by the product of refractive index and the physical thickness of the layer. By a parabolic fit, the shift of an extremum is evaluated and the change in the optical thickness is given as result.

The set-up used consists of a tungsten light source (20 W, 12 V), optical Y-fibre couplers (PMAA, 1 mm core diameter, microparts, Dortmund, Germany) and a diode array spectrometer (MMS, Zeiss, Jena, Germany). Interference spectra were recorded from 450–700 nm.

The temperature of the flow cell was kept constant by a thermoregulator.

Polymer Layers and Analytes

The investigated polymer is a glassy, ultramicroporous polymer (Makrolon® M2400, Bayer AG Leverkusen, Germany). All measurements were performed below the glass transition temperature ($T_g = 145\,°C$), so the glassy polymer chains are restricted in motion and cannot completely homogenize.

All sensitive layers were prepared from solutions of Makrolon in mixtures of chloroform and dichlorobenzene by a spin-coating process. By adjustment of the rotation speed and time the thickness of the layers were varied between 35 nm and 455 nm. Layer thicknesses and refractive indices were determined by spectral ellipsometry. Furthermore the polymer thicknesses were verified by a surface profilometer (Alpha Step 500, Tencor Instruments, Mountain View, USA).

For SPR-measurements, two layers with a thickness of less than 100 nm were prepared on BK7-prisms (refractive index $n_D = 1.51680$, Krombach, Germany) and all other layers were prepared on NPH2-prisms ($n_D = 1.92286$, Ohara, Japan) to fulfil the resonance conditions of the surface plasmons. To prepare layers for the RIfS-measurements, the solutions were spin-coated onto glass-substrates with a Ta_2O_5-layer for reflection enhancement. Si-wafers were used for the ellipsometric measurements.

The analytes (Aldrich Chemicals, Munich, Germany, p.a. quality) were evaporated from Chromosorb (W-AW-DMCS 80/100, Macherey & Nagel, Düren, Germany) by a computer-driven gas mixing station (MKS, Munich, Germany) at 5 °C and the measurements were performed at

20 °C. For reasons of comparability the partial pressure p_i of each analyte was referenced to its saturation pressure p_{i0}. The sensitive layers were exposed to concentrations of the three homologous alcohols between 0 and 0.12 p_i/p_{i0} in eight steps. The measurements were performed in random order and each concentration was measured three times.

Results and Discussion

The interaction behaviour of the homologous alcohols methanol, ethanol and 1-propanol and the ultramicroporous polymer Makrolon was investigated by three different optical methods: spectral ellipsometry, surface plasmon resonance and reflectometric interference spectroscopy.

The interaction kinetic of the smaller analytes methanol and ethanol is fast and reversible. For 1-propanol, the sensitive layers had to be exposed for longer periods to reach a stable signal. The desorption of this analyte is also much slower, but still reversible. As the analytes sorb into the polymer backbone, equilibrium is not reached during the time of exposition.

First the ultramicropores are filled with analyte molecules, thereby mainly the refractive index is changed. The filling of the pores is leading to a Langmuir-type change of the measurand. For the smallest analyte methanol ($0.068\,nm^3$, smaller than the mean pore-volume), most pores are within reach. Langmuir-type behaviour can be observed up to higher concentrations than for ethanol and 1-propanol. Due to the pore-volume distribution and 1-propanol being the largest analyte ($0.125\,nm^3$), it has the least number of pores accessible (see Fig. 1a), meaning that 1-propanol has all its available pores occupied at lower concentration than ethanol ($0.097\,nm^3$) and methanol. Admittedly, the interaction due to its size is the strongest. After nearly all accessible pores are filled, further molecules can sorb into the polymer backbone and cause a swelling of the polymer resulting in a Henry-type isotherm. For methanol, this swelling is the shallowest; it is highest for 1-propanol. This results in a smaller Langmuir-part and a greater Henry-part of the absorption isotherm for this analyte.

The observed sorption behaviour can therefore be described by a combined Langmuir-Henry absorptions isotherm. Equation 1 was fitted to the calibration curves of the three methods and the Henry-Langmuir-behaviour was verified.

As follows, the results for each measuring technique are described to outline its benefits to a comprehensive description of the interaction behaviour of the analytes and the polymeric layers.

In all figures, the mean signal changes and standard deviations are shown. The results of the fitting of Eq. 1 are plotted as lines.

Ellipsometric Measurements

A Makrolon layer of 170 nm was exposed to the alcohols to investigate the influence of the analytes on refractive index and layer thickness. In Fig. 2 the relative changes in the refractive index and relative changes in physical thickness for different concentrations (0 to 0.13 p_i/p_{i0}) of 1-propanol are shown. For lower partial pressure, an abrupt rise is observed. This can be explained by the replacement of air ($n_D = 1$) in the pores by the analyte molecules ($n_D = 1.385$). After the accessible pores are filled, furthermore no significant change in the refractive index can be found with rising concentration of the analyte. For the highest measured concentration even a decrease of the refractive index is observed. This can be explained by a swelling of the polymer by permeation of analyte molecules in the polymer backbone as the refractive index of 1-propanol is lower than that of Makrolon. As shown in Fig. 2, the layer thickness increases with the partial pressure of the analyte (Henry-type isotherm). This substantiates the swelling of the glassy polymer due to intrusion.

For methanol and ethanol similar results were found. The relative changes in the physical thickness were smaller than for 1-propanol for equal concentrations, as the interaction of the smaller analytes with the polymer backbone is weaker.

SPR-Measurements

For the characterization of the interaction behaviour, six ultramicroporous layers of thicknesses between 35 nm and 430 nm have been exposed to the three alcohols. In Fig. 3 the SPR measurements of a 200 nm Makrolon layer with methanol ($n_D = 1.329$), ethanol ($n_D = 1.361$) and

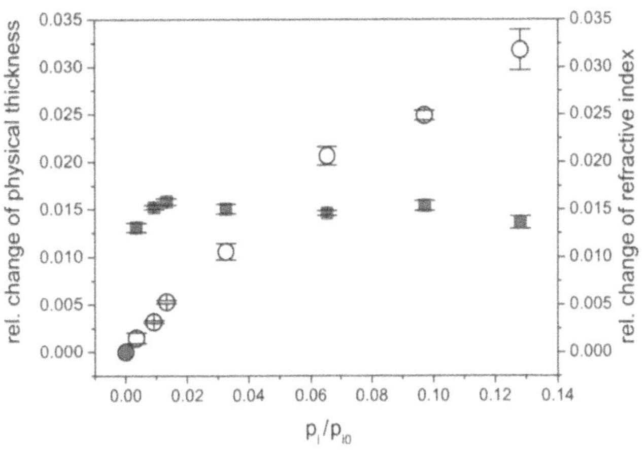

Fig. 2 Relative change in the refractive index (*gray squares*) and relative change in the physical thickness (*open cycles*) of a Makrolon layer of 170 nm during exposition to different concentrations of 1-propanol, measured by spectral ellipsometry

1-propanol ($n_D = 1.385$) are shown. The standard deviations of all measurements are very small, so the interaction of the analytes with the polymer is highly reproducible.

The evanescent field of the surface plasmons decays within some hundred nanometres and is not fully decayed within the 200 nm of this sensitive layer.

As can be seen in Fig. 3, the changes in the sensor signals are increasing with higher concentrations for all analytes. The signals for 1-propanol are higher than for ethanol than for methanol because of the higher refractive index of 1-propanol.

The slope is steeper for the whole concentration range, but the filling of the pores is completed at lower partial pressures, as can be seen in the curvature of the fitted graph. For methanol it is not clearly distinguishable when the filling of the pores is completed.

In Fig. 4 the saturation signals for methanol for the six sensitive layers are plotted as mean and standard deviations. The shifts in the resonance wavelength are increasing with growing layer thicknesses. The signals for 350 nm and 430 nm layers are very similar as the evanescent field decays within some hundred nm and the saturation signals for layers of more than 300 nm shall be almost identical. The Langmuir-part of the graphs is protracted to higher concentrations with increasing thickness of the polymer layers because the total number of pores also rises and diffusion rate also plays an important role. The sorption and desorption kinetics of thinner layers is much faster than for thicker layers. Therefore a compromise between signal height and the ability of fast measurements has to be made for sensing applications.

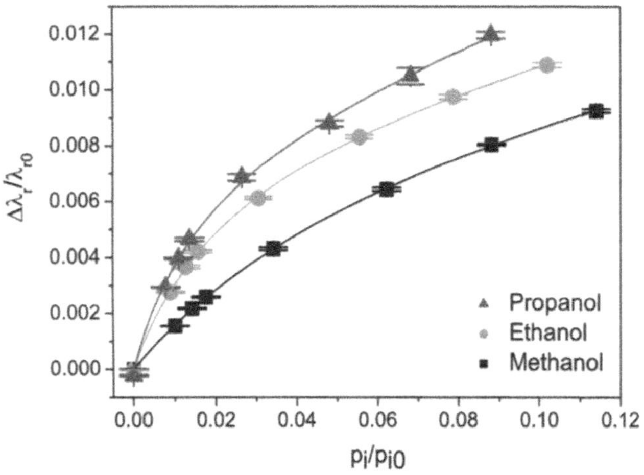

Fig. 3 Shift of the resonance wavelength of the surface plasmons of a layer of 200 nm caused by different concentrations of the three analytes. The sorption isotherms follow the Henry-Langmuir-equation

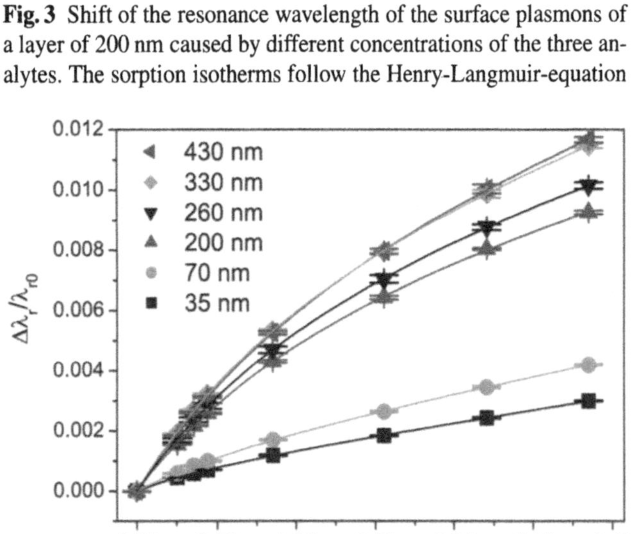

Fig. 4 Eight concentrations of methanol exposed to six different Makrolon layers between 35 and 430 nm. The shift of the SPR resonance wavelength is plotted versus the saturation pressure of the alcohol

RIfS Measurements

The interaction behaviour of five Makrolon layers of different thicknesses between 70 nm and 455 nm and the three homologous alcohols was investigated by reflectometric interference spectroscopy. In Fig. 5 the changes of the optical thickness of a 205 nm layer is plotted versus the concentrations of the three homologous alcohols. For methanol there is a relatively linear increase of the change in optical thickness. For the two greater analytes, the curves are more arcuated and the Henry-Langmuir-behaviour can be seen more clearly.

In Fig. 6 the saturation signals for the five Makrolon layers for methanol are given. The changes of optical thickness depend on the thickness of the sensitive layer, but they are not growing linearly with the layer thickness. For the 70 nm layer the changes are very small and the signal is very low which can be explained by the

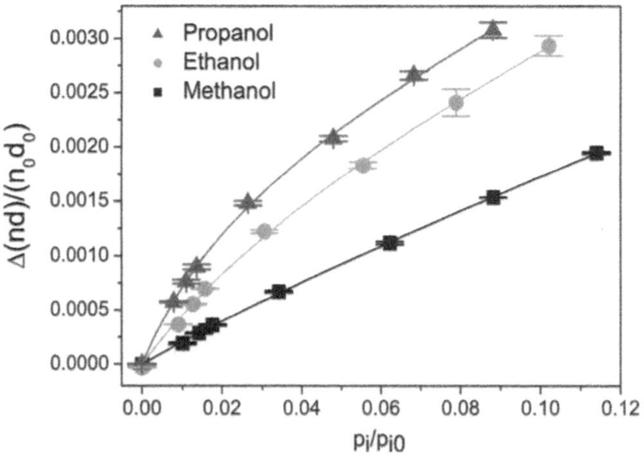

Fig. 5 The change in the optical thickness of a 205 nm thick Makrolon layer plotted against the concentrations of the three homologous alcohols, measured with RIfS

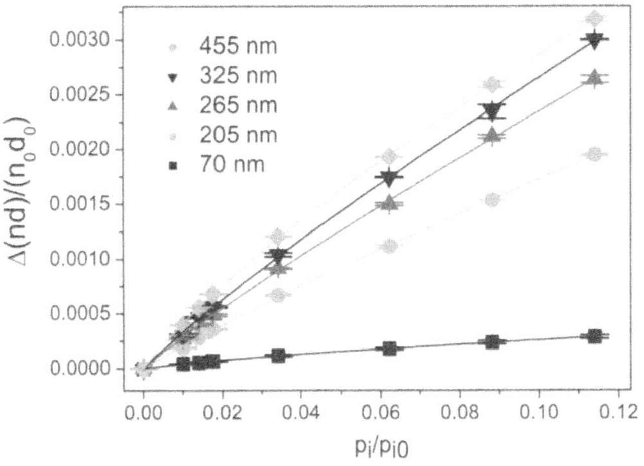

Fig. 6 RIfS-measurements of eight concentrations of methanol exposed to six different Makrolon layers between 35 and 430 nm. The change in the optical thickness is plotted versus the saturation pressure of the alcohol

poor interference pattern of the thin sensitive layer. The relative changes of the optical thickness of the other investigated layers are much higher. The Henry-Langmuir-behaviour of these layers increases with the layer thickness but is less distinctive than in the SPR-measurements. This is due to the fact that both changes in refractive index as well as in layer thickness contribute to the change in the optical thickness. The interaction kinetic of thicker layers is very slow. For practical application a compromise between signal height and kinetic aspects has to be found.

Summary and Conclusion

The 35 nm Makrolon layer shows a relatively linear signal with only a slight curvature in the SPR-measurement. RIfS-measurements with this layer were not possible because such thin layers do not provide a distinctive interference pattern. Even the 70 nm layer provides very small signals for the RIfS-measurements due to the poor interference pattern. For this method, the layer thickness should be at least 100 nm. By the use of thicker layers of e.g. 205–455 nm, good interference patterns can be obtained which result in higher signals. The disadvantages of thicker layers are the slower interaction kinetics which lead to longer sorption and regeneration times.

The layers from 35 nm to 300 nm show an increasing relative signal in the SPR-measurements. The evanescent field of the surface plasmons decays within some hundred nanometres and therefore the signal increases with increasing layer thicknesses, since the evanescent field protrudes less and less out of the layer. The saturation signals of layers thicker than 300 nm are equal and the use of these layers is not reasonable, especially since the interaction kinetics of such thick layers are very slow and the analyte molecules will not interact with the evanescent field until sorbed into the sensitive domain.

The Langmuir-part of the sorption isotherms of RIfS can be observed up to concentrations of approximately 0.02 p_i/p_{i0}. Here the change in layer-thickness has more bearing on the signal. The isotherms of the SPR-measurements are all curved up to higher concentrations.

In any case a compromise between the aspects of signal height and kinetics has to be found. The interaction kinetics of thick layers is very slow but high signal to noise-ratios are found. For thinner layers, the signals are weaker and the fitting of the Henry-Langmuir-equation to the calibration curve is less exact. But for all three methods a good agreement to the sorption model was found.

The optimal layer thickness for the sensor methods depends on the method of observation and the analyte of interest. A good signal to noise ratio has to be obtained while the regeneration should be completed rapidly.

References

1. Gauglitz G (2005) Anal Bioanal Chem 381:141
2. Brecht A, Gauglitz G, Kraus G, Nahm W (1993) Sens Actuators B 11:21
3. Azzam RMA, Bashara NM (1988) Ellipsometry and Polarized Light. North-Holland Physics Publishing, Amsterdam
4. Arwin H (2001) Sens Actuators A 92:43
5. Gauglitz G (1996) in: Baltes H, Göpel W, Hesse J (eds) Sensors Update vol I. Wiley, Weinheim, p 1
6. Liedberg B, Nylander C, Lundström I (1983) Sens Actuators B 4:299
7. Piraud C, Mwarania E, Wylangowski G, Wilkinson J, O'Dwyer K, Schiffrin DJ (1992) Anal Chem 64:651
8. Lakowicz JR (2004) Anal Biochem 324(2):153
9. Brandenburg A, Henninger R (1994) Appl Optics 33:5941
10. Brandenburg A, Hinkov V, Kunz R (1992) in: Göpel W, Hesse J, Zemel JN (eds) Sensors, vol 6. Wiley, Weinheim, p 399
11. Clerc D, Lukosz W (1994) Sens Actuators B 19:581
12. Cush R, Cronin JM, Stewart WJ, Maule CH, Molloy J, Goddard NJ (1993) Biosens Bioelectron 8:347
13. Dieterle F, Belge G, Betsch C, Gauglitz G (2002) Anal Bioanal Chem 374:858
14. Kieser B, Dieterle F, Gauglitz G (2002) Anal Chem 74:4781
15. Busche S, Dieterle F, Kieser B, Gauglitz G (2003) Sens Actuators B 89:192

16. Dieterle F, Busche S, Gauglitz G (2003) Anal Chim Acta 490:71
17. Dieterle F, Busche S, Gauglitz G (2004) Anal Bioanal Chem 380:383
18. Kasper M, Gauglitz G (2004) GIT Labor-Fachzeitschrift 48:447
19. Kasper M, Busche S, Dieterle F, Belge G, Gauglitz G (2004) Meas Sci Tech 15:540
20. Toshima N (1992) Polymers for Gas Seperation. Wiley, Weinheim
21. Michaels AS, Vieth WR, Barrie JA (1963) J Appl Phys 34:13
22. Dlubek G, Clarke A, Fretwell H, Dugdale SB, Alam MA (1996) Phys Status Solidi A 157:351
23. Maier G (1998) Angew Chem 110:3128

Progr Colloid Polym Sci (2006) 132: 23–32
DOI 10.1007/2882_031

Published online: 16 March 2006

P. Müller-Buschbaum

High-resolution Grazing Incidence Small Angle X-ray Scattering: Investigation of Micrometer Sized Structured Polymer Films

Abstract While bulk samples are routinely probed by small angle X-ray scattering (SAXS) and ultra-small angle X-ray scattering (USAXS), the grazing incidence geometry enables the necessary surface sensitivity to probe structured polymer films. Based on examples of polymer blend films of polystyrene and poly-n-butylacrylate as well as polymeric nano-structures of polydimethylsiloxane the present possibilities of high-resolution grazing incidence small angle X-ray scattering (GISAXS) are demonstrated. Three different GISAXS resolutions, denoted with relaxed, high and ultra-high resolution are presented. Thus GISAXS experimentally enables the determination of a large range of length scale varying from the molecular level (nanometer) to several micrometers. This development allows for an overlap with optical techniques combined with the very high sensitivity of the grazing incidence geometry. The impact of an enhanced surface sensitivity is pictured by probing the polymer structures at different depths. In addition, the possibility of determining buried structures is addressed. Depending on the characteristic structure of the thin film a simulation of the two-dimensional GISAXS pattern is important to determine the influence of structural key parameters. Therefore GISAXS pattern simulations are shown to picture sample induced limiting factors of the detection of large scale structures.

Keywords
Grazing incidence ·
Optical microscopy ·
Small-angle scattering ·
Surface scattering ·
Synchrotron radiation ·
Thin film geometry

P. Müller-Buschbaum (✉)
TU München, Physik Department LS E13,
James-Franck-Str. 1, 85747 Garching,
Germany
e-mail: muellerb@ph.tum.de

Introduction

The structure of polymer surfaces and thin polymeric films at the mesoscopic scale is of interest, both for application and basic research [1]. As the size of many technological devices decreases, the natural length scales of many typical polymers such as the radius of gyration, the persistence length, or the domain size in block copolymers, match the feature size and thus the materials are expected to display a new behaviour [2–5]. On the other hand, the tendency towards spontaneous structure formation may stabilize or even generate morphological features of larger size [6–9]. As a consequence, the range of interesting polymeric structures starts on the molecular level and extends up to several micrometers. With the ongoing improvement of sample preparation techniques, these structures may extend to macroscopic surface areas of several tens of square centimetres [10]. Along this line of enlarged structured surface areas, very local spatial probes such as atomic force microscopy (AFM) and transmission electron microscopy (TEM) are of less statistical significance. With AFM and TEM only a few exemplary spots of the surface can be probed. The resulting low statistics include the danger of being mislead by an artificial local structure as for example caused by defects. The use of a strongly averaging experimental technique such as scattering is therefore very advantageous [11–13].

It has been known for many decades that small angle X-ray scattering (SAXS) and ultra-small angle X-ray scattering (USAXS) used in a conventional transmission geometry are powerful tools to investigate electron density fluctuations on a mesoscopic scale. Thus, these non destructive techniques are well suited to study bulk materials; however, are limited when the structure of interest is confined in the surface region of the sample. In order to overcome this limitation and achieve a surface sensitivity, the transmission geometry simply has to be replaced by reflection geometry. By the combination of a grazing incidence geometry and the ideas of SAXS or USAXS, the powerful technique of grazing incidence small angle X-ray scattering (GISAXS) results [13–20]. Furthermore, replacing the transmission by a reflection set-up offers the advantage to decouple the scattering signal of interest from the very intense direct beam. Whereas in USAXS the regime of the smallest wavevectors is shadowed by the beamstop which is urgently required to protect the detector against the direct beam, in GISAXS due to the reflection of the scattered intensity the direct beam is not at all on the detector [21]. Thus geometrical constraints imposed by the use of beamstop are not present and an increased resolution compared to transmission experiment is accessible.

Within the last years, GISAXS experiments receive an increasing attention including the investigation of polymeric films and surface structures. In general, GISAXS does not require a special sample preparation and thus enables kinetic investigations devoted to a structural development as a function of time. Moreover, GISAXS enables the detection of buried structures in a non destructive way. As compared to local probe techniques, GISAXS yields an excellent sampling statistics, which means that it averages over macroscopic regions to provide information down to a nanometer scale. AFM and TEM are still considered as important complementary methods which can be associated to the surface sensitive scattering techniques (i.e. GISAXS) [13]. Both inform about the order of magnitude of the correlation length in the surface structure and provide necessary input for the modelling of GISAXS data [20].

One particular strength of GISAXS is the probing of small sized structures such as micro-phase separation structures in copolymer films. Such molecular structures are easily detected with a relaxed resolution [18–20]. Following the same principles as in the transmission geometry (SAXS to USAXS) larger length scales may be addressed by an increase in the sample detector distance. In accordance to the transmission this sometimes yields a change in the used abbreviation from GISAX to GIUSAXS (or *X-ray reflection ultra small-angle scattering* in pioneering work [21]). However, within this article we restrict to the term GISAXS with the extension of the used resolution, very comparable to other areas of X-ray scattering. Highly collimated X-ray beams enable the extension of GISAXS experiments into the high-resolution regime. By making use of the absence of any beamstop, the resolution can be optimised to the detector pixel size as demonstrated in this article. With the ongoing development towards high-resolution set-ups large sized surface structures such as phase separation structures in polymer blend films are improvingly determinable. In addition, this development enables the overlap with optical techniques combined with the very high sensitivity of the grazing incidence geometry. However, the high demand on collimation requires the use of high flux sources of large scale facilities such as special synchrotron radiation beamlines.

Despite the growing interest in GISAXS and its application to a broad variety of sample systems, it is still an advanced scattering technique and basic articles are very limited in number. Within this article the possibilities of high-resolution GISAXS and its application are focussed. This includes a short theoretical introduction as well as the principles of the experimental set-up required. Selected examples are included to picture this in detail. Within these examples we restrict to polymer blend films and polymeric nano-structures. Both enable to demonstrate all these ideas and blend films mark a class of samples difficult to access with X-rays due to the weak scattering and small contrast between the different blend components. For a comparison, model calculations are shown to picture sample induced limiting factors of the detection of large scale structures.

GISAXS – A Short Introduction

In the region of total external reflection, the incident and exit angle α_i and α_f are small enough to neglect the influence of the periodic crystal structure and to use a description based on one mean refractive index $n_j = 1 - \delta_j + i\beta_j$ [22–24]. Since the dispersion part

$$\delta_j(\boldsymbol{q}, \lambda) = \frac{e^2\lambda^2}{8\pi^2 m_e c^2 \varepsilon_0}\rho_j \frac{\sum_k \left[f_k^0(\boldsymbol{q}, \lambda) + f'(\lambda)\right]}{\sum_k M_k}$$

in case of X-rays is larger than zero, each medium, such as a polymer film and its solid support are optically thinner as compared to the surrounding air or vacuum. In case of polymers the absorption contribution

$$\beta_j(\lambda) = \frac{e^2\lambda^2}{8\pi^2 m_e c^2 \varepsilon_0}\rho_j \frac{\sum_k f''(\lambda)}{\sum_k M_k}$$

is usually very small. Both contributions to the refractive index depend on the speed of light c, the elementary charge e, the electron rest mass m_e, the permittivity constant ε_0, the mass density ρ_j, the wavelength λ, the atomic weight M_k and the dispersion corrections f' and f''. f_k^0 can be approximated by the number of electrons Z_k and the summation is performed over all atoms k of a monomer subunit in the case of polymers under investigation.

Usually the coordinate system is chosen that way, defining the x, y-plane by the sample surface with the x-axis oriented in the direction of the X-ray beam [13,25]. As a consequence, the z-axis is perpendicular to the sample surface (see Fig. 1). In this case

$$q_x = 2\pi(\cos\psi \cos\alpha_f - \cos\alpha_i)/\lambda$$
$$q_y = 2\pi(\sin\psi \cos\alpha_f)/\lambda$$
$$q_z = 2\pi(\sin\alpha_i + \sin\alpha_f)/\lambda$$

denote the components of the wave vector $\boldsymbol{q}$. For specular scattering, with $\alpha_i = \alpha_f$ and $\psi = 0$, the wave vector components are $q_x = q_y = 0$ and $q_z \neq 0$, sampling a depth sensitive information only. With off-specular scattering the lateral component is $q_{||} = (q_x, q_y) \neq 0$ probing the in-plane structure of the sample surface. To enable GISAXS experiments a q_y-resolution is required.

Figure 1 gives a schematic drawing of the basic set-up used in the GISAXS experiments. The two-dimensional detector is only recording the intensity reflected above the sample surface. The direct beam is not recorded with the detector to avoid detector saturation as several orders of magnitude in intensity separate the incoming intensity from the reflected one. In addition the specularly reflected peak (condition $\alpha_f = \alpha_i$) is shielded with a beam stop to

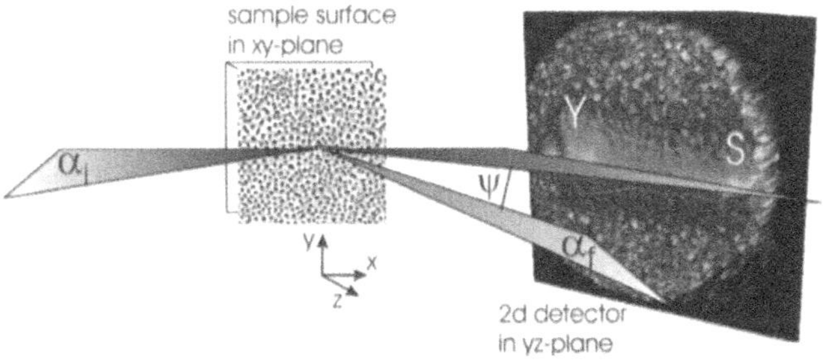

Fig. 1 Schematic top-view of the GISAXS geometry making use of a two-dimensional detector in the yz-plane to picture the scattering geometry. The sample is placed in the xy-plane and consists of a typical polymer blend structure. The incident angle is denoted α_i and the exit angle α_f. The structural information is obtained from a y-cut out of the two-dimensional intensity distribution perpendicular to the scattering plane (xz-plane) indicated by the *blue triangles*. The GISAXS signal is probed as a function of the out-of plane angle ψ. As figuratively shown the diffusely scattered intensity in the scattering plane, the detector scan, exhibits a Yoneda peak (denoted Y) and a specular peak (S) as common features

use the full available flux of the synchrotron beamline because specular and diffuse scattering are as well separated by orders of magnitude in intensity. Figure 1 pictures the anisotropy of the observed scattering pattern recorded in the yz-plane. Different main features are already determined right from the two-dimensional images. The specularly reflected peak (denoted S) in the upper part of the detector and the Yoneda peak (denoted Y) [26] in the middle of the detector are well separated.

Instead of handling the complete two-dimensional intensity distribution, the analysis of the scattering data can be reduced to distinct cuts [13]. Basically two different types of cuts are advantageous [15–17]: Horizontal slices at constant q_z and vertical slices at constant q_y (the naming of horizontal and vertical refers to the sample surface). For an improvement of the statistics the intensity is usually integrated along a slice consisting of a few detector lines only. To model the intensity distribution of a horizontal slice (in literature frequently called out-of plane scan or GISAXS cut), in the framework of the "effective surface"-approximation the differential cross-section using the distorted-wave Born approximation (DWBA) has the easy appearance [16]

$$\frac{d\sigma}{d\Omega} = \frac{C\pi^2}{\lambda^4}\left(1-n^2\right)^2 |T_i|^2 \left|T_f\right|^2 F(\boldsymbol{q}) \propto F(\boldsymbol{q})$$

where C denotes the illuminated surface area, λ the wavelength used, n the refractive index, $T_{i,f}$ the Fresnel transmission functions and $F(q)$ the diffuse scattering factor. In contrast to conventional diffuse scattering experiments the Fresnel transmission functions act only as overall scaling factors, because the incident and exit angle are fixed. As a consequence, the diffuse scattering factor $F(q)$ is directly probed and in case of N identical and centrosymmetrical objects with a random orientation it can be approximated [17]

$$F(\boldsymbol{q}) \propto NP(\boldsymbol{q})S(\boldsymbol{q})$$

to depend on the form factor $P(q)$ of the individual objects and to depend on the structure factor $S(q)$.

However, for more complex samples this easy data analysis is no longer sufficient and instead of one single horizontal slice of a certain width, the full two-dimensional intensity distribution needs to be modelled. This modelling of the GISAXS data is possible with the program IsGISAXS by R. Lazzari [27], which is freely available for non-commercial use. Within the DWBA the program accounts for the four terms involved in the scattering process (associated to different scattering events which involve or not a reflection of either the incident beam or the final beam collected on the detector). Local real space probes such as AFM or TEM can give an important information for setting up the appropriate model with respect to shape and distribution of the polymer structure [20].

Experimental Requirements for High-resolution GISAXS

Despite the strong differences in the intensity between the diffusely scattered intensity carrying the desired information about the lateral sample surface structure and the specularly reflected intensity, the use of a two-dimensional detector is very advantageous for any GISAXS experiment. Because the specular peak, broadened by the experimental resolution function, is only helpful for alignment purpose, in the GISAXS experiment it is shielded with a beamstop [13], which resembles a similarity to the common SAXS experiments in which the direct beam has to be shielded for the same reasons as well. However, whereas in SAXS the intensity carrying information about large bulk sample structures is located very close to this beamstop at small q-values, in GISAXS the Yoneda wing and the specular peak (shielded with a beamstop) can be separated due to the reflection geometry [13]. As a consequence, mainly three different types of GISAXS experiments can be distinguished from the relative position of Yoneda wing (located at the critical angle α_c of the material under investigation) and specular peak (see Fig. 2).

To picture the possibilities of GISAXS within the framework of a statistical analysis, simulations of two-dimensional scattering patterns are presented in Figs. 2, 3 and 4. The simulations were performed using the software IsGISAXS (version 2.5) by R. Lazzari [27]. In the simulations cylindrically shaped polymer objects were assumed. For cylinders of radius R, height H and volume $V = \pi R^2 H$ the form factor

$$P_{\mathrm{cy}}(\boldsymbol{q}, R, H) = 2\pi R^2 H \frac{J_1(q_{\|}R)}{q_{\|}R} \frac{\sin(q_z H/2)}{q_z H/2} \exp(-iq_z H/2)$$

contains the Bessel function of first order $J_1(x)$. A Gaussian type of distribution probability for the parameter R was selected, with the relative width $\sigma R/R$ of the distribution of radii. The arrangement of the cylindrical objects was described by an interference function (structure factor $S(q)$). A 1D-paracrystal, which is a regular one-dimensional lattice with special cumulative disorder inducing a loss of long range order, was chosen. It was parameterised by the position of first neighbour peak and the width w of the distance distribution. Moreover, the decoupling approximation, supposing that the nature of the scatterers and their position are not correlated in such a way that the partial pair correlation functions depend only on the relative positions of the scatterers and not on the class type, was modelled. With respect to the electron densities, polystyrene structures sitting on top of a silicon substrate were taken.

1) In the *first type* of GISAXS experiment, the incident angle fulfils $\alpha_i < \alpha_c$ and consequently on the detector the specular peak appears at $\alpha_f = \alpha_i < \alpha_c$ below the

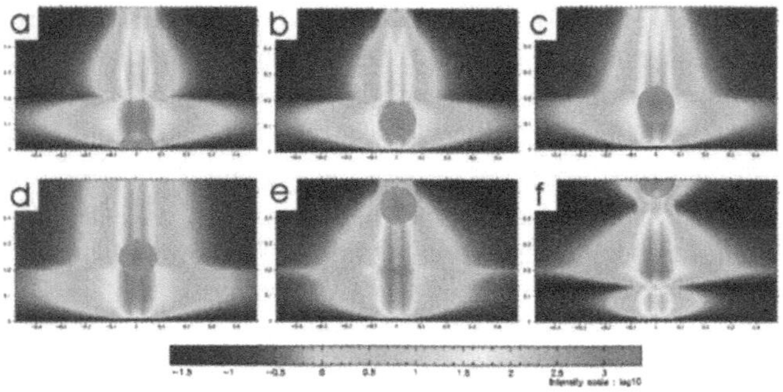

Fig. 2 Simulation of two-dimensional GISAXS scattering pattern assuming an array of cylindrically shaped nano-dots (height $H = 15$ nm, radius $R = 50$ nm, $\sigma R/R = 0.25$), distance $D = 290$ nm and $w = 100$ nm with a variation of the incident angle $\alpha_i =$ **a** 0.01°, **b** 0.12°, **c** 0.18°, **d** 0.25°, **e** 0.45° and **f** 0.55°. The simulated angular range of $0.0° < \alpha_f < 0.55°$ in *vertical direction* and $-0.48° < \psi < 0.48°$ in *horizontal direction* is well comparable with the accessible range using a set-up at the HASYLAB. The position and size of the specular peak is exemplarily shown by the *red spot*

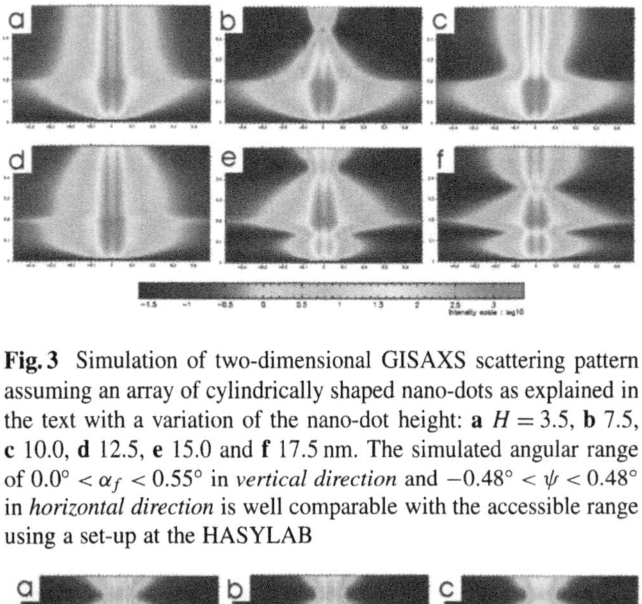

Fig. 3 Simulation of two-dimensional GISAXS scattering pattern assuming an array of cylindrically shaped nano-dots as explained in the text with a variation of the nano-dot height: **a** $H = 3.5$, **b** 7.5, **c** 10.0, **d** 12.5, **e** 15.0 and **f** 17.5 nm. The simulated angular range of $0.0° < \alpha_f < 0.55°$ in *vertical direction* and $-0.48° < \psi < 0.48°$ in *horizontal direction* is well comparable with the accessible range using a set-up at the HASYLAB

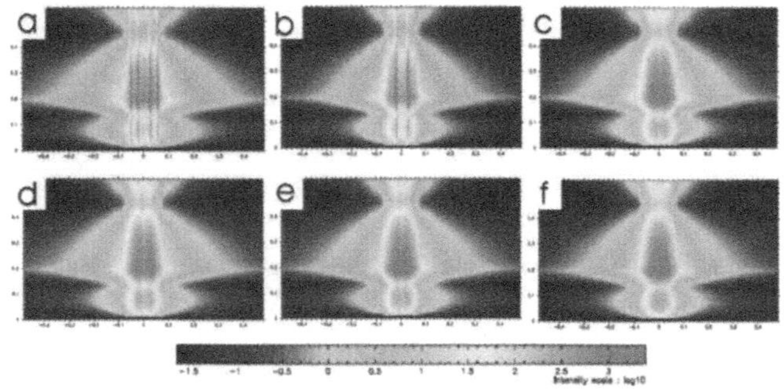

Fig. 4 Simulation of two-dimensional GISAXS scattering pattern assuming an array of cylindrically shaped nano-dots as explained in the text with a variation of the width of the nano-dot distance distribution: **a** $w = 30$, **b** 50, **c** 100, **d** 150, **e** 200 and **f** 300 nm. The simulated angular range of $0.0° < \alpha_f < 0.55°$ in *vertical direction* and $-0.48° < \psi < 0.48°$ in *horizontal direction* is well comparable with the accessible range using a set-up at the HASYLAB

Yoneda wing at $\alpha_f = \alpha_c$ (see Fig. 2a). Depending on the experimental resolution this will enable a separation of both signals along the α_f-axis (or q_z-axis) of the detector and thus a proper separation of the diffuse and the specular scattering, which is essential for the GISAXS analysis (see examples within this article). The extremely small angle of incidence has the advantage of a very high surface sensitivity in combination with a selection of the probed sample depth, such as in the standard grazing incidence diffraction (GID) experiments. But while GID experiments probe relatively large q_y-values, GISAXS experiments focus on significantly smaller q_y-values and thus are much more sensitive to the quality of the beam alignment and the sample perfection (including quality of the sample edges). Therefore this type of GISAXS set-up is only rarely applied with a standard-type beamstop or most likely used in combination with a big linear beamstop oriented along the q_z-axis [28–32]. This type of beamstop, which shields the centre part of the two-dimensional detector, prevents access to high-resolution, whereas as demonstrated within this article the standard-type beamstop (circular) allows for high-resolution experiments.

2) In the *second type* of GISAXS experiment, an incident angle of $\alpha_i = \alpha_c$ (or $\alpha_i \approx \alpha_c$) is selected (see Figs. 2b for α_c of polymer films and 2c for α_c of substrate), to strongly enhance the intensity by making use of the amplification of the Fresnel transmission functions yielding a factor of four. For this special alignment, the Yoneda wing and the specular peak coincide on the detector at $\alpha_f = \alpha_i = \alpha_c$ and GISAXS under these conditions reenacts all the disadvantages of the SAXS experiments. Such as in SAXS, the beamstop necessary to block the uninteresting part of the intensity shadows the region of small q_y-values of the diffusely scattered intensity as well. Therefore the advantage of gaining a higher resolution in GISAXS as compared to SAXS is completely wiped out and no high-resolution is possible. Consequently, data with this type of settings of the GISAXS experiment are not included in this article.

3) In the *third type*, the incident angle fulfils $\alpha_i > \alpha_c$ (see Figs. 2d–f) and the Yoneda wing is placed underneath the specular peak on the detector [13, 19–21]. By increasing the incident angle, the separation between both signals, the Yoneda wing and the specular peak, is easily increased as well (compare position of exemplarily shown specular peak in Figs. 2d–f). With the disadvantage of losing the high surface sensitivity, along this line, the realization of a high-resolution becomes accessible, too. The slit width is adapted to the required resolution in q_y and q_z-direction that way that basically the detector imposes the resolution limit. A very high resolution set-up is possible at very larger distances (larger than 10 m) between sample and detector [21]. The requirements with respect to beam purity and q_y-resolution are significantly higher as compared to low-resolution GISAXS set-ups. As a consequence, typical accumulation times in a high-resolution GISAXS or GIUSAXS set-up are significantly longer as compared to the GISAXS or GID geometry.

By comparing the different simulated two dimensional (2d) scattering patterns shown in Fig. 2, changes in the intensity distribution along the α_f–direction are visible as well. They arise from changes of the surface sensitivity. With increasing incident angle the probed sample surface volume increases in depth. In all simulations shown, the diffuse scattering contribution containing the relevant structural information is plotted. The simulations are based on the distorted wave born approximation (DWBA).

To figure out the influence of the individual parameters on the shape of the diffuse scattering in Figs. 3 and 4 two typical parameters of the nano-dots are varied independently. A fixed incident angle $\alpha_i = 0.55° > \alpha_c$ is chosen (third type of GISAXS experiment). The specular peak containing non of these information is not displayed for clarity. Of course it will be present in a real experiment in the top region of the two-dimensional intensity distribution (compare to Fig. 2f). An array of nano-dots with cylindrical shape is assumed. The diameter of the nano-dots is artificially set to $2R = 100$ nm and the distance between adjacent nano-dots is fixed to $D = 290$ nm. For both, the diameter and the distance a Gaussian type of distribution is modelled.

The simulations shown in Figs. 3 and 4 correspond to a typical scenario in polymer nano-structures. The nano-dots exhibit a rather broad distribution of the diameter $\sigma R/R = 0.25$. As a consequence basically no form factor contribution is visible in the two-dimensional scattering pattern, which is purely dominated by the structure factor contribution. This can be understood as a well defined template with rather well arranged positions but ill defined individual sizes of the nano-dots.

In Fig. 3 the height of the nano-dots is varied to picture its influence on the shape of the Yoneda wing in the q_z-direction. From Figs. 3a to 3f the height increases. With respect to the fixed diameter this increase corresponds to an increase in the aspect ratio from $H/(2R) = 0.035$ to 0.175 and thus the shape of all nano-dots still remains more pancake-like rather than column-like as it is expected for polymer systems. The change in the object height is clearly visible creating an intensity modulation parallel to the vertical axis (α_f-direction) in the region of the Yoneda wing. The structure factor contribution remains visible in the central part of the two-dimensional intensity distribution (two central, vertical, red lines). The drastic changes between Figs. 3d and 3e visualize the high sensitivity of the GISAXS technique, because a change in height only by 12% has a strong effect.

In Fig. 4 the width w of the Gaussian distribution of the nano-dots distance is varied at a fixed height of all nano-dots of 15 nm picturing the visibility of the splitting of the Yoneda wing along the q_y-direction (or *ψ-direction*). From Figs. 4a to 4f the distribution increases in its width. Figure 4a pictures the extreme case of a highly regular lattice type arrangement with an extremely small deviation in the position, with respect to the assumed mean distance of 290 nm. As a consequence the scattering pattern is dominated by the high order of the structure factor (vertical, red lines). With increasing width, the higher orders are damped out as visible in Figs. 4b and 4c. At a distribution width of 200 nm (Fig. 4e) still first order maxima are visible in the central part of the two-dimensional scattering pattern and only at a value of 300 nm (Fig. 4f) the splitting of the Yoneda wing vanishes due to the loss of any lateral order between the nano-dots.

Consequently, the presence of an intensity of the Yoneda peak, split along the ψ-direction is present over a rather broad range of nano-dot distance distributions. It should be noted, that in polymer systems, a very high degree of order as shown in Fig. 4a is only rarely present. In polymer blend systems it is not expected at all.

For the selected experimental examples discussed within the next two sections the conditions of a broad size and distance distribution of the surface structures are well supported by the real space analysis. Thus in the following experimental sections instead of the full measured 2d intensity, only one selected horizontal slice at constant q_z (of the critical angle of the polymer under investigation) will be shown [20]. From a comparison of the Figs. 2–4 it is obvious, that in addition to the full 2d intensity, the split Yoneda peak is visible in a single horizontal slice at constant q_z as well.

Blend Film Structure

Blending different polymers and thereby conserving their individual properties in the blend is an extremely attractive way to obtain new bulk materials [33, 34] or to coat surfaces by functional films [35–40]. Thin films of such blends provide micron-sized surface structures that are well adapted to a large variety of applications where the emphasis is not on nano-structures as, for example, in optics [6, 41, 42]. The resulting structures are mainly determined by the used blending ratio and solvent interaction [43–45].

The model system used consists of blend films of polystyrene (PS) and poly-n-butylacrylate (PnBA) with molecular weights $M_w = 207$ k ($M_w/M_n = 1.02$) and $M_w = 260$ k ($M_w/M_n = 3.78$), respectively. Both components were blended in toluene solution at different weight ratios of PS : PnBA $= 0.1 : 9.9$ to $9 : 1$ and prepared from solution casting onto pre-cleaned silicon (100) substrates. The chosen surface cleaning uses an acid bath of 100 ml of 80% H_2SO_4, 35 ml H_2O_2 and 15 ml deionized water for 15 minutes at 80 °C, a subsequent rinsing in deionized water and a final drying with compressed nitrogen. Varying the solution concentration as well as the amount of solution deposited on the solid support enables the installation of different polymer film thicknesses as shown in Fig. 5b. The smooth drying process in a special designed sample chamber at ambient conditions results in polymer blend films with marked surface structures due to phase separation (the polymer-polymer interaction parameter of PS and PnBA is 0.162 at 20 °C).

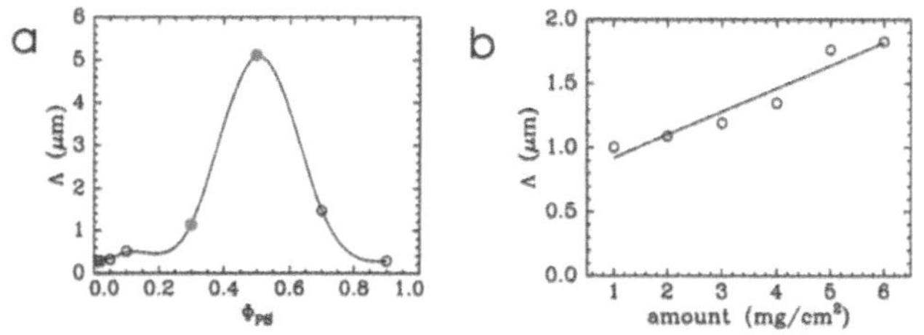

Fig. 5 **a** Most prominent in-plane length Λ as determined from optical micrographs as a function of the weight fraction ϕ_{PS} of the blend component PS for a fixed amount of 3 mg/cm^2 deposited. The *coloured points* resemble samples investigated by GISAXS. **b** Most prominent in-plane length Λ as a function of the amount of solution deposited for a fixed ratio of PS : PnBA $= 3 : 7$

Optical Analysis

The surface structures are analysed with optical microscopy first. Optical micrographs are recorded with different magnifications and the resulting grey-scale pictures are Fourier transformed to enable the detection of most prominent in-plane lengths. In contrast to Fourier transformed height data which are equivalent to the power spectral density function of the height-height correlation function, the Fourier transformed optical micrographs rely on the assumption that different colours (grey values) result from different sample heights [9]. As a consequence the obtainable intensity versus reciprocal length data are not directly comparable with scattering data such as GISAXS cuts. The intensity axis is arbitrary, whereas the reciprocal length axis is exact. Because the preparation technique does not introduce a preferential orientation of the structures, the Fourier transformed exhibit radial symmetry (see Fig. 6a) and statistics are improved by applying a radial average (see Fig. 6b) [44].

The most prominent in-plane length is calculated from the position of the maximum of the intensity via $\Lambda = 1/f$ [44]. As observed in other blend systems as well, the most prominent in-plane length shows a non linear dependence from the weight fraction ϕ_{PS} of the blend component PS. As visible in Fig. 5a Λ exhibits a maximum at a symmetric blending ratio PS : PnBA $= 5 : 5$ and strongly decreases for asymmetric blending ratios. This decrease is accompanied with strong changes in the type of surface structure. As determined by Stenert et al. [46] the majority component forms a matrix embedding the minority component as disperse objects. The size of these objects decreases with decreasing amount of the minority component giving rise to the decrease in the most prominent in-plane length as shown in Fig. 5a.

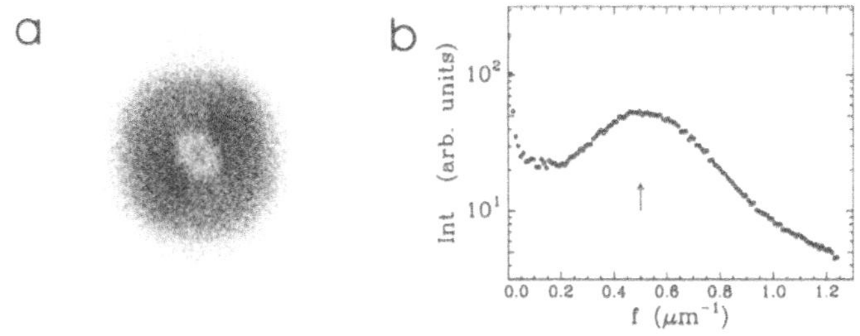

Fig. 6 **a** Example of Fourier transformed of an optical micrograph with high intensity shown in *black* and low intensity in *white*. **b** Resulting intensity versus reciprocal length f after radial averaging. The position of the most prominent in-plane length is pictured with an *arrow*

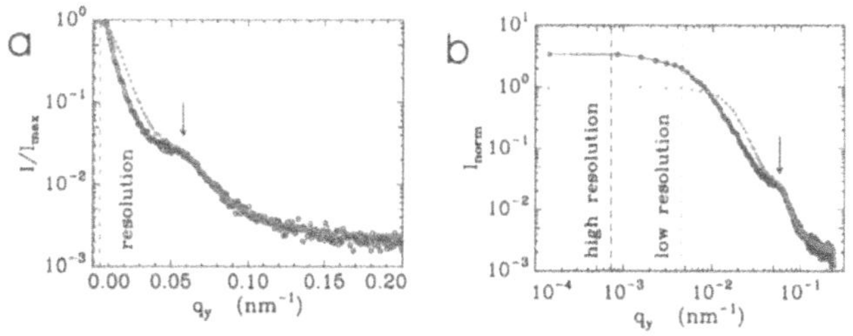

Fig. 7 Example of blend film sample measured under two different experimental resolution conditions: relaxed resolution set-up (*green dots*) and high resolution set-up (*blue open circles*). The corresponding resolution limits are marked by *vertical lines* and the most prominent in-plane length of 110 nm is marked by an *arrow*. **a** Conventional small-angle presentation of the data emphasizing on small length scales and **b** Standard GISAXS presentation of the data with the emphasize on larger lengths. The *solid lines* are fits based on a model described in the text

Comparable to samples prepared by spin-coating [10], the most prominent in-plane length increases with increasing film thickness. Figure 5b shows this behaviour, because the deposited amount is directly proportional to the resulting film thickness. Thus weight fraction and film thickness yield a good tool to prepare samples with different structural sizes which cover a range from 0.1 to 10 μm.

GISAXS

Within the following, we pick-out three different examples to demonstrate different experimental resolutions of a GISAXS experiment. The GISAXS measurements were performed at the BW4 USAXS beamline of the DORIS III storage ring at HASYLAB/DESY in Hamburg [47] using different experimental setting. The selected wavelength was 0.138 nm. The scattered intensity was recorded with a 2d detector which consists of a 512 × 512 to 2048 × 2048 pixel array. Due to the sample-detector distance of ≈ 13 m a high resolution becomes accessible. The beam divergence in and out of the plane of reflection was set by two entrance cross-slits and a guard slit in front of the sample. At one fixed angle of incident the two-dimensional intensity distribution was measured.

Medium Sized Structures. A blending ratio of PS : PnBA = 5 : 5 (weight fraction of the blend component PS $\phi_{PS} = 0.5$) and a small film thickness of 10 nm is chosen, giving rise to structure with PnBA giving the matrix and PS forming disperse objects inside this matrix [46]. In the Fig. 7 the corresponding GISAXS data (horizontal slices at constant q_z) measured with two different experimental resolutions are compared in two different presentations (left: conventional SAXS, right: standard GISAXS plot).

In the so called relaxed-resolution set-up an experimental resolution of $\Delta q_y = 4.47 \times 10^{-3}\,\text{nm}^{-1}$ was realized which enables the detection of lateral lengths up to approximately 1 μm. In Fig. 7 the data measured with this settings are plotted with green dots. As visible by the presence of the shoulder (marked by an arrow) in the decaying intensity, the most prominent in-plane length of 110 nm is easily resolved with this set-up already. It should be noted, that within a standard transmission geometry instead of the applied reflection geometry, the resolution would have been reduced by a factor of 5 and consequently, this set-up resembles the standard USAXS settings at the BW4 beamline.

An improvement of the resolution becomes possible from optimised slit settings and a replacement of the 512 × 512 pixel by a 1024 × 1024 pixel detector. By matching the resolution with the pixel size of 200 μm (yielding $\Delta q_y = 7.24 \times 10^{-4}\,\text{nm}^{-1}$) lateral lengths up to 8 μm become accessible. Data measured with these settings are shown with blue open circles in Fig. 7. In the conventional small-angle presentation (Fig. 7a), which emphasizes on the small length scales, the strong improvement in resolution is less visible as compared to the standard GISAXS presentation (Fig. 7b). Due to the enhanced resolution the shoulder in the decaying intensity is more pronounced mainly due to the fact, that unresolved large length scales which fall into the resolution limit are less smeared in q_y.

Large Sized Structures. With an increased amount of deposited material of 3 mg/cm² and a blending ratio of PS : PnBA = 3 : 7 (weight fraction of the blend component PS $\phi_{PS} = 0.3$) a larger in-plane structure results. A typical optical micrograph is shown in Fig. 8b. Small PS drops are embedded in a PnBA matrix [46]. The statistical analysis of the optical micrographs yields a most prominent in-plane length $\Lambda = 1.13$ μm (shown by the red dot in Fig. 5a).

Therefore with the so called relaxed-resolution set-up this length is inaccessible and the GISAXS data (shown with green open circles in Fig. 8a) exhibits no marked shoulder or peak in the intensity. In contrast with the high resolution set-up this dominant length is still well resolved. The data shown in Fig. 8a (red dots) exhibits a well pronounced peak at a position $q_{y,\text{peak}}$ because in addition to

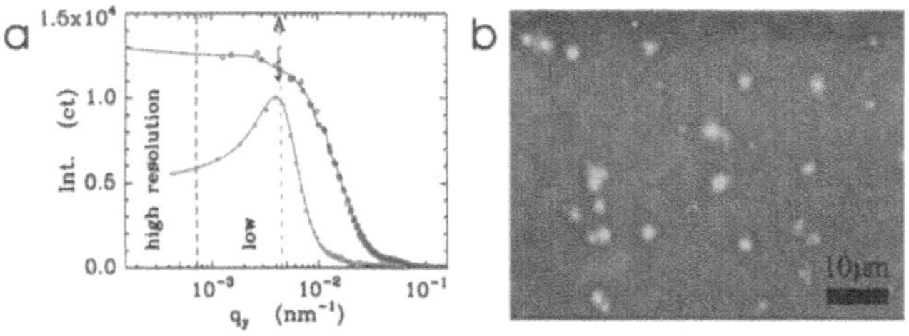

Fig. 8 **a** Example of blend film sample measured under two different experimental resolution conditions: relaxed resolution set-up (*green open circles*) and high resolution set-up (*red crosses*). The corresponding resolution limits are marked by *vertical lines* and the most prominent in-plane length of 1.3 μm is marked by an *arrow*. The *solid lines* are fits based on a model described in the text. **b** Corresponding optical micrograph

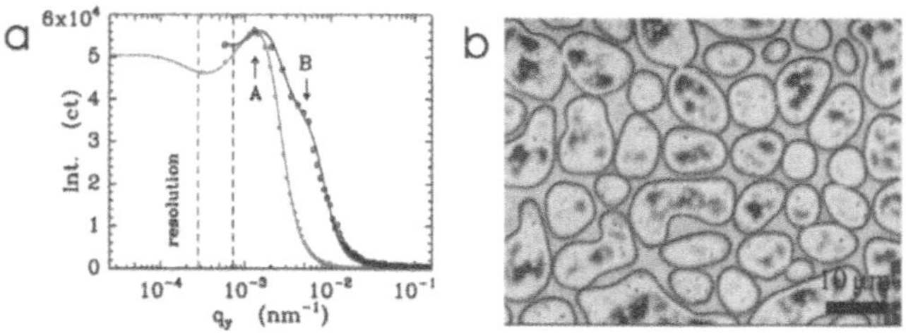

Fig. 9 **a** Example of blend film sample measured under two different experimental resolution conditions: high resolution set-up (*blue open circles*) and ultra-high resolution (*red crosses*). The corresponding resolution limits are marked by *vertical lines* and the most prominent in-plane length of 4.8 μm is marked by an *arrow* "A". The *solid lines* are fits based on a model described in the text. **b** Corresponding optical micrograph

the improved resolution, the incident angle was changed from $\alpha_i > \alpha_c$ to $\alpha_i < \alpha_c$ which enhances the surface sensitivity. As demonstrated in Experimental Requirements for High-resolution GISAXS, the width of the peak displays the width of the distribution function and it can be concluded that the blend structures enlarge in their lateral extension with increasing distance from the sample surface. From the GISAXS data a most prominent in-plane length of 1.3 μm results. Thus it agrees well with the result of the statistical analysis of the optical data, picturing the overlap to the optical regime at high resolution. The slight deviation between both values has to be attributed to the limited possibilities in the statistical analysis of the optical data. As explained above, we have only access to colours and not directly to height or chemical information.

Ultra-large Sized Structures. As shown in Fig. 5a, the increase in the weight fraction of the blend component PS to $\phi_{PS} = 0.5$ gives rise to strong increase in the resulting in-plane structure. The calculated most prominent in-plane length is 5.12 μm and shown in Fig. 5a with the blue circle. The increase coincides with a change in the appearance of the structure in the optical data (see Fig. 9b).

In Fig. 9a GISAXS data measured with the high resolution set-up described in the previous section (blue open circles) are compared with an ultra-high resolution set-up (red crosses). This further increase in resolution marks the actual technical limit with the currently available settings at the BW4 beamline. A further drastic optimisation of the slit settings and the focussing conditions together with a 2d dimensional detector with 78 μm pixel size only are the base of this improvement. The nominal resolution is $\Delta q_y = 2.75 \times 10^{-4}\,\text{nm}^{-1}$ corresponding to 22 μm maximum real space length. This maximum accessible lateral length scale of this ultra-high resolution set-up has been confirmed by the means of Monte-Carlo ray tracing simulation of the beamline BW4 [48]. Due to the limited stability of this ultra-high resolution set-up with time, an experimental validation of this exact value was not possible up to now. However, despite the detailed number, with the ultra-high resolution set-up in-plane structures larger than 10 μm become detectable.

Along the line of the arguments from the previous section, the structure shown in Fig. 9b is close to the resolution limit of the high resolution set-up and easily detected with the improved resolution. The peak in the intensity (marked with "A") corresponds to a most prominent in-plane length of 4.8 μm. It should be noted, that without fitting the data with a simple model as explained in GISAXS – A Short Introduction and Experimental Requirements for High-resolution GISAXS, just from the position of the peak in the high resolution set up (blue open circles in Fig. 9a) a smaller length would be determined, due to the influence of the resolution function. Only in case of a clear separation of the resolution peak and the structural peak, such as measured with the ultra-high resolution set-up for this particular example, the peak position directly gives access to the most prominent in-plane length via a simple Bragg-like analysis and without applying a model fit.

Similar to the previous section in addition to the improved resolution, the incident angle was changed from $\alpha_i > \alpha_c$ to $\alpha_i < \alpha_c$ which enhances the surface sensitivity. Therefore the ultra-high resolution set-up operated at $\alpha_i < \alpha_c$ is highly surface sensitive and probes only the lateral length visible in the optical micrographs, whereas the high-resolution set-up operated at $\alpha_i > \alpha_c$ probes the surface and a surface near volume. It exhibits a second structural feature marked with "B" in Fig. 9a which cannot be related to a surface structure because it is invisible under the surface sensitive conditions. The related real space length is 1.2 μm and pictures the possibility of GISAXS to probe buried structures located inside the film, which are not accessible to optical techniques, as well.

Polymeric Nano-structure

Micrometer sized surface structures in thin polymer blend films are detectable with optical techniques as well and the advantage of the more elaborate technique GISAXS is limited to an improved accuracy in the detection of the most prominent in-plane length and to the access to buried structures. In contrast, polymeric nano-structures consisting of several nanometer sized objects are not resolvable with optical techniques and commonly visualized by AFM [9, 10, 13]. However, the AFM visualization becomes difficult in case these objects are separated by large distances on the order of some micrometers. A possible mechanism for the preparation of this type of surface structures is the deposition from solution by mechanical wiping across the substrate surface.

In the present example polydimethylsiloxane (PDMS) is deposited from an iso-propanol solution onto a glass surface [49]. Due to the small amount of PDMS deposited by this technique, with optical microscopy the sample surface appears structureless. For the optical micrograph shown in Fig. 10b we selected a position which exhibits small dust particles by purpose, to demonstrate that the surface was focused. By optical microscopy the untreated and the nano-structured glass surface are indistinguishable.

As shown in Fig. 10a, GISAXS is able to distinguish between both. The data of the untreated glass are plotted with blue crosses and the data of the nano-structured glass with red circles. Clearly in the GISAXS data of the nano-structured sample a peak in the intensity is visible (marked with an arrow), which is not present in the data of the untreated sample. The corresponding real space length is 7.0 μm. The ultra-high resolution set-up was operated at $\alpha_i < \alpha_c$ because the very high resolution and the highly surface sensitivity are required both to probe this structure.

Summary

Based on examples of polymer blend films and polymeric nano-structures the present possibilities of high-resolution

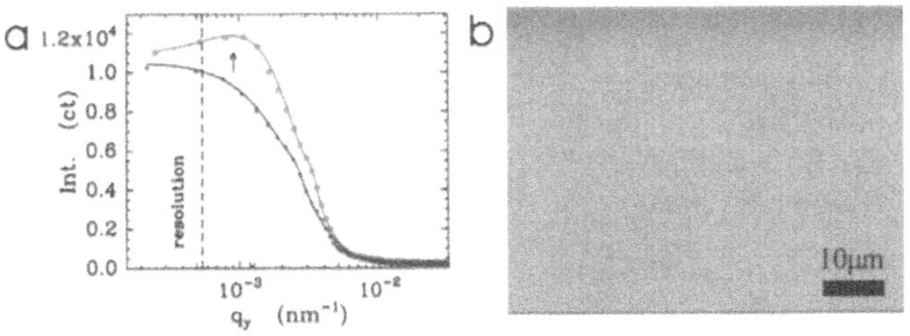

Fig. 10 **a** Example of polymeric nano-structure deposited from solution by mechanical wiping on a glass surface (*red circles*) compared to the untreated glass surface (*blue crosses*) as measured with the ultra-high resolution set-up. The corresponding resolution limit is marked by a *vertical line* and the most prominent in-plane length of 7.0 μm is marked by an *arrow*. The *solid lines* are fits based on a model described in the text. **b** Optical micrograph of the nano-structures

GISAXS are illuminated within this article. Basically three different resolutions have been presented, which for clarity have been denoted within this article by relaxed (structures up to 1 μm), high (up to 8 μm) and ultra-high resolution (more than 10 μm), although even the relaxed resolution GISAXS set-up enables the detection of structures up to one micrometer, which is significantly larger as compared to all USAXS experiments. Thus the switch from a transmission to a reflection geometry admits an increase in resolution due to the absence of the primary beam. Due to technical improvements at the synchrotron beamline BW4 (at HASYLAB, Hamburg) a further increase in resolution was achieved within recent beamtimes. Of course high resolution GISAXS experiments are not limited to this one particular beamline. Requirements are large collimation and sample-detector distances, beam focussing and a small detector pixel size.

As a result of the achieved high resolution, GISAXS experiments expand into the optical regime. This allows for the detection of blend film structures, which have been difficult to detect in the past. Moreover, high-resolution GISAXS can cover the length scale regime, which naturally was only accessible by conventional diffuse scattering within a so called rocking (or transverse) scan in the past. Contrary to the spoiling of the structural signal by the folding with the transmission functions in a rocking scan, with GISAXS object geometry, size distributions and spatial correlations are directly probed.

Nevertheless, GISAXS demands for high intensity sources such as synchrotron laboratories and the increase in resolution is mostly paid by a decrease in X-ray photon flux. Due to the time consuming installation of a ultra-high resolution set-up and its limited time stability (at least at the present conditions) this type of experiments might be still very limited in future. Compared to any AFM, TEM or optical probe, GISAXS is a very elaborate technique and therefore most likely not planned for replacing the real space investigations, rather than complementing them. Advantages as compared to the real space probes are mainly twofold: On the one hand, GISAXS yields an increase in the statistical significance of the detected most prominent in-plane length due to the significantly (in case of AFM) larger probed surface area. It averages over macroscopic regions to provide information on nanometer to micrometer scale. On the other hand, GISAXS gives the access to buried structures. As demonstrated, by tuning the incident angle, surface and film structures are separately detected.

Resulting from its high versatility GISAXS will receive a further increasing attention in the structural investigation of thin film samples.

Acknowledgement This work has benefited strongly from fruitful and enjoyable collaborations with W. Petry and M. Stamm. Support at HASYLAB at the BW4 beamline with respect to improvement of the resolution was provided by S. Cunis, M. Dommach, R. Döhrmann, R. Gehrke and S. V. Roth. Cooperation of E. Bauer, A. Götzendorfer, T. Ittner, S. Loi, C. Lorenz-Haas, E. Maurer, P. Panagiotou, K. Schlögl, T. Titz and P. Volodin in many beamtimes at HASYLAB and stimulating discussions with A. Meyer, C. Papadakis and M. Rauscher is gratefully acknowledged. We thank M. Stenert and F. Bandermann for supplying the PnBA. Financial support was provided by the DFG SPP 1164 "Nano-and Microfluidics" (Mu1487/2) and HASYLAB.

References

1. Fedynyshyn TH (ed) (2002) Advances in Resist Technology and Processing XIX Proceedings of SPIE Vol. 4690
2. Russell TP (2002) Science 297:964
3. Spatz JP, Möller M, Noeske M, Behm RJ, Pietralla M (1997) Macromolecules 30:3874
4. Hamley IW (1998) The Physics of Block Copolymers (Oxford: University Press)
5. Findenegg GH, Herminghaus S (1997) Curr Opin Coll Int Sci 2:301
6. Geoghegan M, Krausch G (2003) Prog Polym Sci 28:261
7. Binder K (1999) Advances in Polymer Science 138:1
8. Wagner AJ, Yeomans JM (1998) Phys Rev Lett 80:1429
9. Müller-Buschbaum P (2003) J Phys Condens Matter 15:R1549
10. Müller-Buschbaum P, Bauer E, Wunnicke O, Stamm M (2005) J Phys Condens Matter 17:S363
11. Stamm M, Schubert DW (1995) Annu Rev Mater Sci 25:325
12. Stamm M (2000) In: Brown W, Mortensen K (eds) Scattering in Polymeric and Colloidal Systems (Amsterdam: Gordon and Breach Science Publisher)
13. Müller-Buschbaum P (2003) Anal Bioanal Chem 376:3
14. Levine JR, Cohen JB, Chung YW (1991) Surf Sci 248:215
15. Salditt T, Metzger TH, Peisl J, Goerigk G (1995) J Phys D: Appl Phys 28:A236
16. Salditt T, Metzger TH, Peisl J, Reinker B, Moske M, Samwer K (1995) Europhys Lett 32:331
17. Naudon A, Babonneau D, Thiaudiere D, Lequien S (2000) Physica B 283:69
18. Smilgies DM, Busch P, Papadakis CM, Posselt D (2002) Synchrotron Rad News 5:35
19. Müller-Buschbaum P, Cubitt R, Petry W (2003) Langmuir 19:7778
20. Müller-Buschbaum P, Hermsdorf N, Roth SV, Wiedersich J, Cunis S, Gehrke R (2004) Spectrochimica Acta Part B: Atomic Spectroscopy 59:1789
21. Müller-Buschbaum P, Casagrande M, Gutmann J, Kuhlmann T, Stamm M, Cunis S, von Krosigk G, Lode U, Gehrke R (1998) Europhys Lett 42:517
22. Born M, Wolf E (1964) Principles of Optics (Oxford: Pergamon Press)
23. James RW (1962) The Optical Principles of the Diffraction of X-Rays (Woodbridge Connecticut: OxBow Press)
24. Tolan M (1999) X-ray scattering from soft-matter thin films (Berlin: Springer tracts in modern physics, Vol. 148)
25. Holy V, Baumbach T (1994) Phys Rev B 49:10668
26. Yoneda Y (1963) Phys Rev 131:2010
27. Lazzari R (2002) J Appl Cryst 35:406
28. Babonneau D, Videnovic IR, Garnier MG, Oelhafen P (2001) Phys Rev B 63:195401
29. Chamard V, Bastie P, Le Bolloch D, Dolino G, Elkaim E, Ferrero C, Lauriat JP, Rieutord F, Thiaudiere D (2001) Phys Rev B 64:245416
30. Gibaud A, Hazra S, Sella C, Laffez P, Desert A, Naudon A, van Tendeloo G (2001) Phys Rev B 63:193407
31. Stemmer S, Li Y, Foran B, Lysaght PS, Streiffer SK, Fuoss P, Seifert S (2003) Appl Phys Lett 83:3141
32. Papadakis CM, Busch P, Posselt D, Smilgies DM (2004) Adv In Solid State Phys 44:327
33. Ryan AJ (2002) Nature Materials 1:8
34. Tanaka H (2001) J Phys Condens Matter 13:4637
35. Walheim S, Böltau M, Mlynek J, Krausch G, Steiner U (1997) Macromolecules 30:4995
36. Walheim S, Ramstein M, Steiner U (1999) Langmuir 15:4828
37. Flebbe T, Dünweg B, Binder K (1996) J Phys II 6:667
38. Karim A, Slawecki TM, Kumar SK, Douglas JF, Satija SK, Han CC, Russell TP, Liu Y, Overney R, Sokolov J, Rafailovich MH (1998) Macromolecules 31:857
39. Krausch G (1995) Mat Sci Eng Rep 14:1
40. Wang J, Composto RJ (2003) Interf Sci 11:237
41. Walheim S, Schäffer E, Mlyneck J, Steiner U (1999) Science 283:520
42. Fasolka MJ, Mayes AM (2001) Annu Rev Mater Res 31:323
43. Gutmann JS, Müller-Buschbaum P, Stamm M (1999) Farady Disc 112:285
44. Müller-Buschbaum P, Gutmann JS, Stamm M (1999) J Macromol Sci B 38:577
45. Müller-Buschbaum P, Gutmann JS, Stamm M (2000) Macromolecules 33:4886
46. Stenert M, Doring A, Bandermann F (2004) e-Polymers 15:1
47. Gehrke R (1992) Rev Sci Instrum 63:455
48. Roth SV in preparation
49. Müller-Buschbaum P, Roth SV, Bauer E, Maurer E, Schlögl K, Gehrke R to be published

Progr Colloid Polym Sci (2006) 132: 33–40
DOI 10.1007/2882_038

Published online: 9 March 2006

Anatoli Serghei
Friedrich Kremer

Unexpected Preparative Effects on the Properties of Thin Polymer Films

Anatoli Serghei (✉) · Friedrich Kremer
Institute for experimental physics I,
University of Leipzig, Linnestr. 5,
04103 Leipzig, Germany
e-mail: serghei@rz.uni-leipzig.de

Abstract Thin polystyrene films are prepared by spin-coating and annealed above the calorimetric glass transition temperature Tg alternatively in vacuum, in a pure nitrogen atmosphere, in the presence of water vapor or in ambient air. It is experimentally shown that these preparative conditions have a pronounced impact on the stability of the samples, on their surface topography and on the molecular mobility. Our results are discussed with respect to actual experiments. For thin polystyrene films annealed in high vacuum and measured in a pure nitrogen atmosphere no T_g reductions are found down to a thickness of 20 nm.

Keywords Polymer mobility · Confined polymers · Thin films

Introduction

The effects of geometrical confinement on the molecular dynamics of polymers together with the search for a deeper understanding of the glass transition phenomenon have stimulated the interest for the investigation of thin polymers films [1–31], a topic of technological importance, as well. A multitude of linear polymers was studied (polystyrene [1–7], polymethylmetacrylate [14–16], polyvinylacetate [16], polyisoprene [17], polydimethylsiloxane [18, 19] etc.), and a variety of experimental techniques was employed: ellipsometry, Brillouin light scattering, dielectric spectroscopy, neutron scattering, X-ray reflectometry. One focus represents the investigation of the polymer mobility in thin films, a factor with manifold manifestations in phenomena like shifts of the glass transition temperature, dewetting, pattern formation, dynamics of polymers near free interfaces. Confinement effects on the glass transition of thin PS films, detected for the first time one decade ago, are still intensively discussed in the literature. While a large body of data reports a reduction of the glass transition temperature T_g in thin PS films [1–7], other studies find no indications of T_g shifts [20–27], even for film thicknesses as small as ~ 3 nm [20]. Unresolved issues of dewetting, another phenomenon related to the mobility of thin polymer films, are also controversial [28–30]. By comparing thin PS films annealed under different conditions, i.e. in vacuum, in a pure nitrogen atmosphere, in the presence of water vapor or in ambient air, we show that these until now unconsidered factors have a pronounced impact on the properties of thin polymer films. Down to a thickness of 20 nm, no reductions of the glass transition temperature T_g are detected by Broadband Dielectric Spectroscopy and capacitive scanning dilatometry in thin polystyrene films annealed in high vacuum and measured in a pure nitrogen atmosphere.

Experimental

Broadband Dielectric Spectroscopy provides a direct experimental access to the molecular relaxations of polymers over a broad frequency and temperature range. It is also especially suitable for the investigation of thin polymer films, because it does not suffer sensitivity loses with decreasing sample amount. This technique does require a special sample preparation for thin films, because of the need to have metal electrodes and good electrical contacts at both interfaces. Spin-coating, one of the most commonly employed methods for the preparation of thin polymer films, has been successfully adapted for dielectric studies [3, 14, 17, 18]. Nevertheless, the resulting sample geometry, essential for a quantitative analysis of the dielectric data, has not been yet thoroughly characterized. This important factor is going to be addressed in this paper, together with other important preparative aspects as well: reproducibility of the sample preparation, thermal stability of thin polymer films, water adsorption, thickness determination, annealing. The polystyrene used in this study for the preparation of thin films was purchased from PPS Polymer Standard Service GmbH. It has a molecular weight of 700 000 g/mol and a polydispersity of 1.04.

Preparation of Thin Polymer Films

The preparation procedure is schematically illustrated in Fig. 1. Glass plates of 1 cm × 1 cm × 1 mm, previously cleaned in an ultra-sound alkaline bath, rinsed in pure acetone and dried under nitrogen flow, were used as support substrates. After cleaning, aluminum electrodes (width 0.5 mm, height 60 nm) were deposited on the glass substrate by thermal evaporation in high vacuum (10^{-6} mbar). Subsequently, thin polymer films were spin-coated from solution at a moderate spinning-rate (2000 rotations/min), to avoid possible chain breaking effects. The film thickness was adjusted by changing the concentration of the polymer in the solution. After spin-coating, the samples were annealed at temperatures above the glass transition (i.e. T_g +50 K) in high vacuum (10^{-6} mbar) for at least 10 hours. The final step of the preparation procedure was the evaporation of the second aluminum counter-electrode on top of the polymer film.

Characterization of the Sample Geometry

Measurements by Atomic Force Microscopy (in tapping-mode, using a standard silicon tip with a radius of ~ 10 nm) of the surface topography of both the lower and the upper interface (the surface of the aluminum electrode and that of the polymer film, respectively) revealed a typical root-mean square roughness of ~ 3 nm or less on micrometric domains (parameter Sq in Fig. 2a,b). The deviations of the real area of the rough interface from its corresponding 2D projection did not exceed 2–3% (parameter Sdr in Fig. 2a,b). The sample geometry for a film thickness of ~ 20 nm is illustrated in Fig. 2c, where the interfaces represent linecuts in the above AFM images. Although the surface of the aluminum electrode might look quite rough in Fig. 2c, it is in fact rather smooth if we consider the difference of three orders of magnitude between the horizontal and the vertical scaling.

The half width of the thickness distribution, calculated using the geometrical profile measured by AFM, was typically ~ 4 nm (Fig. 3), this value representing the main contribution to the error bars of the thickness determination.

The deviations from parallel interfaces induced by the roughness were calculated in the manner illustrated in the inlet of Fig. 4: the tangent of the inclination angle is equal to the first derivative of the geometrical profile $z = z(x)$, where z is the vertical coordinate and x the lateral one. The total deviation δ from parallel surfaces, defined as the difference between the inclination angle of the upper and the lower interface ($\delta = \alpha_{\text{upper}} - \alpha_{\text{lower}}$), turned out to be smaller than one degree, which assured practically uniform electrical fields between the electrodes.

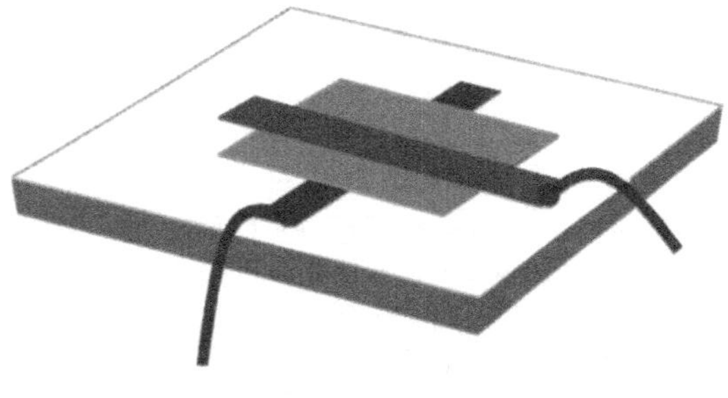

Fig. 1 Scheme illustrating the preparation of thin polymer films for dielectric studies

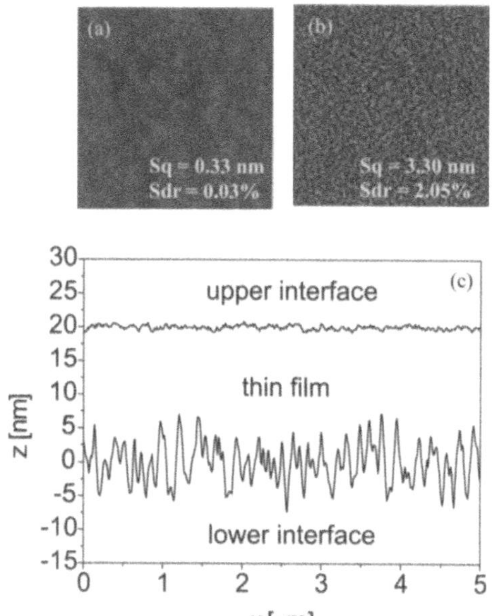

Fig. 2 **a** Surface morphology (by AFM) of a PS film prepared by spin-coating (5 μm × 5 μm); **b** surface morphology (by AFM) of an aluminum electrode prepared by thermal evaporation (5 μm × 5 μm); Sq represents the root-mean-square roughness and Sdr the ration between the real area of the rough surface and its corresponding 2D projection; **c** geometry of a thin film with an average thickness of 20 nm; the interfaces represent linecuts in the AFM images presented above

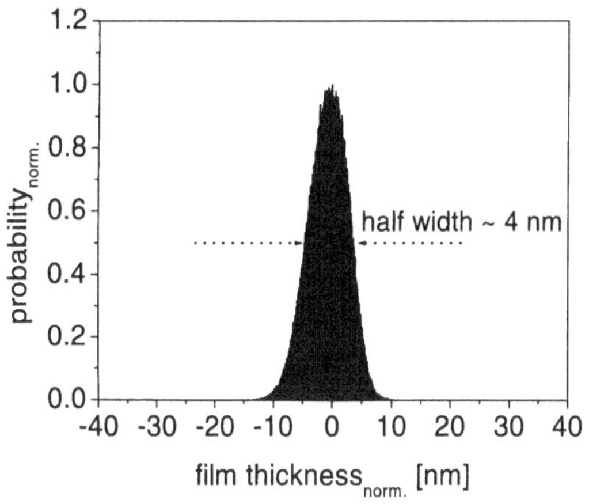

Fig. 3 Normalized thickness distribution for a typical geometrical profile, as determined by AFM

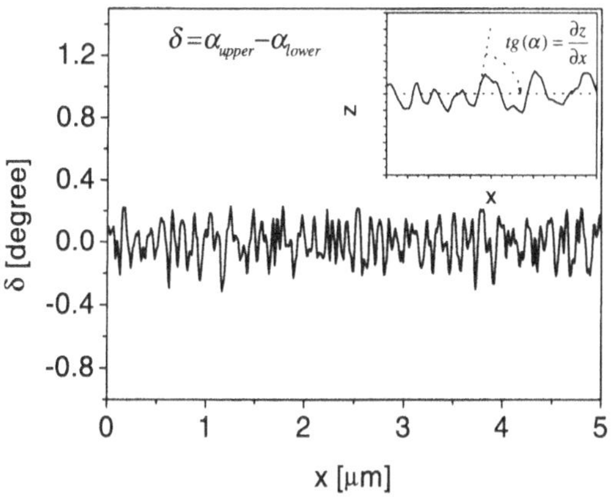

Fig. 4 Angle of deviation δ from parallel interfaces. *Inlet*: the calculation of the inclination angle α from the measured geometrical profile

Thickness Determination by Capacitive Method

Several methods are generally used for the determination of the film thickness: the capacitive method, Atomic Force Microscopy, mechanical nano-profilometry, ellipsometry, X-ray reflectometry, etc. The capacitive method, the easiest one in the case of dielectric investigations, will be discussed in detail in the following. It consists in measuring the real part of the complex sample capacity in a spectral region not affected by dielectric dispersions, i.e. at high frequencies and sufficiently low temperatures. An accurate determination of both the area of the electrodes and the value of ε'_∞ is required, where ε'_∞ is the real part of the dielectric permitivity in the limit of high frequencies. This method relies on the assumption that ε'_∞ is not thickness dependent, which was proven in the literature for some linear polymers [30], but which might be not valid for polymers having special architectures (i.e. hyper-branched polymers, grafted polymers, star-branched polymers, etc.).

The influence of the interfacial roughness on the thickness determination has to be taken into consideration as

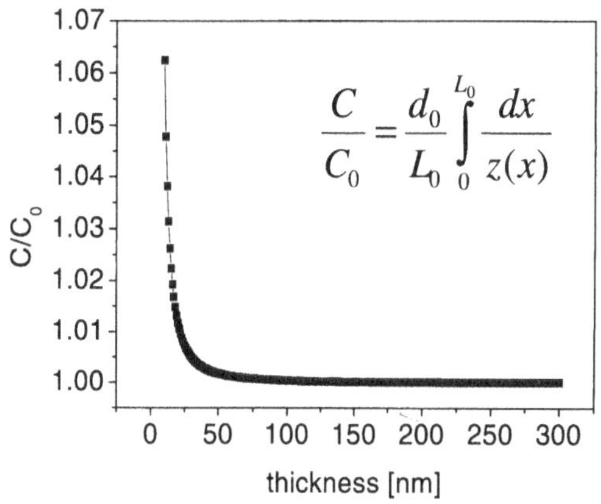

Fig. 5 Calculation of ratio between the sample capacity corresponding to rough interfaces (using a typical geometrical profile, as illustrated in Fig. 2c) and that corresponding to an ideally flat geometry as a function of film thickness. $z = z(x)$ is the vertical surface profile as determined by AFM, and d_0 and L_0 represent the film thickness and the lateral dimension of an ideally flat thin film

well. The ratio between the sample capacity corresponding to rough interfaces (using a typical geometrical profile, as illustrated in Fig. 2c) and that corresponding to an ideally flat geometry can be calculated using the following formula:

$$\frac{C}{C_0} = \frac{d_0}{L_0}\int_0^{L_0}\frac{\mathrm{d}x}{z(x)} \tag{1}$$

where $z = z(x)$ is the vertical surface profile as determined by AFM, and d_0 and L_0 represent the film thickness and the lateral dimension of an ideally flat thin film. Deviations of up to 6% from the value corresponding to an (ideal) flat geometry, for film thicknesses decreasing down to 10 nm (Fig. 5), are calculated using the above formula. This contribution has to be added to the error bars as well, especially for thicknesses smaller than 50 nm.

Reproducibility of the Sample Preparation and Thermal Stability of Thin Polymer Films

The reproducibility of the sample preparation has been carefully checked. An example is given in Fig. 6: measuring two different samples (thin films of hyper-branched polyesters) prepared under identical experimental conditions revealed coincident dielectric spectra.

In terms of their dielectric response, thin polymer films turned out to be stable thermally in a pure nitrogen atmosphere, even when kept for a long time (i.e. 24 hours) at temperatures well-above the glass transition. An example is given in Fig. 7 for a thin PS film of 89 nm: after 24 hours at 180 °C under a flow of pure nitrogen the sample was measured again and no changes in the dielectric response were detected.

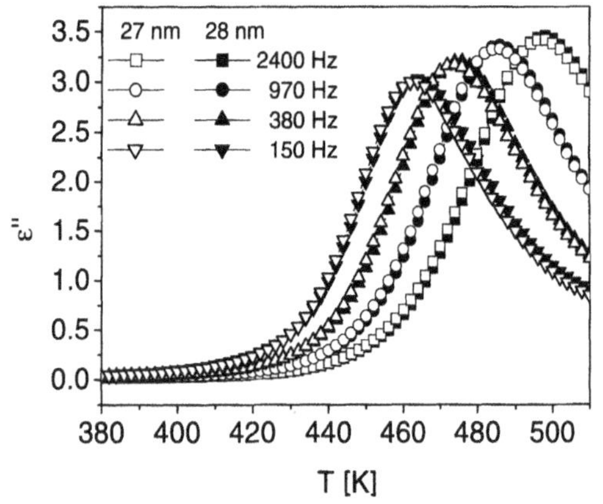

Fig. 6 Dielectric loss vs. temperature at different frequencies, as indicated, for two thin films of hyper-branched polyesters of 27 nm and 28 nm

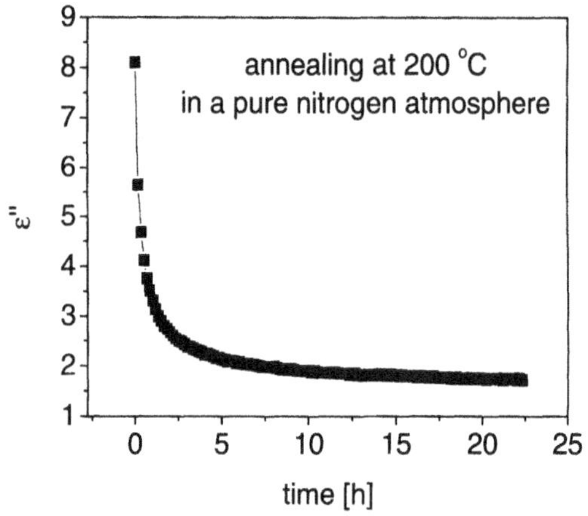

Fig. 8 Time dependence of the dielectric loss at 200 °C and 0.3 Hz in a pure nitrogen atmosphere for a thin film of hyper-branched polyesters of 17 nm

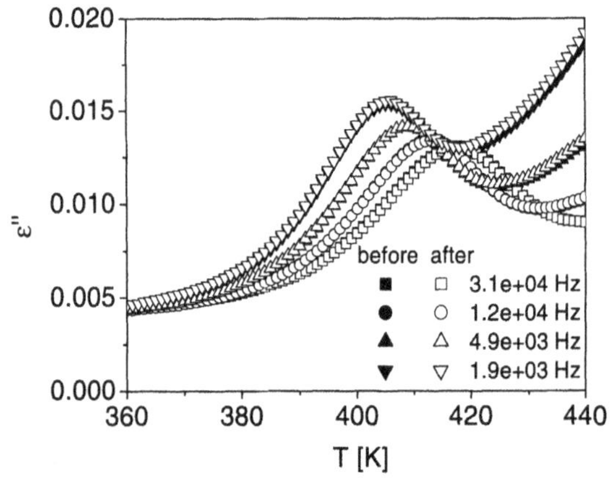

Fig. 7 Dielectric loss vs. temperature at different frequencies, as indicated, for a thin polystyrene film of 89 nm, before and after 24 hours of annealing at 180 °C in a pure nitrogen atmosphere

Annealing

The first annealing of each sample was carried out immediately after spin-coating in order to eliminate the solvent and to equilibrate the thin polymer film. After evaporating the aluminum counter-electrode and prior to dielectric measurements, a second annealing step was performed in a pure nitrogen atmosphere at temperatures above T_g for several hours in the cryostat of the dielectric spectrometer. An estimation of the required time was determined by measuring the dielectric response of the sample during annealing. An example is given in Fig. 8 for a thin film of hyper-branched polyesters (17 nm thickness). The dielectric loss decreased steeply during the first few hours of annealing at 200 °C in a pure nitrogen atmosphere and reached a constant value after ∼ 10 hours. At this point, the measurement sequence could be started, without exposing the sample again to ambient air.

Ambient Effects on the Properties of Thin Polymer Films: Water Adsorption and Oxygen-Induced Chain Scissions

When adsorbed (from ambient air), water molecules might act as plasticizers and alter the dynamics of polymers. Moreover, water has a strong dipole moment and, consequently, dielectric active relaxation processes, which could partially occlude significant parts of the dielectric spectra of interest. Special attention to this effect has to be paid when the dynamics of thin polymer films is investigated, for example in relation to phenomena like the glass transition, dewetting, pattern formation, surface mobility etc.

Even if some polymers might be hydrophobic in the bulk, concerns about possible water adsorption effects in thin films are justified, because of the preponderant role of the interface in confinement. For example, negligible water adsorption is reported in the handbook of polymers for polystyrene in the bulk: 0.05% at 23 °C and 50% relative humidity. In spite of this, strong water adsorption-desorption effects were observed in thin polystyrene films: the dielectric loss decreased with more than one decade when a thin film was measured in a dry nitrogen atmosphere and after 2 hours of annealing at 135 °C (Fig. 9, example for a film thickness of 223 nm). A pronounced decrease of ε' (∼ 20% at 1 Hz) was detected as well.

Conversely, on a time scale of minutes, both the real part of the complex sample capacity (and correspondingly ε') and the dielectric loss increased when a polystyrene thin film (20 nm) was replaced from a dry nitrogen atmosphere and exposed to ambient water vapor at room temperature (Fig. 10).

Even more drastic effects were observed when such an adsorption experiment was performed at temperatures above the glass transition (Fig. 11). A thin PS film of 63 nm was kept at 180 °C in vacuum (∼ 4 mbar). The real part of the complex sample capacity, a parameter sensitive

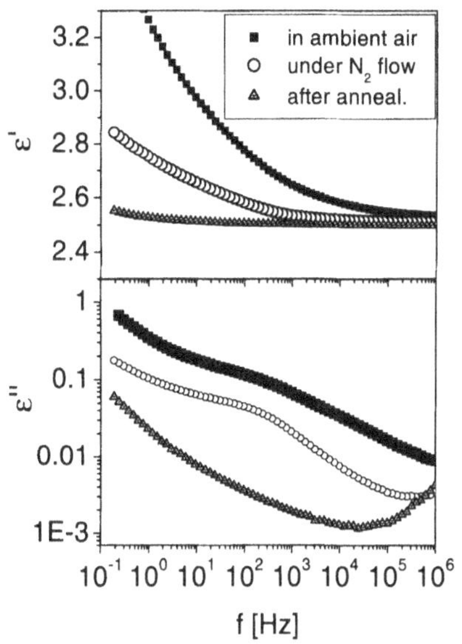

Fig. 9 ε' and ε'' vs. frequency at 25 °C for a thin PS film of 223 nm, in ambient air, under pure nitrogen flow and after annealing 2 hours at 135 °C

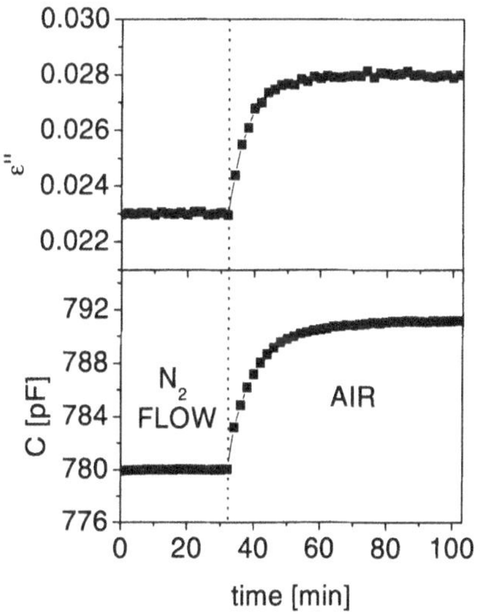

Fig. 10 Time dependence of the dielectric loss and of the real part of the complex capacity at 24 °C and 0.1 MHz for a thin PS film of 20 nm in a pure nitrogen atmosphere and in ambient air

to any changes in the sample geometry and ε', remained constant, proving sample stability under these conditions. After 2–3 hours, Millipore water was injecting in the vacuum chamber through a membrane, keeping the vacuum pump working. Before injection, the water was previously degassed for at least one hour in vacuum. In the presence of water vapor, at the same annealing temperature, a sudden increase of the real part of the complex sample capacity was observed first, due to the water adsorption. After this, the capacity started to increase continuously in time, revealing a gradual change of the sample geometry.

This phenomenon is related to the formation of a specific undulated pattern in thin PS films and it was investigated in detail in [32]. In essence, it emphasizes that the mobility of PS is enhanced in the presence of water vapor, via plasticizing effects or, possibly as well, via an oxidation which might occur due to water dissociation at high temperatures. The enhancement of the polymer mobility was proven by measuring the alpha relaxation process of thin PS films by means of Broadband Dielectric Spectroscopy. The dielectric measurements were performed using a High Resolution Alpha Analyzer (Novocontrol GmbH) in a frequency range covering eight orders of magnitude (10^{-1}–10^{7} Hz) and in a wide temperature interval (320–440 K). The temperature stability was better than 0.1 K. While at 140 °C in a pure nitrogen atmosphere no changes of the dynamics were observed in time for a thin PS film of 84 nm (Fig. 12a), the dynamic glass transition became faster in ambient air (Fig. 12b).

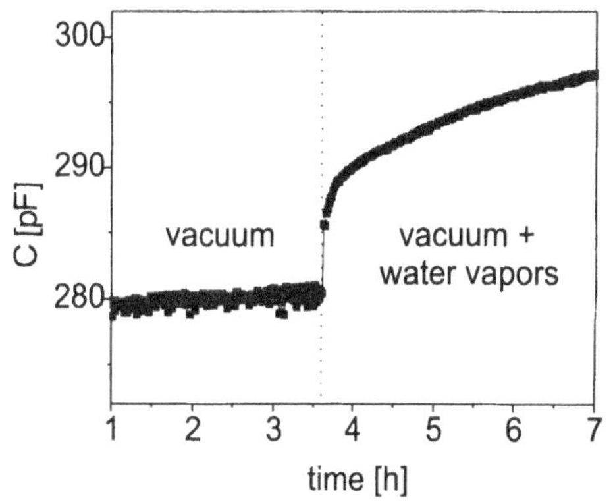

Fig. 11 Time dependence of the real part of the complex capacity at 180 °C and 0.1 MHz for a thin PS film of 63 nm in vacuum (4 mbar) and in the presence of water vapors

Even more pronounced effects were observed when the samples were annealed at higher temperatures, i.e. 180 °C. The sample geometry of a thin PS film of 239 nm remained unmodified during 4 hours of annealing at 180 °C in a dry nitrogen atmosphere (Fig. 13a), while after an identical heat treatment in ambient air large changes were detected (Fig. 13b). An analogous behavior was observed by measuring the real part of the complex sample capacity of a thin PS film (inlet Fig. 13c, example for a film thickness of 63 nm): it remained constant in time at 180 °C in a pure nitrogen atmosphere, while a drastic increase was detected in ambient air. Similarly, after 24 hours at 180 °C in a pure nitrogen atmosphere no changes of the alpha relaxation process were detected, while in air the dynamic glass transition became one order of magnitude faster (Fig. 13c).

The origin of this effect was revealed by IR-spectroscopic measurements on PS films (Fig. 14): after keeping a sample (3 µm thick) 24 hours at 200 °C in air, a new band appeared in the IR-spectrum at 1680 cm^{-1}, which was identified as corresponding to a carbonyl C = O band. This was a clear indication of chain scissions induced in

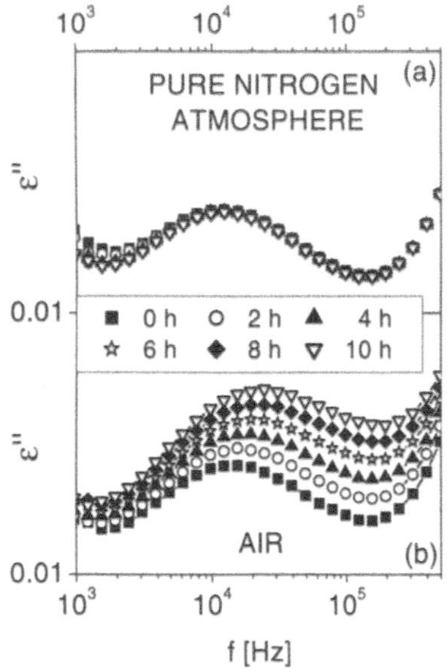

Fig. 12 Dielectric loss vs. frequency at 140 °C for a thin PS film of 84 nm in a pure nitrogen atmosphere (a) and in air (b)

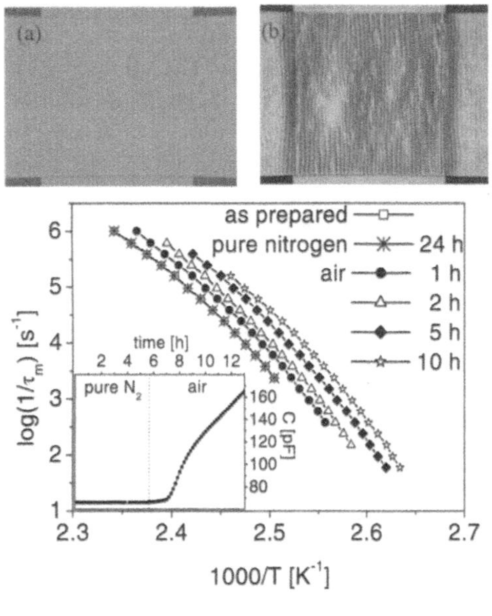

Fig. 13 a Optical image (*top view*) of the sample geometry for a thin PS film of 239 nm after 4 hours at 180 °C in a pure nitrogen atmosphere (~ 1 mm × 1 mm); b the same sample after 4 hours at 180 °C in air (~ 1 mm × 1 mm); c the relaxation time of the dynamic glass transition vs. inverse temperature for a thin PS film of 63 nm after different annealing times at 180 °C in a pure nitrogen atmosphere and in air. *Inlet*: time dependence of the sample capacity at 180 °C and 0.1 MHz in a pure nitrogen atmosphere and in air

the presence of oxygen, causing a decrease of the average molecular weight, which, according to the Flory–Hugens relation, has as a consequence a decrease of the glass transition temperature T_g.

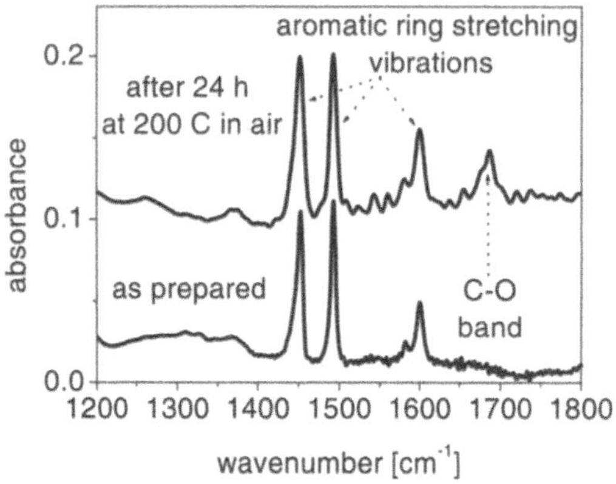

Fig. 14 IR-spectra for a PS film of 3 μm as prepared and after 24 hours at 200 °C in ambient air

Dynamic Glass Transition in Thin Polystyrene Films

For thin polystyrene films annealed for 12 hours at 150 °C in high vacuum (10^{-6} mbar) and measured in a pure nitrogen atmosphere the dynamic glass transition was characterized using two experimental techniques: capacitive scanning dilatometry and Broadband Dielectric Spectroscopy. Data from the first method are presented in Fig. 15a, showing the real part of the complex capacity at 1 MHz as a function of temperature for a thin PS film of 33 nm.

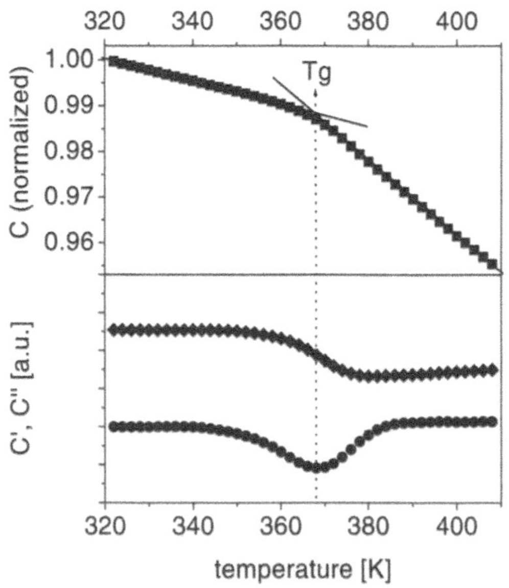

Fig. 15 a Real part of the complex capacity measured at 1 MHz for a thin PS film of 33 nm as a function of temperature; b and c the corresponding first and second numeical derivative, respectively

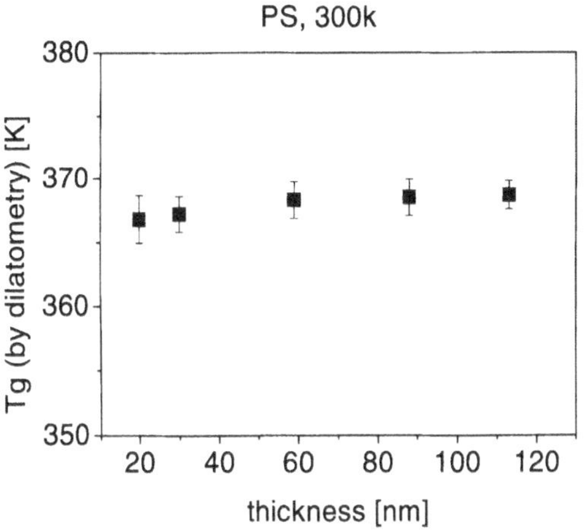

Fig. 16 Glass transition temperature of thin PS films measured by capacitive scanning dilatometry as a function of film thickness

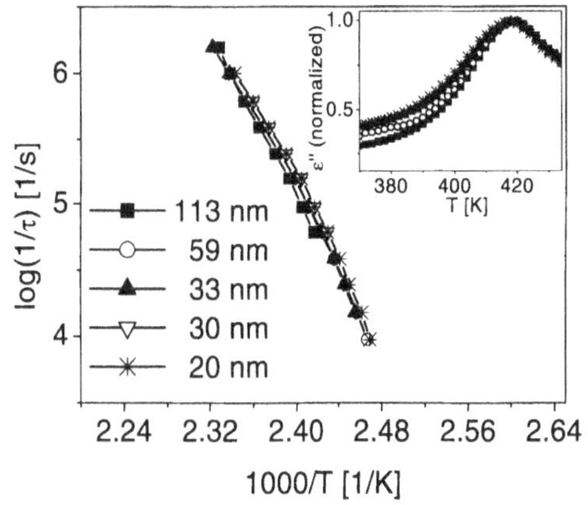

Fig. 17 Relaxation rate of the dynamic glass transition vs. inverse temperature for different film thicknesses, as indicated. *Inlet*: dielectric loss vs. temperature at 31 kHz showing the dynamic glass transition of thin PS films for different film thicknesses, as indicated

The kink observed around 367 K corresponds to a change of the thermal expansion coefficient from a glassy to a liquid-like state and, by that, marks the position of the glass transition temperature. Usually, the T_g is calculated as a intersection point between two linear dependencies. Nevertheless, a more convenient method is the calculation of the first and second numerical derivatives of the experimental data (Fig. 15b,c). In this case, the T_g is defined as the minimum position in the second numerical derivative plot (Fig. 15c). Down to a thickness of 20 nm, no shifts of T_g as determined by capacitive scanning dilatometry were found (Fig. 16).

Similar results were obtained by Broadband Dielectric Spectroscopy (Fig. 17): no shifts in the relaxation time of the dynamic glass transition were detected, even for PS films as thin as 20 nm.

Conclusions

In conclusion, we show that thin polymer films are extremely sensitive to ambient conditions, i.e. to the presence of oxygen and water vapor. Without special experimental precautions, plasticizers effects and chain scission in the presence of oxygen affect the measured parameters. We have shown that these preparative conditions have a pronounced impact on the molecular mobility, on the glass transition temperature T_g, and on the surface topography, after annealing at elevated temperatures. Using two experimental techniques, no shifts of the glass transition in thin polystyrene films were detected down to a thickness of 20 nm for samples annealed in high vacuum and measured in a pure nitrogen atmosphere.

Acknowledgement The authors thank B. Voit, K.-J. Eichhorn and Y. Mikhailova for providing the hyperbranched polyesters.

References

1. Keddie JL, Jones RAL, Corry RA (1994) Europhys Lett 27:59
2. Forrest JA, Dalnoki-Veress K, Dutcher JR (1997) Phys Rev E 56:5705
3. Fukao K, Miyamoto Y (2000) Phys Rev E 61:1743
4. Kawana S, Jones RAL (2001) Phys Rev E 63:021501
5. Forrest JA (2002) Eur Phys J E 8:261
6. Ellison CJ, Torkelson JM (2003) Nature Mat 2:695
7. Fryer DS, Nealey PF, Pablo JJ (2000) Macromolecules 33:6439
8. Herminghaus S, Jacobs K, Seemann K (2001) Eur Phys J E 5:531
9. de Gennes PG (2000) Eur Phys J E 2:201
10. Torres JA, Nealey PF, Pablo JJ (2000) Phys Rev Lett 85:3221
11. Long D, Lequeux F (2001) Eur Phys J E 4:371
12. Anderson PW (1995) Science 267:1615
13. Binder K, Baschnagel J, Kob W, Paul W (1999) Physics World 12:54
14. Hartmann L, Gorbatschow W, Hauwede J, Kremer F (2002) Eur Phys J E 8:145
15. Grohens Y, Hamon L, Reiter G, Soldera A, Holl Y (2002) Eur Phys J E 8:217
16. Fukao K, Uno S, Miyamoto Y, Hoshino A, Miyaji H (2001) Phys Rev E 64:051807
17. Serghei A, Kremer F (2003) Phys Rev Lett 91:165702
18. Hartmann L, Kremer F, Pouret P, Léger L (2003) J Chem Phys 118:6052
19. Kremer F, Hartmann L, Serghei A, Pouret P, Léger L (2003) Eur Phys J E 12:139
20. Efremov MYu, Olson EA, Zhang M, Zhang Z, Allen LH (2003) Phys Rev Lett 91:085703
21. Efremov MYu, Olson EA, Zhang M, Zhang Z, Allen LH (2004) 37:4607

22. Kim H, Rühm A, Lurio LB, Basu JK, Lal J, Lumma D, Mochrie SGJ, Sinha SK (2003) Phys Rev Lett 90:068302
23. Weber R, Zimmermann K-M, Tolan M, Stettner J, Press W, Seeck OH, Erichsen J, Zaporojtchenko V, Strunskus T, Faupelet F (2001) Phys Rev E 64:061508
24. Ge S, Pu Y, Zhang W, Rafailovich M, Sokolov J, Buenviaje C, Buckmaster R, Overney (2000) Phys Rev Lett 85:2340
25. Liu Y, Russel TP, Samant MG, Stohr J, Brown HR, Cossy-Favre A, Diaz J (1997) Macromolecules 30:7768
26. Xie L, deMaggio GB, Frieze WE, DeVries J, Gidley DW, Hristov HA, Yee AF (1995) Phys Rev Lett 74:4947
27. Lupascu V, Huth H, Schick C, Wübbenhorst M (2005) Thermodinamica Acta 432:222
28. Thiele U (2003) Eur Phys J E 12:409
29. Tsui OKC, Wang YG, Zhao H, Du B (2003) Eur Phys J E 12:417
30. Green PF, Ganesan V (2003) Eur Phys J E 12:449
31. Wallace WE, Beck Tan NC, Wu WL, Satija S (1998) J Chem Phys 108:3798
32. Serghei A, Huth H, Schellenberger M, Schick C, Kremer F (2005) Phys Rev E 71:061801

Progr Colloid Polym Sci (2006) 132: 41–47
DOI 10.1007/2882_030

Published online: 16 March 2006

Michael Osterhold

Characterizing Physical Properties of Coating Surfaces – Mar Resistance and Surface Structure

Michael Osterhold (✉)
DuPont Performance Coatings
GmbH & Co. KG, Christbusch 25,
D-42285 Wuppertal, Germany
e-mail:
michael.osterhold@deu.dupont.com

Abstract Mar resistance and surface structure (appearance) are two of the most important physical properties to characterize the mechanical and optical quality of a coating system. In this paper the relevant measuring techniques are discussed using typical examples reflecting the current status of application in the automotive and coatings industry. Especially the techniques to determine the mar resistance were recently improved (single scratch tests) giving additional results in characterizing the mechanical behavior of coatings. In addition, surface structure measurements on coatings for plastics are shown. These investigations are of particular interest due to the fact that plastics are more and more used as exterior car parts.

Keywords Coatings · Scratch · Mar Resistance · Surface Structure · Appearance

Introduction

In addition to good flow (resulting in surface structure) and high gloss, resistance to mechanical damage – by stone chippings and scratching – is particularly important in order to obtain a high-quality appearance for clearcoats. The brushes and dirt in a car wash, for example, produce scratches measuring only a few micrometers in width and up to several hundred nanometers in depth.

With this background several measuring methods are currently being discussed in the automotive and paint industry with the objective to obtain a clear characterization of the scratch resistance of clearcoats. Especially procedures that create a single scratch have recently been developed (micro or nano scratch method) [1–8]. These methods are different from more practically oriented procedures that are based on relatively simple methods to try to test or even come close to reality (e.g. car wash brush method).

One of the best suited methods to determine the surface structure (smoothness of the surface) is based on mechanical profilometry. In the past, it was used mainly to examine the influence of sheet substrate structures on the paint surface [9–19]. In order to simulate the visual impression obtained from optical inspection of surface structures, the German company Byk-Gardner developed the so-called "wave-scan DOI". Here, the measuring principle is based on the modulation of the light of a small laser diode reflected by the surface structures of the sample. In this paper some methods for characterizing surface structure were applied to coatings for plastics.

Characterization of the Mar Resistance

Micro-Scratch Experiments

In micro-scratch experiments single scratches of characteristic and realistic phenotype can be generated. The indentation depth depends on the applied force and the indentation body (indenter). Indentation depth is usually in the range of 1 μm and smaller. A diamond indenter (radius at the peak 1–2 μm) is pushed onto the sample surface applying a defined force to generate the scratch while the sample is moved in a linear direction at constant velocity underneath the indenter (see Fig. 1). The applied normal force

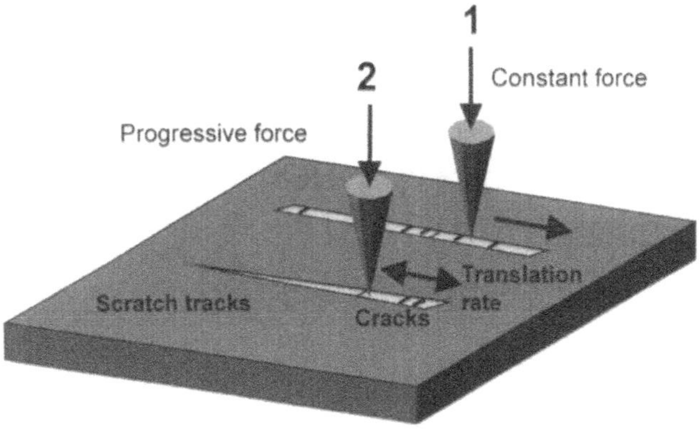

Fig. 1 The principle of single-scratch tests

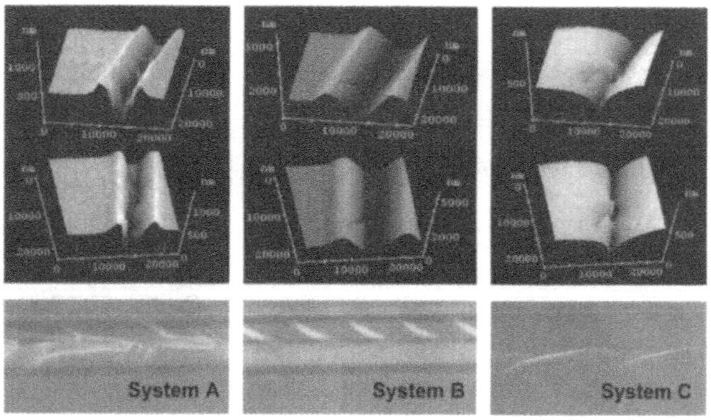

Fig. 2 AFM (*top*) and microscopical (light microscope) images (*bottom*) of cracks in different coating systems (A acrylate/melamine, B acrylate/isocyanate, C elast. melamine system)

can be constant (1) or progressive (2) during the scratching procedure.

Basically it is possible to measure the tangential forces and the indentation depth during the scratching procedure. Deformations or damages can be observed with a microscope or additionally with a video camera. The profile of the generated scratches can be measured using AFM technology. In combination with the parameters measured during the scratching procedure it is possible to calculate physical parameters which allow conclusions about the elastic and plastic deformation behavior and the fracture behavior.

Tests with Progressive Load. In this context a method was developed in the DuPont Marshall Lab. (Philadelphia) that generates and evaluates a single scratch on a surface (nanoscratch method). This method is used for tests along with development work for clearcoats and it is constantly being improved.

A commercial instrument by the Swiss company CSEM (now CSM) was tested in the course of a project at the Research Institute for Pigments and Paints (Forschungsinstitut für Pigmente und Lacke e.V. (FPL)). This instrument is based on the method developed at the DuPont Marshall Lab. Objective of the project – where manufacturers of paints, raw materials and automobiles worked together – was the evaluation of the CSEM nano-scratchtester regarding reproducibility, accuracy and applicability under the aspect of a realistic determination of scratch resistance.

The key point in this method is the determination of the critical load where first irreversible cracks or fractures are generated and which therefore indicates the transition from plastic deformation to significant/lasting damages. For that, the normal load is constantly increased and indentation depth and tangential load are simultaneously recorded. The transition from plastic deformation to the fracture range is indicated e.g. by unsteadiness or fluctuations in the detected load flow and the indentation depth. This transition range can also be evaluated by additional optical or AFM analysis. As a typical dimension besides the critical load it is common to determine the residual indentation depth after scratching – typically at a normal load of 5 mN – in the range of plastic deformation.

A scratch experiment consits of three main steps (experimental procedure):

- Prescan with scanning the surface at minimum load to determine the level zero.
- Increasing the load up to maximum load to detect the critical load = fracture threshold.
- Postscan at minimum load to determine the residual depth.

Depending on the clearcoat system different values for critical load and scratch profile can be observed – especially in the transition range. Figure 2 shows the transition ranges of three so-called public domain samples (measurements by FPL).

The top pictures show AFM pictures of the scratch ridge itself where cracks are clearly visible. The bottom

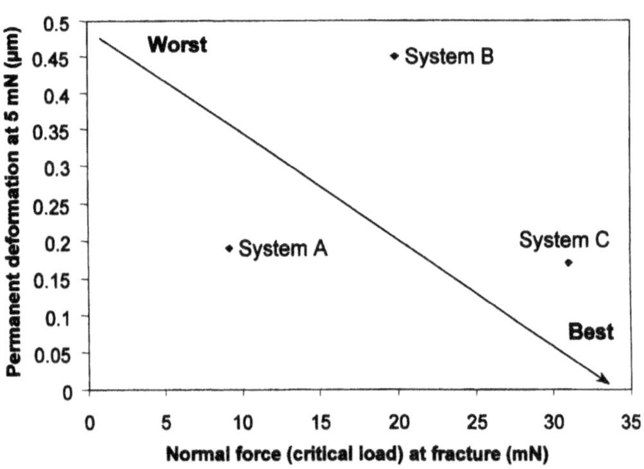

Fig. 3 Permanent deformation (scratch depth) as a function of critical load

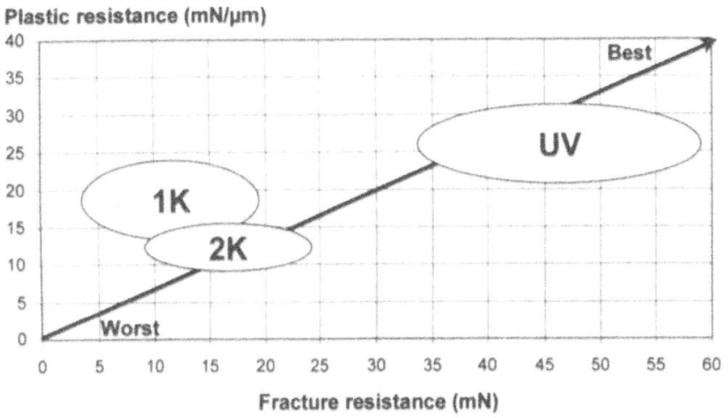

Fig. 4 Plastic resistance and fracture resistance (= critical load) for typical coating systems

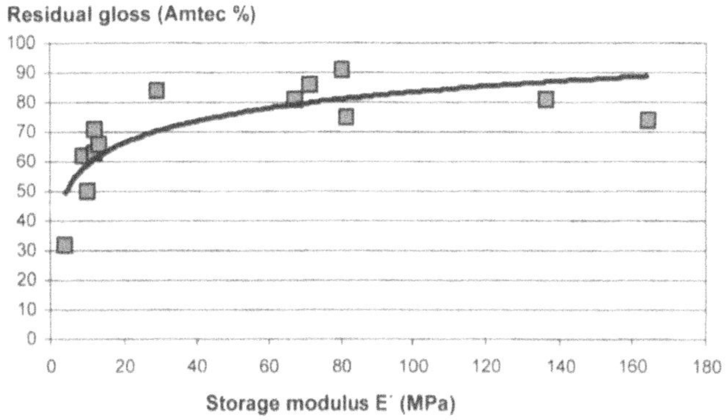

Fig. 5 Correlation of scratch resistance (Amtec values) and cross-linking density (storage modulus E´)

pictures are taken with a microscope at a magnification factor of app. 1000.

Critical load and indentation depth at 5 mN are the parameters printed as a function of each other to evaluate the scratch resistance (see Fig. 3). A good scratch resistance is the result of high values for the critical load when the values for residual indentation depth are low.

If the determined indentation depth at 5 mN is used to calculate the ratio of load (5 mN) and depth as a measure for plastic resistance, the result for typical 1 K, 2 K and UV systems is the correlation [20] shown in Fig. 4. It has to be noted that in comparison with Fig. 3 the reciprocal value of the indentation depth is actually plotted on the *y*-axis. Therefore, a good scratch resistance is the result of high values for the critical load and also high values for the plastic resistance.

Dynamic Mechanical Analysis

For certain clearcoat systems a partial healing of scratches can be observed on the time scale. In literature this is known as the reflow effect [21]. Thermal relaxation phenomena may be used for a physical explanation of this effect. In connection with scratch resistance the cross-linking density of clearcoats is also a decisive factor. Meanwhile, dynamic mechanical analysis (DMA) has been established as a method to determine cross-linking density [21–23].

Several ranges can be characterized when observing the behavior of the storage modulus E' as a function of temperature. The glass transition range is followed by the rubber elastic range, which can be more or less distinct depending on the degree of cross-linking. In this range the value of the storage module E' is connected with the cross-linking density. The value of E' at the local minimum is often chosen as a quantitative measure for cross-linking density. The clearcoats were examined by submitting free films to a tension test. A DMA 7 instrument by Perkin Elmer was used for these tests. In this context, measurements of paint systems are described e.g. in [23].

Figure 5 shows a comparison of the results of DMA analysis and values obtained in car wash simulations.

The clearcoat to be tested (applied on a standard metal sheet) is moved back and forth 10 times under a rotating car wash brush. The brush is sprayed with washing water during the cleaning procedure. Because the metal sheets used for the test are clean, a defined amount of quartz powder is added to the washing water as a replacement for street dirt (Amtec method). A gloss measurement in 20°-geometry is used to evaluate the scratch resistance. The initial gloss and the gloss after the cleaning procedure are measured. The percentage of residual gloss with regard to the initial gloss is a measure for scratch resistance. High values indicate good scratch resistance [8].

A clear correlation between E' as a measure for cross-linking density and the value for scratching can be observed. The quality of the scratching level (high residual gloss) increases with increasing cross-linking density [24].

Determination of the Surface Structure

Methods

Mechanical Surface Characterization. The surface profiles were measured by mechanical profilometry using a Hommeltester (Hommel, Germany). For all measurements, a dual-skid tracing system with a skid distance of 10 mm or a so-called datum system both with a diamond tip radius of 5 µm were used. The vertical resolution of this mechanical profilometric system is approx. 0.01 µm. The surface profiles were recorded over a scan length of 48 or 15 mm and the detected signals were stored in a computer after digitizing. The evaluation of mechanical profile measurements according to typical roughness parameters – e.g. average roughness Ra – gives an integrated information about the surface structure. In comparison with roughness parameters, Fourier techniques (FFT) yield a more detailed characterization of the surface structure. It is also possible to detect the surface structure pseudo-3-dimensional. The sample is moved by a precise positioning table for a small distance between two line scans. A cut off wavelength of 8 mm for a scan length of 48 mm was used to separate between roughness and waviness profile (cut off 2.5 mm for 15 mm scan length).

In a so-called autopower spectrum, the intensity (square of amplitudes) of the calculated sine/cosine waves representing the surface profile is plotted versus the corresponding wave number (reciprocal wavelength).

For wavelengths from 10 to 1 mm (integral 1, long waviness) and from 1 to 0.1 mm (integral 2, short waviness), the intensities of the autopower spectra are added up and used for further evaluation of the surface structures.

Optical Surface Characterization.

wave-scan. To simulate the visual impression obtained from optical inspection of surface structures, the German company Byk-Gardner developed the so-called "wave-scan DOI". The measuring principle is based on the detection of the modulation of the light of a small laser diode reflected by the surface structures of the sample. The laser light shines on the surface at an angle of 60°, and the reflected light is detected at the gloss angle (60° opposite). During the measurement, the "wave-scan" is moved across the sample surface over a scan length of approx. 10 cm, with a data point being recorded every 0.027 mm. The signal is divided into 5 wavelength ranges in the range of 0.1 to 30 mm and computed by mathematical filtering. For each of the 5 ranges a characteristic value (Wa 0.1–0.3 mm, Wb 0.3–1.0 mm, Wc 1.0–3.0 mm, Wd 3.0–10 mm, We 10–30 mm) as well as typical wave-scan-values longwave (LW, approx. 1–10 mm), shortwave (SW, approx. 0.3–1 mm) were calculated. Low wave-scan-values means a smooth surface structure. Additionally a LED light source is installed in the wave-scan DOI and illuminates the surface under 20° after passing an aperture. The scattered light is detected and a so-called dullness value (du, < 0.1 mm) is measured. By using the three values of the short wave range Wa, Wb and du a DOI value is calculated. Here a high value means a smooth, shiny surface.

Investigations on Plastics Coatings

Investigations on plastics coatings are of special interest to achieve a similar structure development on a painted car not depending on the different substrate materials (steel or plastics) used. Against this background parameters influencing the surface of plastics parts are becoming recently more important.

Measurements on Reinforced Plastics (Structure Effect). Polymeric substrates (Poly-butylterephthalate PBT) with different amount of glass fibers were painted with a typical coating system for plastics. The complete coatings were measured by wave-scan and mechanical profilometry [25]. Optical and mechanical measurements (Ra) show both the same trend, an increasing measuring value of structure for an increasing amount of glass fiber (GF) in the PBT (Fig. 6). The same correlation was found by using the integral-values of the FFT (Fig. 7).

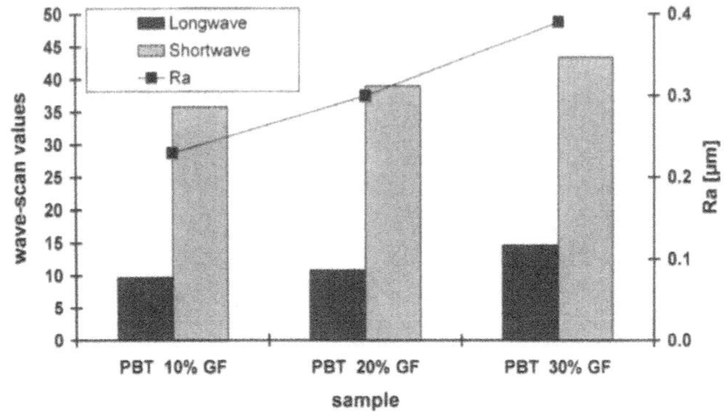

Fig. 6 Measuring values of the coatings in dependence on the fiber content

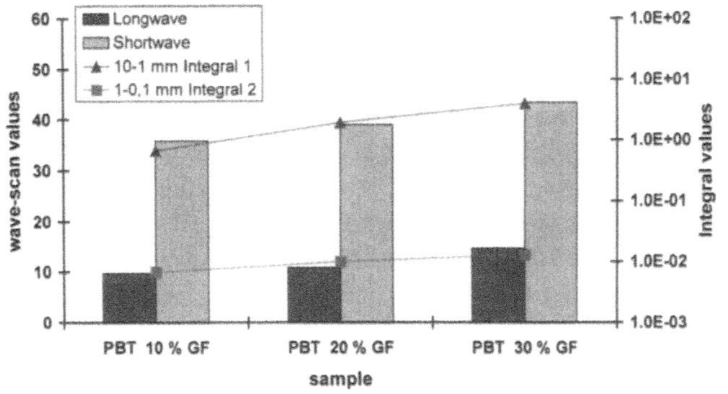

Fig. 7 Comparison FFT (arb. units) and wave-scan values

In addition to the effect of the amount of fiber content also a dependence on the type of the used fiber can be observed in a different example (Fig. 8, [26]). In this investigation it was possible to carry out the wave-scan measurements directly on the unpainted surfaces.

This effect is also demonstrated in Fig. 9 showing a comparison of 2 substrates with high and low surface structure. The amplitude of the structures is reduced, but the basic structure of the substrate is partially transferred through each single coating layer and influences finally the topcoat appearance. These results are very important due to the effect, that the substrate structure clearly influences the final coating structure. Based on surface structure measurements a preselection of plastics substrates can be done to obtain an optimal structure (smoothness) of the coating system applied.

Temperature Effect on Reinforced Plastics. Often, fibers and polymer matrix have different coefficients of thermal expansion. After painting the liquid coating flows

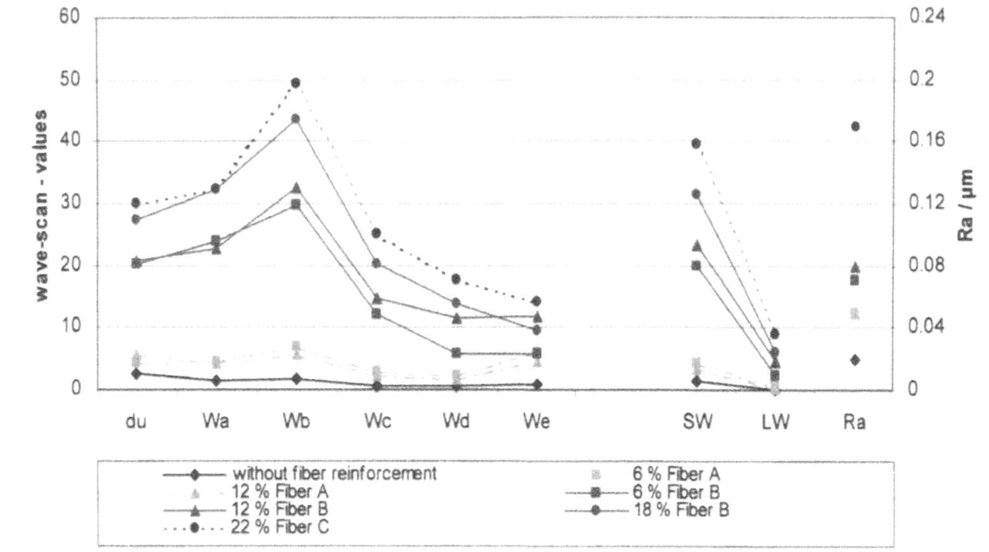

Fig. 8 Measurement of substrate structure (same polymer matrix)

off to a smooth surface equalizing structures of the substrate. During the drying process the polymer matrix is expanding more than the fibers. After cross-linking the solid coating is a duroplastic and not longer able to flow. Directly after taking the plastics parts out of the oven the coated surface looks well. But after cooling down an inverse image of the warm phase is seen as topcoat structure.

This physical behavior of the plastics is very important due to the fact that in practice the steel car body and the plastics exterior parts are connected at different steps of the painting process (and therefore at different temperatures loads, off-line, on-line, in-line).

Figure 10 shows this effect. At the left side the topography and a line scan of the uncoated fiber reinforced material is shown. At the right side of Fig. 10 the roughness of the clearcoat, coated directly on the substrate and dried under varying conditions is represented. The higher drying temperature causes a more distinct surface structure.

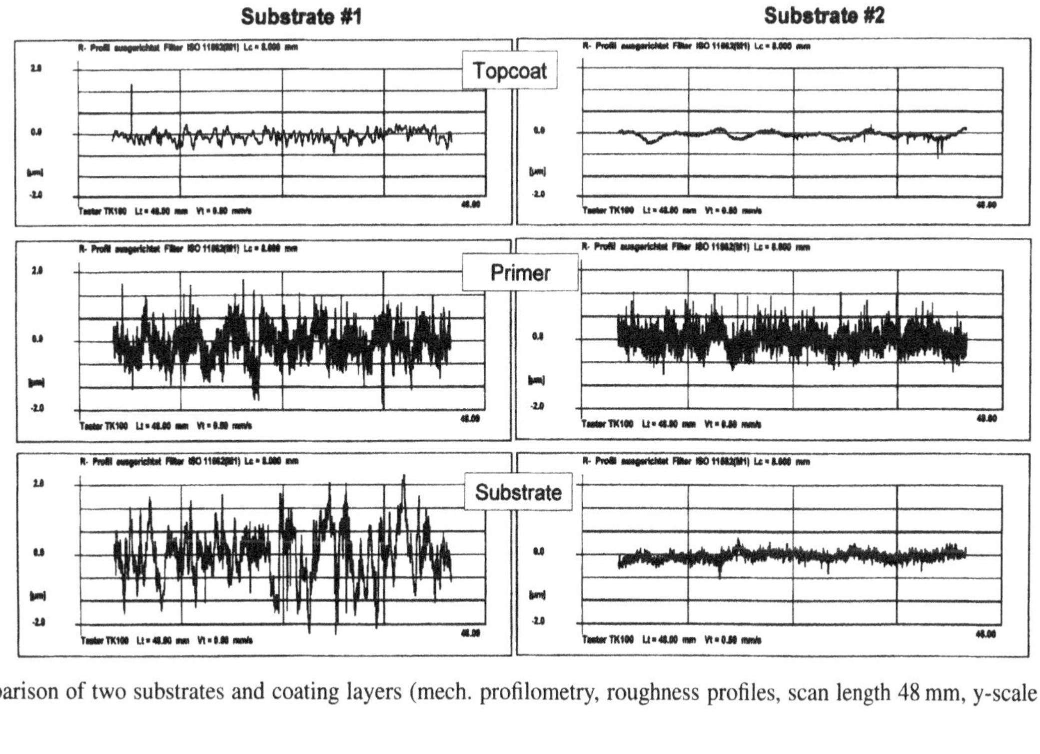

Fig. 9 Comparison of two substrates and coating layers (mech. profilometry, roughness profiles, scan length 48 mm, y-scale ±2 µm)

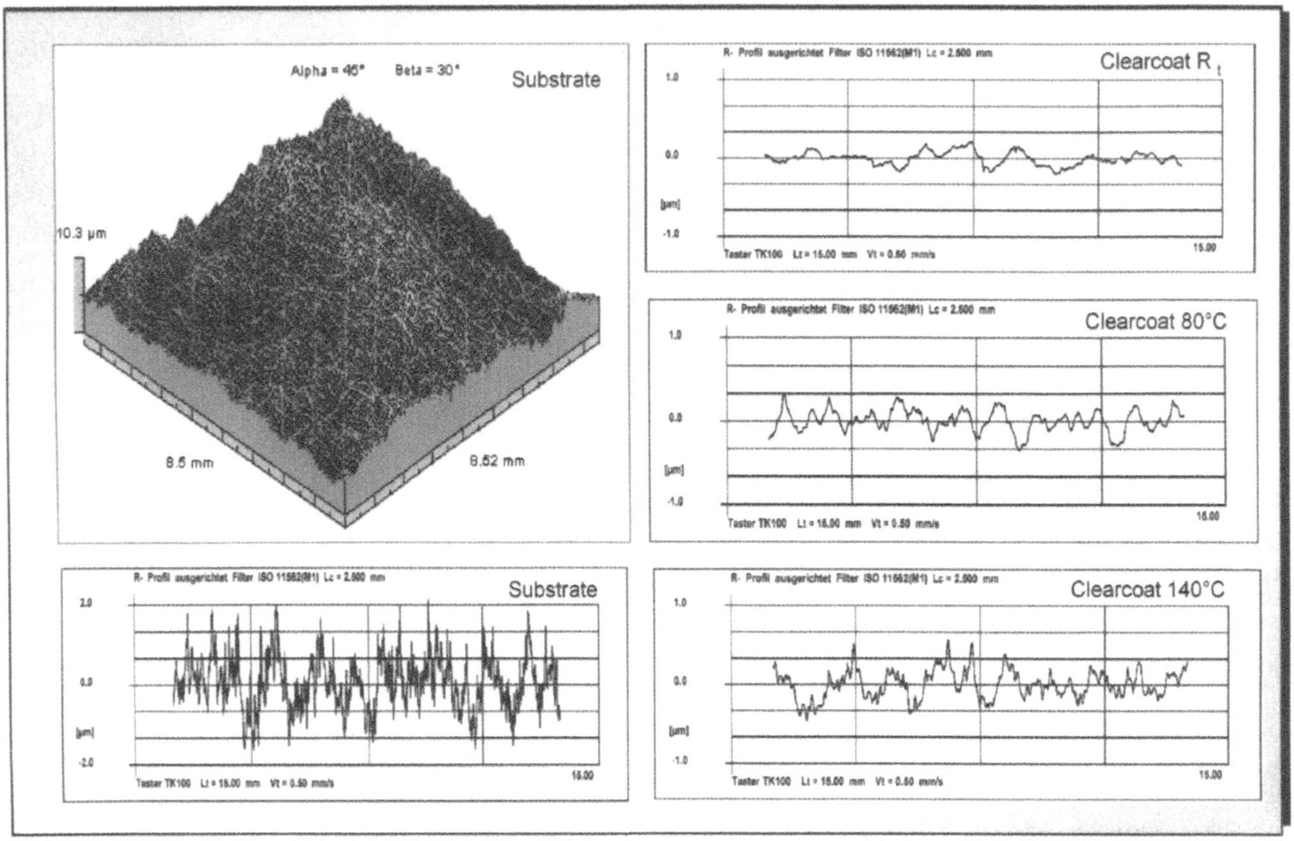

Fig. 10 Structure development of clearcoat at varying drying temperature; *left*: 3-D substrate, *bottom*: roughness profile; *right*: roughness profile clearcoat coated directly on substrate at RT, 80 °C and 140 °C (scan length 15 mm, y-scale ± 2 µm *left*, ± 1 µm *right*)

Summary

In this paper the recently developed techniques to characterize the mar resistance of coating systems were presented. The techniques base on methods that create a single scratch onto a surface. Characteristic values like the critical load as a measure for the transition from plastic behaviour to brittle fracture can be determined and used to rank different clearcoat systems and to compare these results with other physical properties. In the field of mar resistance the cross-linking density of the coating film determined via DMA plays an important role.

To characterize the overall appearance of a coating system mainly two different techniques are used in the automotive and paint industry. The mechanical profilometry yields a detailed information of the topography so that substrate influences and other effects on the final coating appearance can be described. These methods, in the past mainly used for coated steel substrates, were demonstrated showing typical examples of coatings for plastics. The optical method (wave-scan) yielded similar results.

References

1. Blackmann GS, Lin L, Matheson RR (1999) ACS Symposium Series: Microstructure and Tribology of Polymer Surfaces
2. Lin L, Blackman GS, Matheson RR (2000) Prog Org Coat 40:85
3. Adamsons K, Blackmann G, Gregorovich B, Lin L, Matheson R (1998) Prog Org Coat 34:64
4. Courter JL (1999) Proc. 5th Nürnberg Congress, Nuremberg, Germany, 351
5. Klinke E, Eisenbach CD (2001) Proc. 6th Nürnberg Congress, Nuremberg, Germany, 249
6. Lin L, Blackman GS, Matheson RR (2001) Materials Science and Engineering A317:163
7. Wagner G, Osterhold M (1999) Mat.-wiss. u. Werkstofftech. 30:617
8. Osterhold M, Wagner G (2002) Prog Org Coat 45:365
9. Boyd DW (1987) Proc. XIII. Int. Conf. Org. Coat. Sci. Techn., Athens, 59
10. Fister F, Dingerdissen N, Hartmann C (1987) Proc. XIII. Int. Conf. Org. Coat. Sci. Techn., Athens, 113
11. Armbruster K, Breucker M (1989) farbe + lack 95:896

12. Nakajima T, Yoshida Y, Miyoshi Y, Azami T (1991) Proc. XVII. Int. Conf. Org. Coat. Sci. Techn., Athens, 227
13. Geier W, Osterhold M, Timm J (1993) Metalloberfläche 47:30
14. Bastawros AF, Speer JG, Zerafa G, Krupitzer RP, SAE Technical Paper Series 930032
15. Timm J, Armbruster K, Osterhold M, Hotz W (1994) Bänder, Bleche, Rohre, 35, No. 9:110
16. Osterhold M, Hotz W, Timm J, Armbruster K (1994) Bänder, Bleche, Rohre, 35, No. 10:44
17. Osterhold M (1996) Prog Org Coat 27:195
18. Osterhold M (1998) Mat.-wiss. u. Werkstofftech. 29:131
19. Deutscher O, Armbruster K (2003), Proc. 3rd Stahl-Symposium, Düsseldorf, Germany
20. Dössel KF (2001) SURCAR, Cannes, France
21. Gräwe R, Schlesing W, Osterhold M, Flosbach C, Adler HJ (1999) Europ. Coat. Journal, No. 3:80
22. Schlesing W (1993) farbe + lack, 99: 918
23. Schlesing W, Buhk M, Osterhold M (2004) Prog Org Coat 49:197
24. Frigge E (2000) farbe + lack 106, No. 7:78
25. Stegen H, Buhk M, Armbruster K (2002) Paper TAW Seminar "Kunststofflackierung – Schwerpunkt Automobilindustrie", Wuppertal, Germany
26. Armbruster K, Stegen H (2004) Proc. DFO Congress "Kunststofflackierung", Aachen, Germany, 78

Progr Colloid Polym Sci (2006) 132: 48–53
DOI 10.1007/2882_041

Published online: 8 March 2006

Victoria Albrecht
Andreas Janke
Astrid Drechsler
Gert Schubert
Ernő Németh
Frank Simon

Visualization of Charge Domains on Polymer Surfaces

Victoria Albrecht · Andreas Janke ·
Astrid Drechsler · Frank Simon (✉)
Leibniz Institute of Polymer Research
Dresden, Hohe Straße 6, 01069 Dresden,
Germany
e-mail: frsimon@ipfdd.de

Gert Schubert · Ernő Németh
Institute of Mechanical Process and
Reprocess Engineering, Freiberg
University of Mining and Technology,
Agricolastraße 1, 09596 Freiberg, Germany

Ernő Németh
SKZ Süddeutsches Kunststoff-Zentrum,
Friedrich-Bergius-Ring 22,
97076 Würzburg, Germany

Abstract The tribo-electric charging of polymers is widely known and applied in many technical processes. However, the molecular and electronic reasons for this phenomenon are not fully understood. In a former work we showed that the tribo-electric charging of polymer particles can be described by their electron pair acceptor and donator properties. But polymer charging cannot be fully explained on a molecular level. During the collision of two particles expanded areas come into contact and an enormous number of molecules is involved in the charge formation process. In the present study, the lateral distribution of charges and their dissipation over the time were investigated using two electric force microscopy techniques (*surface potential imaging mode* and *electric force imaging mode*). The recorded images show that oppositely charged domains exist on polymer surfaces side by side. The dissipation of deposited charges depends on the electronic properties of the polymer. Also water adsorbed from the surrounding atmosphere or incorporated in the polymer bulk has a great influence on the charge stability.

Keywords Atomic force microscopy · Charge dissipation · Charge domains · Electric charges on polymer surfaces · Electric field imaging mode · Surface potential imaging mode

Introduction

Most polymers have a low electrical conductivity. One commonly known and often observed physical phenomenon is the electrical charging of those insulator materials. In the 18th century, the electrical charging of amber was an object of scientific interest and public performances showed the electrostatic effects. Hence, the Greek name of amber ʹΗΛΕΚΤΡΟΝ (electron) is the origin of all names related to electron, electricity, electronics and so on. Today, the electrical charging is used in innumerable technical processes and technologies, like electro-photography [1], electrostatic filtration as well as flocculation [2, 3] and electrostatic coating by means of dye particles [4]. Technical processes were developed to separate sodium and potassium chloride crystals in a scale of thousands of tons per year [5]. And in our daily work we do not want to miss the electrostatic phenomena for copying and printing processes [6]. In contrast to the common publicity and wide use of electrical charging effects, the electronic and molecular reasons of the observed phenomenon are largely unknown.

In his phenomenological study [1], Brück ordered a number of polymers in a row describing their affinity to become positively or negatively charged when rubbed with each other. Brück's row and other comparable rows first demonstrated the fundamental relationship between the chemical structure of a polymer and its tribo-electrical charging properties. In a former work [2] we followed that idea and showed that the tribo-electrical charging of two polymer species charged in a fluidized bed clearly depends on the affinity of the polymer surface to take up electron pairs (electron pair acceptor properties) or provide electron pairs (electron pair donator properties). In our study [3] we observed that particles of the same polymer species can be oppositely charged (bi-polar charging) in a fluidized bed where all sides were lead with the polymer to prevent polymer particle/metal contacts. Therefore it can be concluded that functional groups of the polymer may act both as an electron pair donator and an electron pair acceptor producing positive and negative charges on the particle surface.

In reality, the charging of polymer particles cannot be described only by interactions between single molecules or functional groups. The interacting surface must be considered as an expanded area containing an enormous number of molecules with the same chemical constitution and the same functional groups. The non-conductivity of the polymer prevents a flux and a charge compensation between the ionised sites. In our study [3] we found, however, a non-negligible influence of the ambient humidity on the tribo-electric charging of polymer surfaces.

In order to gain an insight into the surface charge distribution and stability, particular atomic force microscopy (AFM) techniques, the *surface potential imaging* (SPM) and *electric force imaging* (EFM) mode were developed. Already in 1986 the principles of imaging electrostatic charges with an AFM were described [4, 5]. Stern and coworkers [6, 7] used electrochemically etched Ni wires as AFM tips to study "one of the oldest unresolved problems in physics ... the mechanism of charge exchange between contacting surfaces when at least one of them is insulating". They imaged charges on polycarbonate (PC) deposited by contacting the polymer with the conductive Ni tip on which a voltage was applied. A tribo-electric charging was observed by slowly approaching a poly(methyl methacrylate) surface to the tip without applying a voltage [7, 8]. Morita and Sugawara [9] studied the time evolution of deposited charges on silicon samples with a defined thin silicon oxide layer under ambient conditions by EFM measurements with atomic resolution. They conjectured that the contact charging and dissipation mechanism of positive charges involves only one stage, the charge carrier's diffusion on the silicon oxide surface. In contrast, the dissipation of negative charges showed a more complex three-stage mechanism. The SPM technique (also known as Kelvin force microscopy [9]) was employed to detect dipoles on buried nano-particle/polymer interfaces and record their potential distribution with accurate lateral and energetic resolution [10]. In this way it was able to quantify electronic anisotropies and charge transfer processes in semi-conductors [11].

The aim of the present paper is to explain the bipolar charging of chemically identical polymers observed in tribo-charging experiments [3] by showing the coexistence of negatively and positively charged domains side by side on the polymer surface. Using AFM surface potential imaging, the distribution and strength of these charge domains are visualized on various polymer surfaces. The stability of charges and the influence of environmental humidity is investigated by electric force imaging.

Materials and Methods

Materials

All of the measurements described in this paper were carried out with polymers containing a minimum amount of additives, impurities and other accompanying substances (Table 1).

In order to remove water traces from the polymer surface and the bulk phase all polymer samples were carefully dried in a vacuum oven at 50 °C and a slight vacuum at least five days.

For all experiments employing the atomic force microscopy (AFM) polymer melt films were prepared from the granules. One polymer granule was put on a silicon wafer and heated up to the melting point. Then, the polymer melt was pressed against a second silicon wafer using a press, usually employed to prepare samples for infrared spectroscopy. After cooling, the silicon wafers were carefully removed and the polymer film was mounted on a steel sample puck by a piece of electrically conductive double-sided adhesive tape. To detect surface impurities all melt layers were analysed by means of X-ray photoelectron spectroscopy (XPS). Only samples without any surface impurities and a surface composition equal to that of the granules were used for the AFM measurements.

Atomic Force Microscopy (AFM)

All AFM measurements were carried out by means of a NanoScope IIIa – Multimode (Veeco, USA). The instrument was equipped with an electrically conducting cantilever with a silicon tip (Pointprobe EFM sensor, NanoSensors, Wetzlar, Germany). An overall metallic

Table 1 Polymers used for charging experiments

Polymer	Abbreviation	Trade name
Polyethylene	PE	Mirathen AL 21 FA
Polypropylene	PP	Daplen PP KS 10
Polyamide-12	PA	Vestamid L 1600
Poly(vinyl chloride)	PVC	Decelith 23002
Poly(ethylene terephthalate)	PET	Polyclear T 86
Poly(oxymethylene)	POM	Ultraform N 232003
Poly(methyl methacrylate)	PMMA	Plexiglas G7 clear 10100

coating (PtIr5) on both sides of the cantilever increased the electrical conductivity of the tip. The tip radius was lower than 25 nm. The spring constant was 2.8 N/m and the resonance frequency 75 kHz. The sample mounted on a steel sample holder was grounded, the sample acted as a dielectric in a capacitor.

An environmental chamber ensured that temperature and humidity were constant during the measurements.

All polymer films were used as prepared, i.e. the samples were not mechanically rubbed. Samples that were contacted with other polymer sheets and afterwards separated, showed highly electrically charged surfaces which cannot be studied by the AFM techniques mentioned below.

Surface Potential Imaging Mode (SPM)

The *SPM* technique allows measurements of the local sample surface potential. The NanoScope recorded two passes. In the first pass, the sample surface topography was obtained by the standard tapping mode. The surface potential was measured during the second pass carried out in the lift mode (lift height was 100 nm). Here the cantilever's vibration is turned off and an oscillating voltage $U_{AC} \cos \omega t$ is applied directly to the cantilever tip. This creates an oscillating electrostatic force F at the frequency ω on the cantilever:

$$F = \left(\frac{dC}{dz}\right) U_{AC} U_{DC} \qquad (1)$$

where (dC/dz) is the vertical derivative of the tip/sample capacitance, U_{DC} the d.c. voltage between the tip and the sample, and U_{AC} is the oscillating voltage applied to the cantilever tip.

An external d.c. voltage (U_{tip}) applied on the tip compensates the voltage U_{DC} between the tip and the sample until the oscillation amplitude becomes zero ($F = 0$). At this point the external voltage U_{tip} will be the same as the unknown surface potential.

The voltage U_{tip} is analyzed to construct a voltage image of the surface. The surface topography recorded in the first pass is involved to compensate topographic effects.

Electric Force Imaging Mode (EFM)

The EFM technique was applied to detect the electric field gradient's distribution produced by electrical charges on the sample surfaces. As explained above, the measurements were carried out by two passes where the first pass records the sample surface topography. In the second pass (lift height was 100 nm) the cantilever vibration stimulated by a piezoelectric element was kept near the cantilever's resonant frequency and a d.c. voltage was applied to the cantilever. As response on an additional force gradient (e.g. due to the interaction of the charged tip and a charge domain on the sample surface) the resonant frequency changes. Attractive forces (generated by oppositely charged tip and surface) reduce the cantilever resonant frequency, and repulsive forces (generated by tip and surface charges having the same sign) increase the resonant frequency. The frequency shift can be used to construct an image showing the lateral distribution of surface charges.

We used the EFM technique to study the stability of charge domains on polymer surfaces. After recording the sample surface topography the cantilever tip with an applied d.c. voltage of +12 V or −12 V touched the sample surface for 120 s. During this period charges were transferred to the surface. A series of passes in the lift mode were carried out to detect the evolution of the lateral charge distribution on the sample surface over the time. The applied d.c. voltage on the tip was −10 V.

Results and Discussion

Visualization of Potential Distribution on Polymer Surfaces

The SPM image in Fig. 1a shows a typical example of the lateral distribution of the surface potential for a polystyrene melt film.

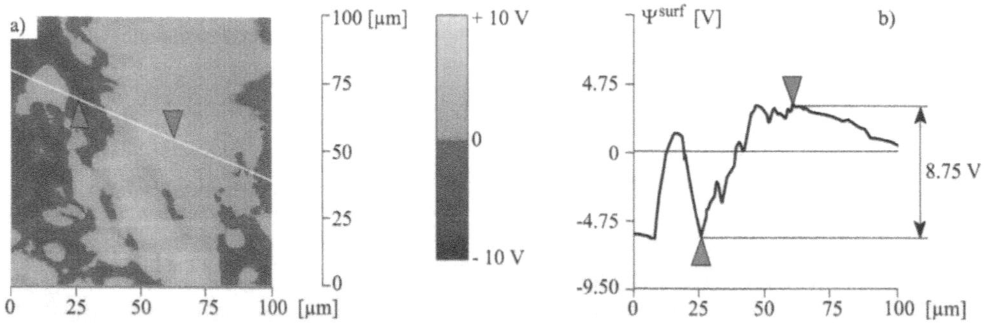

Fig. 1 Distribution of the surface potential (Ψ^{surf}) on polymer surfaces visualized by means of the SPM technique for PS (**a**) and line scan (**b**) along the white line in the SPM image

The positively and negatively charged domains exist side by side like a patchwork. The immediate neighbourhood of positively and negatively charged domains, indicated by more or less sharp domain borders can lead to considerable electrical field strengths. Figure 1b shows an arbitrarily selected line-scan through the charge patterned polystyrene surface. Along the line-scan the field strength E was calculated to $E \approx 2.4\,\mathrm{kV\,cm^{-1}}$. The high value results from the short distance of the two oppositely charged domains. However, the surface potential ($\Psi_{surf} \approx 8.75\,\mathrm{V}$) was too low to ionise molecules in the contacting gas phase that could cause an electrical breakdown.

During the collision of two polymer grains areas with a size of a few square micrometers come into contact. In the contact zone an electron transfer can take place as described in [2] (Fig. 2a). In case of the collision of two chemically different polymer particles one of them preferably interacts as electron pair donator and the other species as electron pair acceptor. In this way laterally expanded charge domains can be formed (Fig. 2b). The low electrical conductivity of the polymer bulk and surface prevents a rapid charge dissipation, hence the formed domain structure seems to be permanently stable.

However, in case of the contacting of two polymer samples of the same chemical constitution the discussed electron transfer does not fully explain the existence of laterally expanded charge domains nor the bi-polar charging as the chemically identical functional surface groups should not predominately or selectively act as electron pair donator or electron pair acceptor. A pronounced tribo-electric charging is only observed for polymer species consisting of rather polar molecules. These polar molecules have pronounced electron pair donator or acceptor properties. The existence of a strong electron pair donator centre in a molecule creates an electron pair acceptor centre in its immediate neighbourhood and an electron pair acceptor centre generates a donator centre [2]. However, the tendency to act as potential electron pair donator or acceptor will not be different for those interacting particles.

The formation of charge domains showed that the fundamental electron transfer process involves a high number of surface sites. The tribo-electric charging of polymers has to be considered as a collective phenomenon.

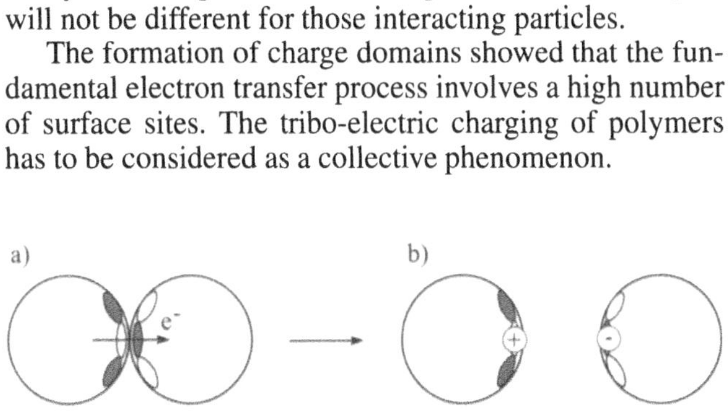

Fig. 2 Model concept of the contact charging of polymer grains. **a** Contact between an electron pair donator domain (*empty dots*) of the particle 1 and an electron pair acceptor domain (*grey dots*) of the particle 2. Charge transfers (e^-) are taking place. After separation the two particles (**b**), a positively charged (⊕) and a negative charged (⊖) domain, remain on the particles surface

Stability of Charge Domains on Polymer Surfaces

Because of the low surface and bulk conductivity of most polymer materials the stability of charges on polymer surfaces should be stable over the time. However, the polymer surface is an interface where other substances can come into contact with charged polymers and these substances can provide charge carriers influencing the time-stability of the charges. Especially traces of water can support the discharging of polymer surfaces because the water molecules auto-dissociate forming hydronium ions (H_3O^+) and hydroxyl ions (OH^-). Brønsted acidic or basic surface groups may be involved in the ion formation and provide additional charge carriers or traps to neutralize charges. The presence of water molecules does not only depend on the environment that is in contact to the polymer surface. Enormous amounts of water can also be contained in the polymer bulk [12] where the water molecules are more or less mobile and able to transport ions. As shown in the work by Morita and Sugawara [13] EFM is suitable to deposit charges on insulator surfaces and study their behavior over the time.

Figure 3 shows four series of EFM images recorded on different polymer melt film species. The first image of each series shows the non-charged melt film surface (initial state). Employing a conductive silicon cantilever charges were deposited on the polymer melt film surfaces. Immediately after the charge deposition the time was set at $t = 0$.

After various time spans t the distribution of the deposited charges was recorded. PE (Fig. 3a) and PP (not shown in Fig. 3) show a high stability of the deposited charges. Over a time of a few hours no change in the charge image was observed. PE and PP are rather non-polar polymers without pronounced electron pair acceptor or donator properties [2]. The low surface polarity does not allow a stabilization of adsorbed water films on the surface. Also, the bulk phase does not take up water [12]. Hence, no charge flux can take place and a charge dissipation cannot be observed.

According to the charge stability PMMA shows a similar behaviour (Fig. 3c). Deposited charges seem very stable over a long time. However, it is well known that PMMA is able to adsorb and incorporate considerable amounts of water [12]. It is assumed that the water molecules are immobilized in the network of the PMMA's bulk phase and cannot act as mobile charge carriers. PVC is a polymer with a rather low hydrophilic character comparable to the other polyolefins, and water films should not be stable on the PVC surface. But PVC exhibits a high electron pair acceptor strength as well as a considerable value of electron pair donator strength [2]. From this point of view PVC can be considered as a polar polymer. The SPM images show pronounced oppositely charged domains on its surface. Charges deposited on the PVC surface do not seem to be stable. Figure 3b shows a lateral

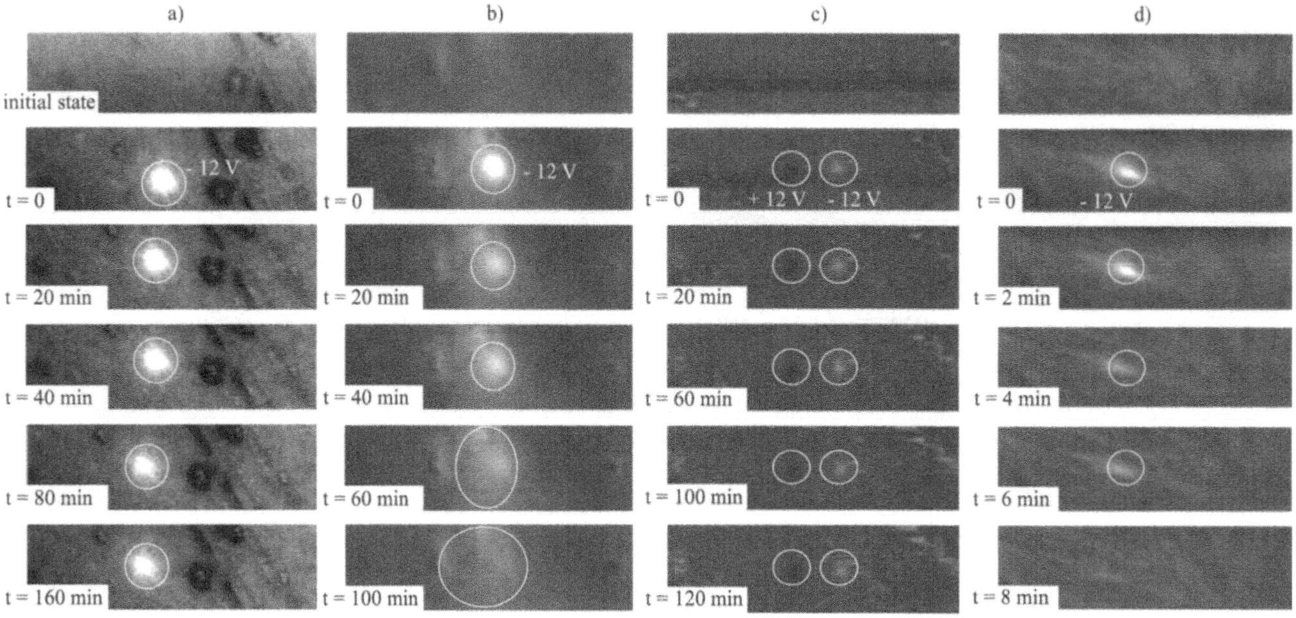

Fig. 3 Series of EFM images taken for PE (**a**), PVC (**b**), PMMA (**c**), and PA (**d**). The images show the *initial states* without deposited charges, the states immediately after deposition of charges by the silicon cantilever ($t = 0$) and after different time spans ($t = t_i$). The positions of the deposited charges are indicated by a *white circle*. Negative charges appear pale, positive charges are *dark*. All measurements were carried out at a relative humidity of $\varphi = 30\%$

spreading of charges over the time. A conclusive evidence for this phenomenon cannot be given. Light irradiation of PVC causes a substitution of chlorine by oxygen on the PVC surface. Maybe traces of the carboxylic groups formed in this way can act as adsorption centres for water molecules or provide hydronium ions which can act as charge carriers.

PA is another polar polymer. It is able to take up and incorporate tidy amounts of water [12]. Figure 3d shows a quick spreading of the deposited charges within a few minutes. It can be clearly seen that the charge flux follows the grain boundary of the PA's spherulites where water molecules can move. Obviously, in the case of PA water seems responsible for the charge transport that involves the discharging of the PA surface.

The influence of water on the dissipation of surface charges can be studied by adjusting the environmental humidity which controls the water amount offered for adsorption. The influence of humidity on the charge stability is illustrated in Fig. 4. For three polar polymer samples the

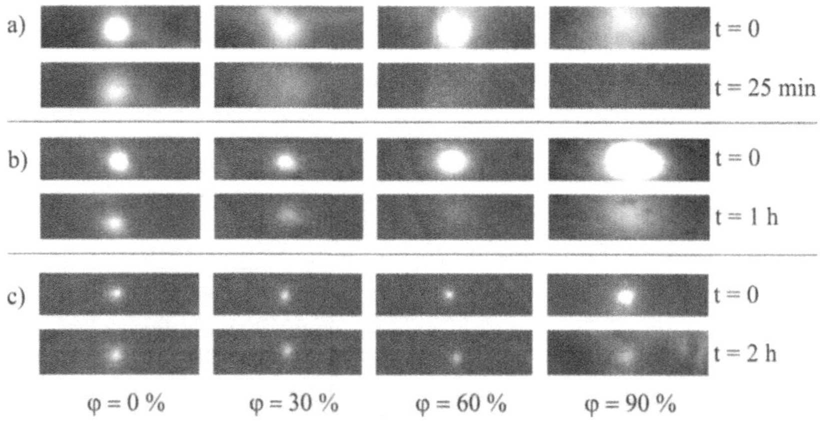

Fig. 4 Series of EFM images taken for PET (**a**), PVC (**b**), and PMMA (**c**) in dependence on the relative humidity (φ). After deposition of charges by the silicon cantilever ($t = 0$) the samples were stored in an atmosphere of constant humidity for different time spans ($t = t_i$). In all images negative charge domains appear pale

charge stability was measured in dependence on the time and the exactly adjusted humidity φ.

Figure 4 shows for all three polymer samples that in the case of an environment containing a relative high amount of water charges immediately spread on the surfaces after their deposition. In case of PMMA the spreading seems more moderate, but it can be clearly detected. The result corresponds with findings of the time-dependent measurements shown in Fig. 3c and the assumption of water which is fix incorporated in the PMMA-network. The image of the PET sample taken at $\varphi = 30\%$ indicates that the charges rapidly spread. After a storage of the PET sample in wet atmosphere ($\varphi = 60\%$ and 90%) all charges were dissipated within only 25 min. For PVC the same course is observed, however over a longer time span and the image recorded at $\varphi = 90\%$ indicates that a complete dissipation of charges does not take place. Here it is necessary to note that the flux of charges starts immediately after the deposition of the first charges on the PVC surface (an expanded pale area can be seen in Fig. 4b). That means, during the charge deposition time a significantly higher number of charges were deposited.

Conclusions

The tribo-electric charging of polymers is a collective phenomenon of numerous electron transfer processes in the contact zone of two colliding particles. The SPM technique allows to visualize the charge distribution on polymer surfaces. It was shown that oppositely charge domains can stably exist side by side although considerable filed strengths are present.

The dissipation of deposited charges was studied by means of EFM measurements. Depending on the polymer's polarity the rate of charge dissipation can be very different what indicates different mechanisms of discharging. Traces of water which can be adsorbed from the surrounding air or incorporated in the polymer bulk have a strong influence on charge dissipation. Furthermore, kinetic studies quantifying the discharging mechanisms will be published in a separate paper [14].

Acknowledgement The financial support from the Deutsche Forschungsgemeinschaft (DFG-Sonderforschungsbereich 285) is acknowledged.

References

1. Brück R (1981) Kunststoffe 71:234
2. Németh E, Schubert G, Albrecht V, Simon F (2003) J Electrostatics 58:3
3. Németh E, Schubert G, Albrecht V, Simon F (2005) Aufbereitungstechnik 45:35
4. Binnig G, Quate CF, Gerber C: Phys Rev Lett 56:930
5. Martin Y, Williams CC, Wickramasinghe HK (1987) J Appl Phys 61:4723
6. Stern JE, Terris BD, Mamin HJ, Rugar D (1988) Appl Phys Lett 53:2717
7. Terris BD, Stern JE, Rugar D, Mamin HJ (1989) Phys Rev Lett 63:2669
8. Terris BD, Stern JE, Rugar D, Mamin HJ (1990) J Vac Sci Technol A8:374
9. Zerweck U, Loppacher C, Otto T, Grafström S, Eng LM (2005) Phys Rev B71:125424
10. Eng LM, Grafström S, Hellmann I, Loppacher C, Otto T, Renger J, Schlaphof F, Seidel J, Zerweck U (2004) In: Meyendorf N, Baaklini GY, Michel B (eds) SPIE Proceedings Testing, Reliability, and Application of Micro- and Nano-Material Systems II, p 21–35
11. Loppacher C, Zerweck U, Teich S, Beyreuther E, Otto T, Grafström S, Eng LM (2005) Nanotechnology 16:S1
12. Albrecht V, Simon F, Schubert G, Németh E (2002) Chemie-Ingenieur-Technik 74:1425
13. Morita S, Sugawara Y (2001) Thin Solid Films 393:310
14. Albrecht V, Janke A, Schubert G, Németh E, Simon F, Kinetics of discharging polymer samples in dependence on humidity. J Electrostatics, in preparation

Progr Colloid Polym Sci (2006) 132: 54–61
DOI 10.1007/2882_032

Published online: 9 March 2006

Susanne Temmel
Wolfgang Kern
Thomas Luxbacher

Zeta Potential of Photochemically Modified Polymer Surfaces

Susanne Temmel
Polymer Competence Center Leoben GmbH, A-8700 Leoben, Austria

Wolfgang Kern (✉)
Institute for Chemistry and Technology of Organic Materials, Graz University of Technology, A-8010 Graz, Austria
e-mail: kern@ictos.tugraz.at

Thomas Luxbacher
Anton Paar GmbH, A-8054 Graz, Austria

Abstract The present work is focused on the introduction of sulfonic acid groups ($-SO_3H$) onto low-density polyethylene (LDPE) surfaces by photosulfonation. The generation of sulfonic acid groups at the polyethylene surfaces and the degree of photosulfonation were examined by FTIR-spectroscopy, contact angle testing as well as zeta potential measurements. The contact angle θ of water decreased from $\theta = 99°$ (pristine LDPE) to about $\theta = 30°$ (photosulfonated LDPE) and then remained constant. From contact angle data with different test liquids a significant increase in the polar component γ^P of the surface tension γ was evidenced, while the dispersive component γ^D remained almost constant. Zeta potential measurements were performed for the investigation of electrokinetic effects at the solid/liquid interface. The ζ-potential of the modified LDPE surfaces shifted to less negative values with increasing UV irradiation time corresponding to a higher hydrophilicity of the photosulfonated polyethylene surfaces. Concomitantly the isoelectric point was shifted to lower pH values, which indicates an increasing amount of acidic $-SO_3H$ groups present at the sample surface. However, strongly photosulfonated LDPE surfaces became partly soluble in aqueous media which limits the amount of $-SO_3H$ groups present at the modified LDPE surface.
To reduce these effects, LDPE samples were cross-linked by e-beam irradiation and then subjected to the photosulfonation process. Compared to standard LDPE, crosslinked LDPE displayed a higher content of $-SO_3H$ groups and higher surface polarity after photosulfonation. This was evidenced both by zeta potential and contact angle measurements. It is thus demonstrated that sample pre-treatment by crosslinking provides more stable surfaces which maintain their polar properties during water contact. This is explained by a lower amount of extractable components as a result of radiation-induced network formation.

Keywords Contact angle · Cross-linked LDPE · Photosulfonation · Polyethylene · Surface modification · Zeta potential

Introduction

The use of polymers in various industrial applications has increased dramatically during the past few decades [1, 2]. Despite the variety of many plastics, their use is often limited due to their surface properties, although their mechanical properties may be well-suited. Therefore the surface of a polymer has to be modified to obtain the desired surface properties, such as hydrophilicity, conductivity or chemical composition without altering the bulk properties.

Several techniques including corona discharge [1], plasma treatment [1, 3, 4], flame treatment [1], and irradiation with UV light in the presence of a UV sensitive gas [5–8] have been developed to modify the polymer surface. The principle of those surface treatment technologies is to introduce polar groups onto the polymer surface. This provides significant improvement of wettability, paintability, biocompatibility and also adhesion to other polymers or metals.

The growing demand for detergents led to the development in processes for introducing sulfonic acid groups $-SO_3H$ (sulfonation) in organic molecules. This results in changing the physical and chemical properties of molecules as the solubility in water, detergency, wettability, emulsifying power, and foaming power, which is of great practical interest. In the industry the most important technique is sulfonation using oleum (SO_3 in sulfuric acid) [9, 10]. The introduction of sulfonyloxy groups in alkylbenzenes, olefins and fatty alcohols can also be achieved using sulfur dioxide. However, the sulfonation of saturated aliphatic hydrocarbons is not possible by this method due to absence of active sites. Alkanesulfonates can be generated by two photochemical processes: sulfochlorination and sulfoxidation [11]. In both methods a gaseous mixture of SO_2 and an oxidant is used. In the sulfochlorination chlorine is used as the oxidizing agent, whereas in the sulfoxidation oxygen is the oxidant. The overall process for photoinduced sulfoxidation ("photosulfonation") is represented in Eq. 1 [12–15].

$$RH + SO_2 + \frac{1}{2}O_2 \xrightarrow{h\nu} RSO_3H \qquad (1)$$

The application of photosulfonation reactions to introduce $-SO_3H$ groups onto low-density polyethylene (LDPE) surfaces has already been described in the literature [16–20]. Both the reaction pathways and the optimization of reaction conditions (e.g. the volume ratio SO_2/air) have been discussed.

The aim of the present study was a closer investigation of the properties of LDPE surfaces subjected to photosulfonation processes. Until today no zeta potential studies had been carried out on photosulfonated LDPE. In the present investigation such modified LDPE surfaces were characterized by FTIR spectra, contact angle testing and zeta potential measurements were carried out. To improve the stability of modified LDPE surfaces, polyethylene samples were crosslinked by e-beam irradiation and subsequently photosulfonated. The surface properties of crosslinked LDPE are compared to those obtained with standard LDPE. Such data are useful to exploit the potential of surface sulfonation processes both for laboratory and technical purposes.

Experimental Section

Materials. Low density polyethylene (LDPE) foils (thickness: 37 μm) were purchased from a commercial supplier. Some LDPE samples were cross-linked using electron beam irradiation (dose: 250 kGy) under inert gas conditions. SO_2 of > 99.98% purity as well as compressed air were received from Linde AG.

FTIR Spectroscopy. Infrared spectra were recorded on a Perkin Elmer Spectrum One FTIR spectrometer equipped with a Deuterated Triglycine Sulfate (DTGS) detector and KBr optics. Data collection was performed at 4 cm^{-1} spectral resolution in the region of 4000–450 cm^{-1} and averaged over 5 scans. All samples were measured in transmission mode.

Contact angle measurements. The surface tension γ of the sample surfaces was determined by measuring the contact angle with a Drop Shape Analysis System DSA100 (Krüss GmbH, Hamburg, Germany) using water and diiodomethane as test liquids (drop volume ~ 20 μL). Based on the Owens-Wendt method [21], the surface tension γ as well as the dispersive and polar components (γ^D and γ^P) were evaluated. The literature values for the surface tension components of the test liquids used are given in Table 1 [22]. The contact angles were obtained by means of the sessile drop method and they were measured within two seconds. The reproducibility was within 2°.

Zeta potential measurements. The zeta potential ζ of the sample surfaces was determined by the streaming potential method, using a clamping cell connected with an EKA electrokinetic analyzer (Anton Paar GmbH, Graz, Austria). In this cell the sample is pressed against a PMMA spacer with seven channels. The measurements were performed with a KCl electrolyte solution (10^{-3} M, 500 mL). The pH was adjusted to about 10 by adding NaOH (0.1 M,

Table 1 Surface tensions γ, dispersive components γ^D and polar components γ^P of water and diiodomethane. Data taken from Ström [22]

	γ [mN/m]	γ^D [mN/m]	γ^P [mN/m]	Density [g/cm]	Viscosity η [mPa s]	Temperature [°C]
Water	72.8	21.8	51	0.998	1.002	25
Diiodomethane	50.8	50.8	0	3.325	2.762	20

2.5 mL) and was then decreased stepwise (0.3–0.4 units) by titration with HCl (0.1 M), until pH = 3 was reached. Using the Clamping Cell a pressure ramp from 0 to 400 mbar was employed to force the electrolyte solution through the cell. Streaming potentials were converted to zeta potentials using the Helmholtz-Smoluchowski equation [23] and the Fairbrother-Mastin approach [24]. Each value of the zeta potential at a given pH value represents an average value over at least five individual measurements.

Crosslinked polyethylene. For additional investigations, LDPE samples were crosslinked by e-beam radiation with a dose of 250 kGy. In order to remove residual soluble components from the cross-linked LDPE samples, a suitable extraction process was developed [16]. After being clamped in a frame, the polyethylene foils were extracted in m-xylene for 24 hours at a temperature of 80 °C. The addition of a small amount of BHT (2,6-di-tert-butyl-4-methylphenol), an antioxidant, prevented unwanted oxidation of the LDPE foils. Afterwards the samples were dried carefully at room temperature for 20 hours, and then dried in vacuo (20 mbar, 48 hours at 40 °C) to constant weight.

Irradiation setup. The LDPE samples to be irradiated were placed in a reaction chamber made of stainless steel (type 360Ti) equipped with a quartz window pasted into the top cover, a gas-inlet and a gas-outlet. During the irradiation experiments, a gas stream composed of sulfur dioxide and compressed air at a volume ratio of 1 : 1 was passed through the reaction chamber. The composition of the gas mixture was controlled with two flow meters. The total flow was about 5 litres per hour in all cases. Irradiation was carried out with a medium pressure mercury lamp (Heraeus, 1300 W). For all experiments, an etched (i.e. corrugated) quartz plate was positioned as a diffusor element between the lamp and the reaction chamber in order to provide a more uniform light intensity over the whole sample area. The distance between the lamp and the sample surface was 7.2 cm. The radiation emitted by the medium pressure Hg lamp was determined by a UV spectroradiometer from Solatell® measuring UV light in the range from 230 nm to 465 nm. The spectral distribution of the radiation is represented in Fig. 1. The light intensity integrated over the whole spectral range (230–465 nm) was about 70 mW/cm^2 (at the sample surface).

Irradiation procedure. The LDPE samples (size 40 mm × 20 mm, thickness 37 μm) were purified in an ultrasonic bath (10 minutes, 30 °C) using a mixture of acetone/water and then dried in vacuo (20 mbar, 1 hour at 40 °C). The samples were positioned in the reaction chamber, which was then purged with a mixture of SO_2 and synthetic air (1 : 1, volume) for about 10 minutes for conditioning. Subsequently, the polyethylene samples were irradiated with the Hg lamp for different periods of time (1 to 5 minutes) maintaining the gas flow. Finally the chamber was purged for additional 3 minutes before opening the reaction chamber. After removal from the chamber the modified LDPE samples were extracted in an ultrasonic bath (10 minutes, 30 °C) using a mixture of acetone/water (1 : 1) and dried in vacuo (20 mbar, 1 hour at 40 °C) over again. This extraction ensured the removal of soluble reaction products as well as excess SO_2 and H_2SO_4.

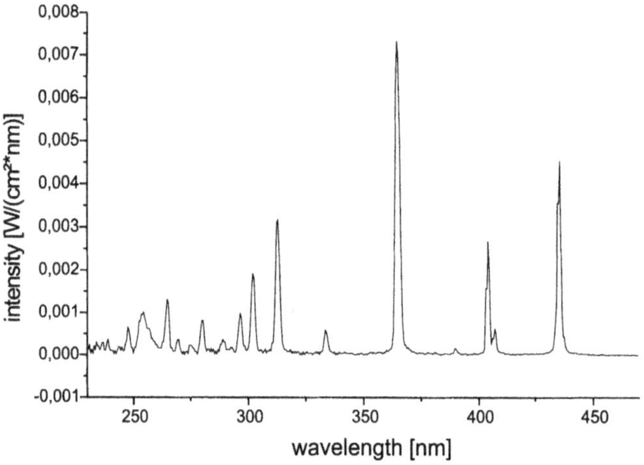

Fig. 1 Emission spectrum of the Hg-lamp as recorded with the spectroradiometer. The power density was measured at the sample position

Results and Discussion

Non-crosslinked LDPE

Although the photosulfonation of polyethylene has already been described in the recent literature [16–20], in the following a full characterization of LDPE surfaced modified by photosulfonation will be presented. First, the identity and the purity of our pristine LDPE samples were probed by FTIR spectroscopy. A characteristic IR-spectrum is

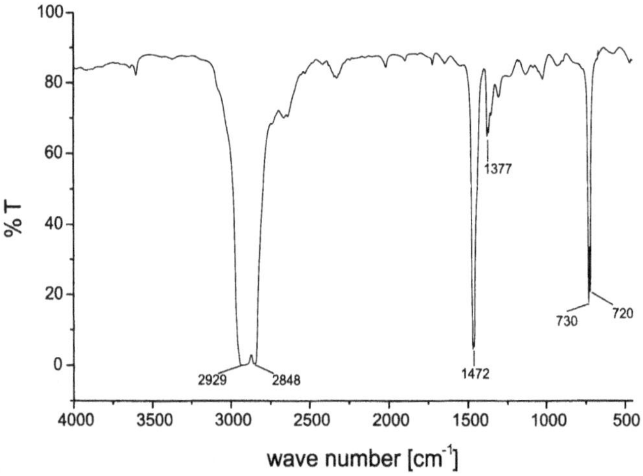

Fig. 2 FTIR spectrum of the unmodified LDPE sample

shown in Fig. 2. The typical IR bands of LDPE are observed around 2929 and 2848 cm^{-1} (C – H valence vibration), 1472 and 1377 cm^{-1} (C – H deformation vibration), 730 and 720 cm^{-1} ($-CH_2$-rocking vibration).

Figure 3 shows an FTIR spectrum of LDPE after the photomodification process with SO_2 and O_2. Besides the typical IR bands of LDPE, new bands around 1170 and 1037 cm^{-1} were detected after UV irradiation (3 minutes) in the presence of sulfur dioxide and air at a volume ratio of 1 : 1. The peak at 1170 cm^{-1} can be assigned to the asymmetric vibration mode of SO_3 units, whereas the signal at 1037 cm^{-1} originates from the symmetric stretching vibration of SO_3 units. These two bands are typical of sulfonic acids and are getting more intense with increasing irradiation time until they remain constant after four minutes of photosulfonation. The signal around 1720 cm^{-1} may be attributed to aliphatic aldehydes, which seem to be formed as a by-product during photosulfonation.

The contact angle of water on non-crosslinked polyethylene decreased from 99.2° (unmodified LDPE) to 30.3° after an UV irradiation time of 5 minutes (photosulfonated LDPE, see Fig. 4). The increased wettability is attributed to the introduction of polar sulfur acid groups on the surface of polyethylene. The measured contact angles and the surface tension data (γ, γ^D and γ^P) of the unmodified and sulfonated samples are listed in Table 2. The contact angle θ then remained constant after an UV irradiation time of about four minutes. Thus it appears that a saturation of the surface with sulfonic acid groups is achieved after this period. As expected from the increasing wettability, the value of the surface tension γ increases at increasing irradiation time, see Fig. 5. The polar surface tension component γ^P rises significantly, whereas the dispersive component γ^D remains virtually constant. To illustrate the improved wetting behaviour, Fig. 6 shows micrographs of water droplets placed on the surface of pristine and photosulfonated LDPE.

The ζ potential of the samples obtained using the Clamping Cell is no absolute value since the grooved PMMA spacer of the measuring cell contributes one measured surface while the other is the sample surface. Any-

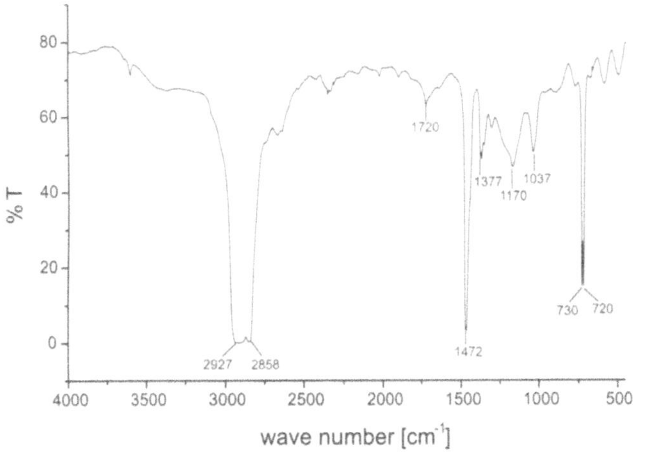

Fig. 3 FTIR spectrum of a LDPE sample after UV irradiation in the presence of SO_2 and O_2 and subsequent extraction with acetone/water.

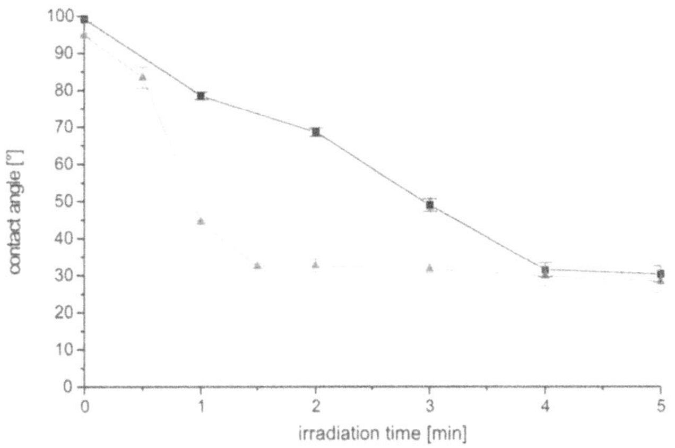

Fig. 4 Contact angle θ of water on non-crosslinked (■) and crosslinked LDPE (▲) as a function of the UV irradiation time under SO_2/air atmosphere

Table 2 Contact angle and surface tension data for non-crosslinked LDPE subjected to photosulfonation

Irradiation time [min]	Contact angle [°], water	Contact angle [°], diiodomethane	γ [mJ/m²]	γ^D [mJ/m²]	γ^P [mJ/m²]	Surface polarity [%]
0	99.2	55.4	35.85	35.72	0.13	0.36
1	78.5	52.2	38.57	33.04	5.53	14.34
2	68.6	51	43.69	33.71	9.98	22.84
3	48.9	46.1	56.68	36.43	20.26	35.74
4	31.5	45.1	66.87	36.97	29.90	44.71
5	30.3	44.9	67.50	37.06	30.44	45.10

γ ... surface tension
γ^D ... dispersive component
γ^P ... polar component

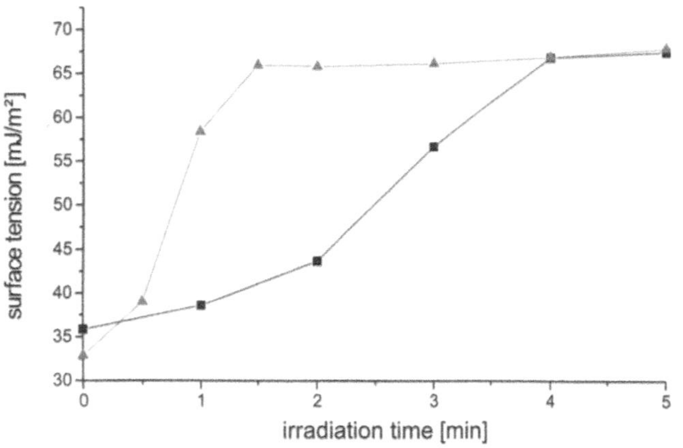

Fig. 5 Photosulfonation of LDPE. Surface tension γ versus UV irradiation time for non-crosslinked (■) and crosslinked (▲) polyethylene

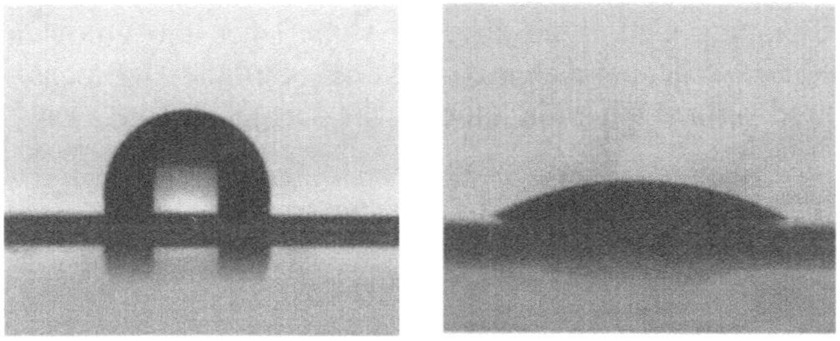

Fig. 6 Photo of a water droplet placed on the surface of unmodified (*left*) and sulfonated LDPE irradiated for 5 minutes (*right*). The contact angle decreases from 99° to 30°

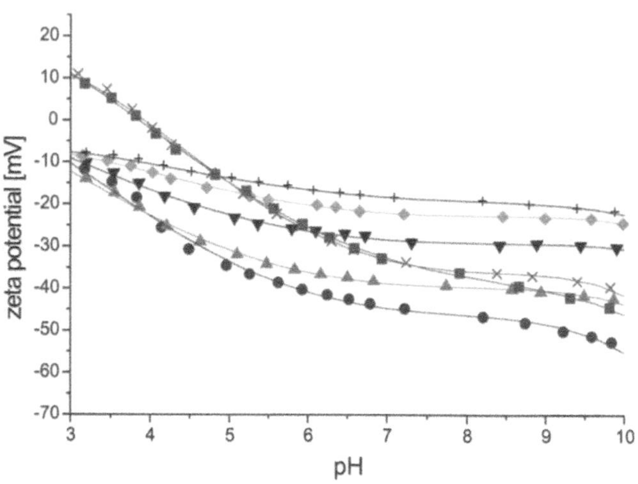

Fig. 7 Non-crosslinked LDPE: Zeta potential versus pH curves of unmodified LDPE (■) and photosulfonated LDPE (• 1 min; ▲ 2 min; ▼ 3 min; ◇ 4 min; + 5 min of UV irradiation). (×) PMMA reference

way, changes in the zeta potential of LDPE depending on the degree of surface modification can be detected and interpreted quite well. Figure 7 displays the zeta potential versus pH curves of unmodified and irradiated LDPE samples (non-crosslinked). The repeatability of all zeta potential measurements shown in this paper was within 2 mV. The ζ potential versus pH curve of unmodified polyethylene is very similar to the curve of the PMMA reference. The isoelectric point (IEP) around pH = 3.9 and the shape of the curve are typical of polymers bearing no dissociating groups. Due to the hydrophobic character of the unmodified polyethylene, preferred absorption of the chloride anions is observed which gives a negative zeta potential in the range of pH = 4–10. Compared to the untreated LDPE all modified samples showed a marked shift of the isoelectric point towards lower pH and a plateau occurring at high pH. The shift of the isoelectric point can be explained by the presence of acidic groups (sulfonic acid groups) on the surface and – as expected – this change in IEP is more pronounced at increasing irradiation time. The plateau of the zeta potential curves at higher pH values is characteristic of acidic groups in general. We observed that the zeta potential shifts to less negative values (at high pH) as a result of the photosulfonation. This is remarkable, since one would expect a trend to more negative values when strongly acidic $-SO_3H$ units are attached to the surface. This finding can be explained as follows: on the one hand, the sulfonic acid groups are deprotonated at high pH values, which would shift the zeta potential to more negative values. On the other hand, the increasing hydrophilicity of the modified LDPE leads to increasing water adsorption and also swelling during contact with the electrolyte solution. Consequently the adsorption of chloride anions to the surface is decreased which results in a shift of the zeta potential to less negative values. When the second effect exceeds the first one, the overall result is a shift of the zeta potential towards less negative values. Indeed, a similar result has been reported in the literature for the sulfonation of polypropylene [25]. As shown in Fig. 7, the photosulfonation of polyethylene seems to be almost finished after four minutes of UV processing. It is found that the changes in the zeta potential (and the IEP) correspond well with the variation in surface tension.

In another part of this study we wished to see the effects of post-modification treatments on the properties of the modified LDPE surface. Polyethylene samples were photosulfonated for different periods of time. Afterwards they were subjected to an after-treatment by conditioning in an electrolyte solution (aqueous KCl, 10^{-3} M) for 48 hours and then characterized by zeta potential measurements. This conditioning process resulted in a shift of ζ to even less negative values (see Fig. 8). This finding may be explained by the swelling of the polymer samples (water adsorption) in water that causes a shift of the shear plane of the electrochemical double layer into the liquid phase. This effect demonstrates that storage conditions and pre-conditioning may exert a pronounced influence on the zeta potential recorded for surface-modified polymers. Phenomena of this kind have already been described in previous literature [26, 27].

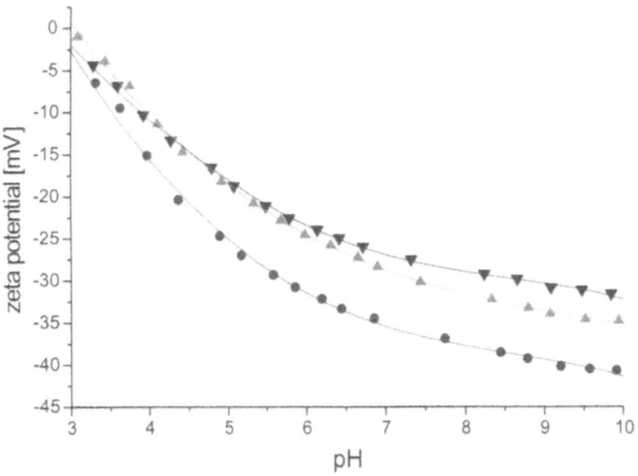

Fig. 8 Photosulfonation of non-crosslinked LDPE (● 1 min; ▲ 2 min; ▼ 3 min of UV irradiation). Zeta-potential versus pH curves recorded after 48 hours of conditioning in an electrolyte solution

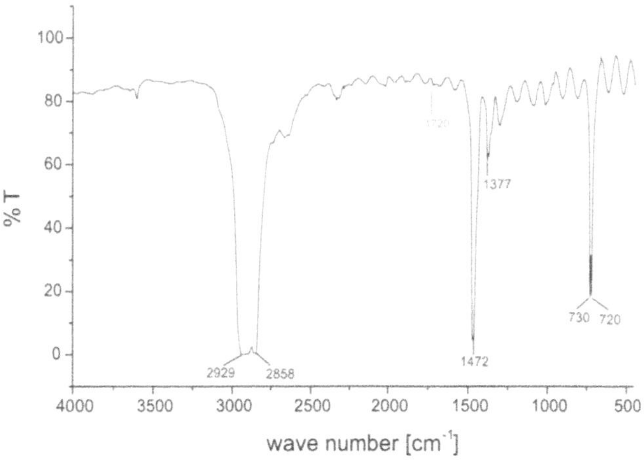

Fig. 9 FTIR spectrum of the e-beam cross-linked LDPE sample (prior to photosulfonation)

Crosslinked LDPE

LDPE samples were crosslinked by e-beam irradiation under inert gas conditions and then extracted exhaustively with m-xylene. Prior to photosulfonation, the cross-linked polyethylene samples were also analyzed with respect to the identity and purity by FTIR-spectroscopy. The FTIR spectrum (see Fig. 9) is virtually identical with that of not cross-linked LDPE and no oxidation products (e.g. carbonyl groups, $\sim 1720\,cm^{-1}$) could be detected.

The initial contact angle of crosslinked LDPE ($\theta = 93.8°$) was slightly lower than the contact angle of non-crosslinked LDPE ($\theta = 99.2°$) which may be a result of some oxidation products present at the surface after the crosslinking process. However, the degree of oxidation at the polymer surface appears to be quite low as no carbonyl species were observed by FTIR spectroscopy (less sensitive).

The contact angle (water/air/sample) of cross-linked polyethylene decreased from 93.8° (unmodified LDPE) to 26.6° after photosulfonation for five minutes. The measured contact angles and the surface tensions γ, γ^P and γ^D of the unmodified and the photosulfonated samples are listed in Table 3. The changes in the contact angle θ and the surface tension γ are represented graphically in Figs. 4 and 5, respectively. For crosslinked LDPE the saturation of the surface with sulfonic acid groups $-SO_3H$ is already achieved after an UV irradiation time of about 90 s. This is explained by a stabilization of the surface layers by the crosslinking pre-treatment. Network formation prevents the extraction of sulfonated polyethylene chains during the post-exposure extraction process with acetone/H_2O. As a result, a higher amount of sulfonic acid groups is maintained at the surface at a given period of photosulfonation.

Table 3 Contact angle and surface tension data for crosslinked LDPE subjected to photosulfonation

Irradiation time [min]	Contact angle [°], water	Contact angle [°], diiodomethane	γ [mJ/m^2]	γ^D [mJ/m^2]	γ^P [mJ/m^2]	Surface polarity [%]
0	94.8	54.1	32.89	31.94	0.95	2.89
0.5	83.4	47	39.04	35.93	3.11	7.97
1	44.5	49.6	58.43	34.49	23.94	40.97
1.5	32.5	46.6	65.98	36.13	29.85	45.24
2	32.6	47	65.86	35.94	29.92	45.43
3	31.7	47.7	66.18	35.55	30.63	46.28
4	29.8	48.9	66.91	34.87	32.04	47.89
5	28.3	48.2	67.9	35.28	32.52	47.89

γ ... surface tension
γ^D ... dispersive components
γ^P ... polar components

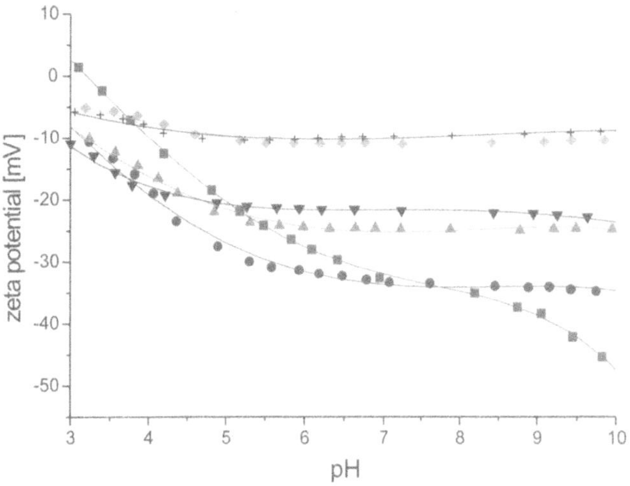

Fig. 10 Zeta potential versus pH curves of cross-linked LDPE prior to (■) and after photosulfonation (● 1 min; ▲ 2 min; ▼ 3 min; ◇ 4 min; + 5 min of UV irradiation)

This is most evident from a comparison of the data obtained for crosslinked and non-crosslinked LDPE surfaces, see Figs. 4 and 5.

The zeta potential versus pH curves of unmodified and photosulfonated LDPE (e-beam crosslinked) are illustrated in Fig. 10. Compared to Fig. 7, a larger amount of sulfonic acid groups seems to be present on the LDPE surface. The shift of the zeta potential to less negative values (at high pH) is an indication for a higher amount of sulfonic acid groups present at the surface of the crosslinked LDPE sample. The zeta potential data are in perfect accordance with the results from contact angle measurements: insolubilization of the polymer by network formation results in a stable surface containing higher amounts of polar groups after photo-induced surface modification. Compared to non-crosslinked LDPE (Fig. 7) the IEP of the unmodified crosslinked sample is somewhat lower (~ 3.2). This can be again explained by the formation of oxidation products during the cross-linking process.

Conclusions

From these investigations the following conclusions can be drawn:

The introduction of sulfonic acid groups onto low-density polyethylene (LDPE) surface was achieved by UV irradiation in the presence of sulfur dioxide and air (photosulfonation). The wettability of the modified polyethylene samples increased with increasing irradiation time but remained constant after prolonged UV irradiation, which was evidenced by contact angle measurements. The zeta potential of the modified LDPE surfaces shifted to less negative values with increasing UV irradiation time, which is explained by an increase of the hydrophilicity of the polyethylene surface. Concomitantly the isoelectric point was shifted to lower pH values, which indicates an increasing amount of sulfonic acid groups present at the sample surface. Furthermore it has been demonstrated that zeta potential measurement is a very sensitive technique to characterize the degree of surface modification of polymers since the contact angle measurements show a plateau after prolonged UV irradiation, whereas the changes of ζ potential-pH-profiles are low, but still noticeable.

Post-modification treatments like storage conditions (air/water) and preconditioning may exert a pronounced influence on the zeta potential obtained for surface modified polymers. Sulfonated LDPE samples, which were subjected to an after-treatment by conditioning in an electrolyte solution, showed a shift of the zeta potential to even less negative values. This may be explained by the swelling of the polymer samples, which occur during storage in aqueous media.

A crosslinking pre-treatment of LDPE prevents the extraction of sulfonated polyethylene chains during the post-exposure extraction process with acetone/H_2O. This is caused by a radiation-induced network formation. As a result, a higher amount of sulfonic acid groups is maintained at the surface at a given period of sulfonation. This was evidenced both by zeta potential and contact angle measurements.

Acknowledgement This study was performed at the Polymer Competence Center Leoben GmbH (PCCL, Austria) within the framework of the K_{plus}-program of the Austrian Ministry of Traffic, Innovation and Technology with contributions of Graz University of Technology (TU Graz), Anton Paar GmbH (Graz) and KEKELIT Kunststoffwerk GmbH (Linz). PCCL is funded by the Austrian Government and the State Governments of Styria and Upper Austria. Further thanks to ARC Seibersdorf research GmbH (Dr. J. Wendrinsky) for performing the e-beam irradiation experiments.

References

1. Chan C-M (1994) Polymer Surface Modification and Characterization. Hanser, Munich
2. Chan C-M, Ko T-M, Hiraoka H (1996) Surf Sci Rep 24:1
3. Armbruster K, Osterhold M (1990) Kunststoffe 80:1241
4. Giroux TA, Cooper SL (1991) J Appl Polym Sci 43:145
5. Bevington JC, Ratti L (1972) Eur Polym J 8:1105
6. Niino H, Yabe A (1993) Appl Phy Lett 63:3527
7. Meyer U, Kern W, Ebel MF, Svagara R (1999) Macromol Rapid Commun 20:515
8. Okoshi M, Murahara M, Toyoda K (1992) J Mater Res 7:1912
9. Garbassi F, Morra M, Ochiello E (1998) Polymer Surfaces – From

Physics to Technology. Wiley, Chichester
10. Siqueira-Petri DF, Wenz G, Schunk P, Schimmel T, Bruns M, Dichtl MA (1999) Colloid Polym Sci 277:673
11. Braun M, Maurette MT, Oliveros E (1991) Photochemical Technology. Wiley, Chichester
12. Orthner L (1950) Angew Chem 62:302
13. Graf R (1952) Justus Liebig Ann Chem 578:50
14. Asinger F, Geisler G, Eckholdt H (1956) Chem Ber 89:1037
15. Harting H (1975) Chem Ztg 99:175
16. Feiertag P (2000) Oberflächenmodifizierung von Polymeren durch Photosulfonierung. Diploma Thesis, Graz University of Technology
17. Feiertag P, Kavc T, Meyer U, Gsoels I, Kern W, Rom I, Hofer F (2001) Synthetic Metals 121:1371
18. Kavc T, Kern W, Ebel MF, Svagera R, Pölt P (2000) Chem Mater 12:1053
19. Kern W, Kavc T (2001) Patent application EP 890227.2
20. Kaneko M, Sato H (2004) Macromol Chem Phys 205:173
21. Owens DK, Wendt RC (1969) J Appl Polym Sci 13:1741
22. Ström G, Frederiksson M, Stenius P. J Coll Interf Sci 10, 119/2:352
23. Smoluchowski M (1921) Handbook of Electricity and Magnetism Vol. 2. Barth, Leipzig
24. Fairbrother F, Mastin H (1924) J Chem Soc 75:2318
25. Stakne K, Smole MS, Kleinschek KS, Jaroschuk A, Ribitsch V (2003) J Mat Sci 38:2167
26. Bismarck A, Mohanty AK, Aranberri-Askargorta I, Czapla S, Misra M, Hinrichsen G, Springer J (2001) Green Chemistry 3:100
27. Wuertz C, Bismarck A, Springer J, Königer R (1999) Prog Org Coat 37:117

Progr Colloid Polym Sci (2006) 132: 62–71
DOI 10.1007/2882_028

Published online: 10 March 2006

J. Friedrich
G. Kühn
R. Mix
K. Hoffmann
U. Resch-Genger

Tailoring of Polymer Surfaces with Monotype Functional Groups of Variable Density Using Chemical and Plasma Chemical Processes

J. Friedrich (✉) · G. Kühn · R. Mix ·
K. Hoffmann · U. Resch-Genger
Bundesanstalt für Materialforschung und -prüfung, 12200 Berlin, Germany
e-mail: joerg.friedrich@bam.de

Abstract Polymer surfaces were modified in low-pressure glow discharge plasmas for introduction of monotype functional groups of different type and density. For this purpose three ways are selected, (**i**) oxygen plasma treatment followed by wet-chemical reduction of O functional groups to OH groups, (**ii**) plasma bromination for introducing C – Br groups and (**iii**) coating by deposition of thin plasma (co-) polymerized layers of functional groups-bearing monomers with OH, NH_2, COOH, epoxy etc. functionalities. Subsequently, these groups were used as anchoring points for chemical grafting of spacer molecules, oligomers, prepolymers, fluorescent labels, ionic and nucleic acid residues, employing different chemical routes. The yield in monosort functional groups at polymer surfaces ranged from 10–14 (process **i**), 20–40 (process **ii**) and 18–31 groups per 100 C atoms (process **iii**) as measured by XPS after derivatization. The consumption of functional groups amounted to 40–90% of all functionalities present at the surface and depended on the dimensions of grafted molecules. For infinitely variably tuning the number of functional groups process **iii** was performed as copolymerization of a functional group-carrying comonomer with a non-functionalized ("chain-extending") comonomer.

Keywords
Introduction of functional groups · Plasma modification · Reactions at polymer surfaces

Introduction

The aim of this work was to introduce monotype functional groups at polymer surfaces for subsequent chemical grafting of a variety of functional molecules, oligomers and polymers. The controlled introduction of functional groups onto most often chemically inert polymer surfaces can be preferentially performed using plasmachemical processes. Such low-temperature glow discharge plasmas are well suited to attack the polymer surface, forming thereby reactive sites, which present binding sites for functional group-forming atoms or radical fragments from plasma. First attempts were made by Hollahan who used NH_3 plasma for introducing primary amino groups onto polypropylene surfaces and thus using these surfaces for interaction with heparin [1]. Later on it could be shown that the reaction of ammonia plasma with polymer surfaces always lead to the formation of co-existing N-functionalities of different types among them NH_2 groups not exceed a threshold concentration greater than 1–3 NH_2 groups per 100 C also for a total N-concentrations $\geq 20N/100$ C [2]. Attempts to unbalance this product distribution using a NH_3/H_2 gas mixture remained unsuccessful [3]. Using the oxygen plasma also a broad variety of different O-functional groups is produced at polymer surfaces [4], amongst them the concentrations of OH groups typically amounting to 2–4 OH groups/100 C. The formation of OH groups demands the preceding abstraction of H from the polymer. Since the binding energies of C – H and C – C bonds are energetically similar a simultaneous polymer chain scission is plausible. Thus, a polymer rearrangement is often observed. Also in this case an improvement in selectivity of the formation of monotype of O-functional groups is not realistic, otherwise the O-functionalization passes into polymer degradation and etching. Therefore, only a two-step process can be considered that counterbalances the unsatisfying result of the O_2 plasma modification by using a wet-chemical reduction process. This was first presented by Nuzzo and Smolinsky [5] and later perfected [6]. In this way a concentration of 10–14 OH groups/100 C was realized (process **i**) with a maximum selectivity of about 60% within all O-functionalities [7].

An elegant one-step process was introduced by our group consisting of the polymer surface bromination by $HCBr_3$ (process **ii**) [8]. Tuning of the Br-concentration was achieved by varying the treatment time. Thus, 20–40 covalently bonded Br-groups/100 C could be introduced onto the surface, with selectivities near 100%. Only traces of Br-anions were found as side-product.

A more complicated but very variable process (**iii**) was the pulsed plasma-initiated polymerization or copolymerization. Here, the desired monotype functional groups are provided by the monomer, which are polymerized in the pulsed plasma. The art in producing such 50 nm thick monotype functionalized polymer coatings lays in carrying out the plasma process under as mild conditions as possible to avoid fragmentation of monomers and to preserve and remain the functional groups of the respective monomers.

More recently, application of the continuous-wave plasma mode allowed polymerization of any evaporable organic substance. Alone this ability showed the complete different polymerization mechanism in plasmas in comparison to classic polymerizations. Accordingly, the plasma polymerization was characterized as "fragmentation-polyrecombination" process or as "atomic" polymerization [9, 10]. It is to be expected that using such a continuous-wave plasma polymerization the retention of functional groups during the process is low within irregular polymer structures caused by the complete fragmentation and random recombination processes. However, this process could be significantly improved by introduction of the pulsed plasma technique consisting of short plasma pulses (0.1–1 ms), required for monomer activation, and long plasma-off periods (1–100 ms) needed for a pure chemical chain propagation instead of the random recombination of fragments. Together with minimizing the plasma power-input the monomer fragmentation could be suppressed and the number of retained functional groups in the deposited plasma polymer layer should be maximum in comparison to the number of functional groups originally present in the monomer molecules. 60–95% of the functional groups could be recovered at the polymer surface [11].

The variety of monomers suited for this "plasma-initiated" polymerization is similar to that of ordinary polymer chemistry. Thus, allyl alcohol, allylamine or acrylic acid produce OH, NH_2 or COOH groups, however, with the anomaly that under low pressure conditions the degradative chain transfer is absent, so that it is possible to produce high-molecular weight allyl polymers. The same reaction principle is valid for the copolymerization used to adjust the number of functional groups. However, also in this case, the rules of classic copolymer chemistry are valid for combining compatible pairs of comonomers. Nevertheless yet, to a smaller content, the known disadvantages of all plasma techniques remain, such as occurrence of trapped radical sites, auto-oxidation, crosslinking and irradiation defects etc. [12].

The accordingly generated functional groups can than be subject to further derivatization and grafting, for example using dioles and diamines (Br groups), silanes and isocyanates (OH groups), aldehydes (NH_2 groups) or alcohols (COOH groups), respectively. In this paper different strategies for controlled functionalization of polymer surfaces are presented including subsequent grafting routes for the introduction of functional molecules, spacers, biomolecules, and prepolymers.

Experimental

Polymer Substrates

Polypropylene (PP) foils with a thickness of 100 µm (goodfellow, U.K.) or 300 µm (Ciba Geigy Switzerland) as well as polyethylene (PE, Lupolen, BASF) foil (100 µm) were ultrasonically cleaned in a diethyl ether bath for 15 min. The monomers used, i.e. acrylic acid, allyl alcohol and allylamine (all > 99%), were purchased from Merck KGaA, Darmstadt, Germany and distilled before use. The monomers, ethylene and butadiene-1,3, supplied by Messer-Griesheim GmbH, Germany, were used as received.

Plasma Functionalization of Polymer Surfaces

Plasma treatments (O_2 and $HCBR_3$ plasmas) were performed in a cylindrical plasma reactor of 50 dm^3 volume. The design of the plasma reactor has been described in detail in [13]. The reactor was equipped with a radio-frequency (r.f., 13.56 Mhz) generator with an automatic matching unit and an r.f. bar antenna (length: 35 cm). The substrate foil was mounted on a continuously rotating, grounded steel cylinder (ø = 10 cm, length = 35 cm, rotating frequency = 0.5–1 s^{-1}) at the distance of 10 cm to the r.f. powered electrode. Surface functionalization of PP foils in the O_2 plasma was performed in the continuous-wave (cw) mode. The power input was varied between

$P = 10–100$ W. Mass flow controllers for gases and a standard pressure of 6 Pa were adjusted. For production of an oil-free high vacuum a turbomolecular pump was applied. Using aggressive gases or vapours a cryo trap was interposed between reactor and pump.

Deposition of Functional Groups Bearing Plasma Polymer Layers

Plasma polymer layers were deposited in the same reactor as described before. However, in this case, the pulsed plasma mode was applied. The duty cycle of pulsing was adjusted generally to 0.1 and the pulse frequency to 10^3 Hz. The power input was varied between $P =$ 100–300 W. Mass flow controllers for gases and vapours, a heated gas/vapour distribution in the chamber, and control of pressure and monomer flow by varying the speed of the turbomolecular pump were used. The gas flow was adjusted to 75–125 sccm and the pressure was varied between 10 to 26 Pa depending on the respective polymerization or copolymerization process. The deposition rate was measured by a quartz microbalance.

Surface Analysis

The XPS (X-ray Photoelectron Spectroscopy) and IR analyses have been described in detail elsewhere [14]. Here, only some important facts are summarized. The XPS data acquisition was performed with a SAGE 150 Spectrometer (Specs, Berlin, Germany) using a non-monochromatized MgK_α or AlK_α radiation with 12.5 kV and 250 W settings at a pressure $\approx 10^{-7}$ Pa in the analysis chamber. XPS spectra were acquired in the constant analyzer energy (CAE) mode at 90° take-off angle. Peak analysis was performed using the peak fit routine from Specs.

The FTIR spectra were recorded with a NEXUS instrument (Nicolet, USA) using the ATR (Attenuated Total Reflectance, 45° angle of incidence) technique with a diamond or a Ge cell ("Golden Gate", Specac, Kent, UK). The IR signal results from a near-surface layer of the polymer film. The information depth depends on the material used as ATR crystal and amounts to about < 1.5 µm using Germanium and 2.5 µm applying diamond.

Metal Evaporation and Peel Test

For measuring the Al-polymer (PP) peel strengths the plasma polymerization was performed using the previously described reactor. Then, the plasma polymer coated polymer samples were transferred into a separate electron beam metallizer (Auto 306, Edwards, UK). The thickness of deposited aluminium layers was adjusted to 150–200 nm using a quartz microbalance. The metal peeling technique for Al-PP and Al-PTFE systems followed DuPont's preparation and peeling procedure. The Al side of the composite was fixed on a double-faced scotch tape, which was pressed onto a steel plate. Then, a 90° peel test was carried out at a peel speed of 25 mm/min for all Al-polymer systems.

Spectroscopic Characterization of Surface Functionalities by Means of XPS Probing

Derivatization of OH Groups. Samples with OH groups at the surface were exposed to trifluoroacetic anhydride (TFAA) vapour for 10–15 min after evacuation to 10^3 Pa in order to transform the hydroxyl groups into a fluorinated derivate that can much better detected and quantified:

$$|{-}OH + (CF_3 - CO)_2O$$
$$\rightarrow |{-}O - CO - CF_3 + CF_3 - COOH$$

Afterwards these samples were degassed at a pressure of 10^{-5} Pa to remove free trifluoroacetic acid [15]. The derivatization of OH groups was proved to be highly selective (ca. $> 90\%$ [16]). F1s peak intensities in the XP spectrum as well as $> C=O$ (1785 cm^{-1}) and CF_3 (1223 cm^{-1}) stretching vibrations were used to evaluate the number of surface OH groups. The yield of this reaction is about 90%.

Derivatization of NH_2 Groups. Primary amino groups react with pentafluorobenzaldehyde (PFBA) by formation of an schiff's base:

$$|{-}NH_2 + OHC - \text{Phenyl} - F_5 \rightarrow |{-}N = C - \text{Phenyl} - F_5$$

Similar to the OH derivatization, labelling of the NH_2 groups was performed as a gas-phase reaction. The samples were exposed to PFBA vapour at 45 °C for 2 h and subsequently degassed at a pressure of 10^{-5} pa for 8 h to remove non-consumed PFBA. This labelling reaction reaches yields in the range of 80% [17].

Derivatization of COOH Groups. Carboxylic groups were identified by labelling with vapours of trifluoroethanol in the presence of di-*tert*-butyldiimide and pyridine over 8 to 20 h [18]:

$$|{-}COOH + CF_3 - CH_2 - OH +$$
$$(CH_3)_3C - N = C = N - C(CH_3)_3 \rightarrow$$
$$|{-}COO - CH_2 - CF_3 +$$
$$(CH_3)_3C - NH - CO - NH - C(CH_3)_3$$

Afterwards the samples were degassed at a pressure of 10^{-5} Pa to remove non-reacted components. The yield of this reaction is about 90% (see also [19]).

Reduction and Grafting Techniques

General Method. All solvents used were of analytical reagent grade. Toluene and THF were refluxed under N_2, and over sodium and benzophenone. Pyridine was distilled and stored over KOH pellets. Water was doubly distilled before use. All the silanes were vacuum-distilled and all reactions were performed under N_2 atmosphere. Reactions on polypropylene films were normally carried out with pieces of 5 × 5 cm or for industrial applications with foils of 35 × 24 cm.

Reduction of O_2 Plasma Treated Polypropylene Film. The oxygen plasma treated foils were stirred under N_2 at room temperature in 12 ml dry THF and 3 ml of 1 M diborane solution for 18 hours. The foils were removed and immersed in an alkaline H_2O_2 and THF for 2 hours. Then, the foils were washed with THF, water (thrice) and with methanol. The modified foils were dried and stored in a desiccator. The reduction with $LiAlH_4$ was performed in ether for 2 hours. Vitride (SDMA-Sodium dihydrobis(2-methoxyethoxy)aluminate) was processed in toluene for 2 hours and then the reaction product was hydrolized with NaOH.

Glycolate Spacer. For grafting of glycolate spacers PP foils with C – Br groups were dipped into a glycolate solution. For this purpose dried glycolates were dissolved in dried tetrahydrofuran and mixed with sodium at 50 °C employing NaI as catalyst. Then, the plasma-brominated PP foil was rotated in the glycolate solution for 20 h. Washing and extracting processes to remove by-products completed the reaction. The derivatization with diamine spacers was achieved using a similar procedure as with glycolates.

$$|{-}Br + Na[O - CH_2 - CH_2]_6 - OH \rightarrow$$
$$|{-}[O - CH_2 - CH_2]_6 - OH + NaBr \quad \text{(Williamson)}$$
$$|{-}Br + Na[NH - CH_2 - CH_2]_6 - NH_2 \rightarrow$$
$$|{-}[NH - CH_2 - CH_2]_6 - NH_2 + NaBr \quad \text{(Gabriel-like)}$$

Undecenylsilane Spacer. The undecenyl chloride was prepared by chlorination of undecenyl alcohol using carbon tetrachloride and triphenyl phosphine, in 80% yield. After purification of undecenyl chloride, it was converted into a Grignard reagent and was added to a solution of $SiCl_4$ in dry benzene. This afforded 10-undecenyltrichlorosilane in a 75% yield. NMR and IR spectroscopies characterized all intermediates and final products.

Adsorption of Undecenyl Trichlorosilane onto the Polypropylene Film. Chemisorption of undecenyltrichloro-silane was performed at ambient temperature in stirred solution of undecenyltrichlorosilane (concentration –0.02 M) in a solvent mixture of 80% *n*-hexadecane + 12% carbon tetrachloride + 8% chloroform. All the solvents were passed through columns of alumina and chloroform was stabilized with absolute ethanol (1 vol %). The polypropylene film was immersed in the above solution for 1–2 minutes and washed with carbon tetrachloride and water and then extracted with 1 : 1 v/v chloroform and ethanol for 15 minutes in a Soxhlet apparatus [20].

$$|{-}OH + Cl_3Si - (CH_2)_9 - CH = CH_2 \rightarrow$$
$$|{-}O - Si(OH)_2 - (CH_2)_9 - CH = CH_2$$

Hydroboration of Vinyl Groups. The 10-undecenyltrichlorosilane coated polypropylene substrate was immersed into a solution of diborane in THF (1M, Aldrich) and was dipped for 1 minute into an alkaline H_2O_2 solution. The film was rinsed with distilled water and dried.

$$|{-}O - Si(OH)_2 - (CH_2)_9 - CH = CH_2 + B_2H_6/H_2O_2 \rightarrow$$
$$|{-}O - Si(OH)_2 - (CH_2)_{11} - OH$$

Adsorption of γ-Aminopropyltrimethoxysilane onto the Polypropylene Film. The reaction with γ-aminopropyltrimethoxysilane was performed using 2% v/v toluene solution under N_2 atmosphere for 16 hours. The film was successively washed with toluene, chloroform, and acetone. Then the film was washed with water for half an hour. Finally, methanol washing was performed and the film was dried.

Introduction of the Glutaraldehyde Spacer. The γ-aminopropyltrimethoxysilane modified surface was treated with 5% v/v aqueous glutaraldehyde solution. The reaction was performed at ambient temperature for 24 hours. The film was washed with water and dried.

Alanine or Cysteine Grafting. The PP surface modified with γ-aminopropyltrimethoxysilane and glutaraldehyde was incubated in a solution consisting of 1.0 g of alanine or cysteine dissolved in 10 ml of doubly distilled water for a period of 3 hours. The, the film was washed with water and methanol and dried.

Reactions of Isocyanates with OH and NH_2 Groups. Under exclusion of humidity, isocyanate-containing monomers react slowly with OH groups at polymer surfaces. The reaction was catalyzed by dibutyltin dilaureate. Thus, a covalent linkage by formation of urethane groups was reached. Using diisocyanates and hydrolyzing the non-consumed isocyanate groups the formation of spacer bonded amino groups is possible. Using surfacial NH_2 groups instead of OH groups isocyanates react under formation of urea linkages.

Results

Monotype Functionalization

Reduction of Oxygen Plasma Introduced Oxygen-containing Groups on PP and PE Surfaces to OH Groups by Diborane. Using the oxygen r.f. pulsed-plasma modification the attachment of oxygen containing functional groups onto the surface of PP could be demonstrated by a series of ATR-FTIR spectra (Fig. 1). A strong carbonyl band ($\approx 1700\,\text{cm}^{-1}$), alcoholic structures ($\approx$ 1100–1200 cm^{-1}), a broad OH peak and adsorbed water related features ($\approx$ 3100–3500 cm^{-1}) appear. Surfaces with an O/C ratio of about 0.2 were used for further chemical processing.

The XP spectra of PP of different stages of processing are shown in Fig. 2, e.g. for as received, after low-pressure oxygen plasma pretreatment, diborane reduction and derivatization with TFAA. The plasma treatment produces a strong O1s peak. After TFAA derivatization of plasma-generated OH groups only a small F1s peak at a binding energy of 689 eV is detected at the PP surface indicating low concentration of OH groups after O_2 plasma treatment.

This low concentration of OH groups is not surprising because the oxygen plasma cannot deliver additional hydrogen for the formation of OH groups. Such hydrogen sources may be abstracted hydrogen from the polymer backbone and hydrogen from dissociation of desorbed water from the walls of the reactor. Comparing the F1s peaks in the survey scans with and without diborane reduction and derivatization of formed OH groups (cf. Fig. 2) it is obviously that the diborane reduction strongly increases the yield of OH groups from about 2–4 OH/100 C without further chemical post-plasma treatment to 11–13 OH/100 C under optimal conditions of the reduction with diborane. These conditions were the hydrolysis with NaOH and the oxidation of $C=C$ bonds in the presence of H_2O_2.

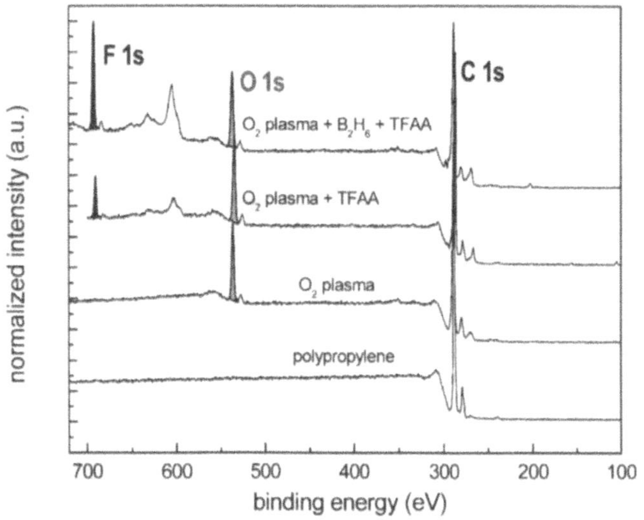

Fig. 2 XP spectra after different treatment steps of pp foils

Comparison of Reduction between Diborane and $LiAlH_4$. Using $LiAlH_4$ as reducing agent $C-O-C$ bonds can be additionally converted to OH groups. Accordingly, the yield in OH groups is thus slightly increased (Fig. 3). However, the $C-O-C$ scission are unwished, leads to a degradation of the polymer backbone and a decrease in molecular weight. Thus, it cannot be excluded that a number of newly formed OH groups are removed together with polymer fragments during wet-chemical processing. Similar results were obtained using vitride as reducing agent.

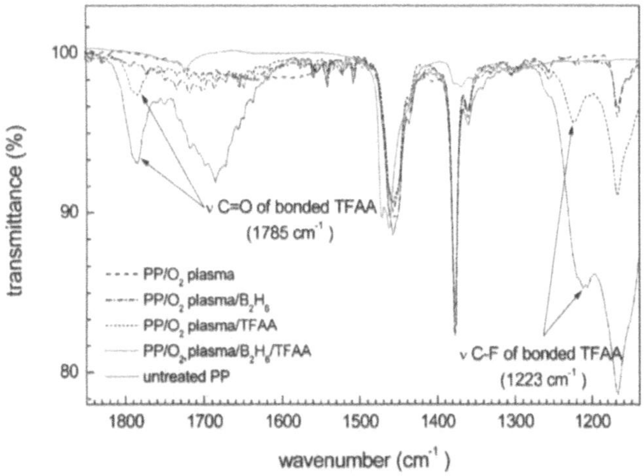

Fig. 1 ATR-FTIR absorbance spectra after different treatment steps of PP foils using a Ge crystal and 45° angle of incidence

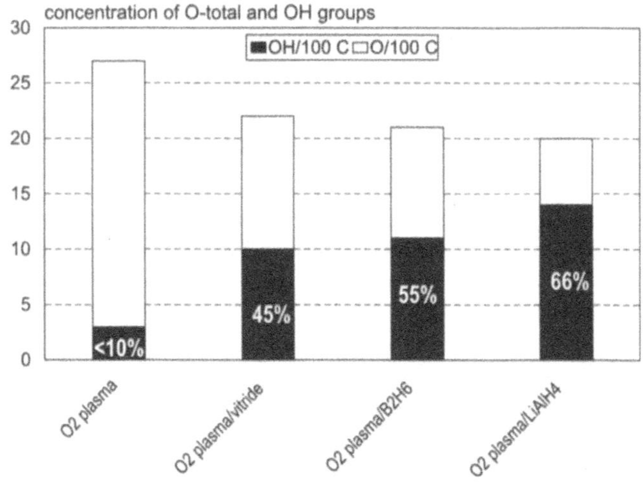

Fig. 3 Yield of OH groups for different types of wet-chemical reduction of O functional groups produced by O_2 plasma pretreatment

Bromination. Bromination of polymer surface employing the bromoform plasma presents a highly selective method of controlled surface functionalization. Only traces of Br anions were detected as negligible side-products by means of XPS and, under adverse conditions, the deposition of polymer layers was observed as measured with a microbalance. The XPS measured introduction of Br at polymer surfaces for two types of plasma treatment is illustrated in Fig. 4.

The percentage of surface bromination could be adjusted between 0 and 45 Br per 100 C by variation of treatment time. Exposure to air causes a post-plasma addition of oxygen.

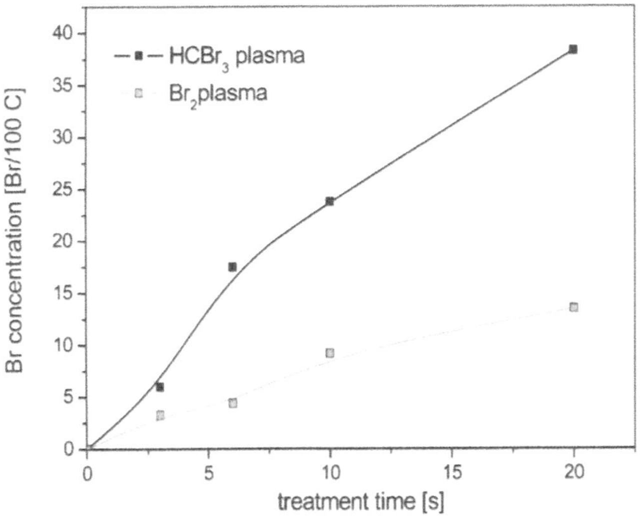

Fig. 4 Introduction of Br vs. treatment time

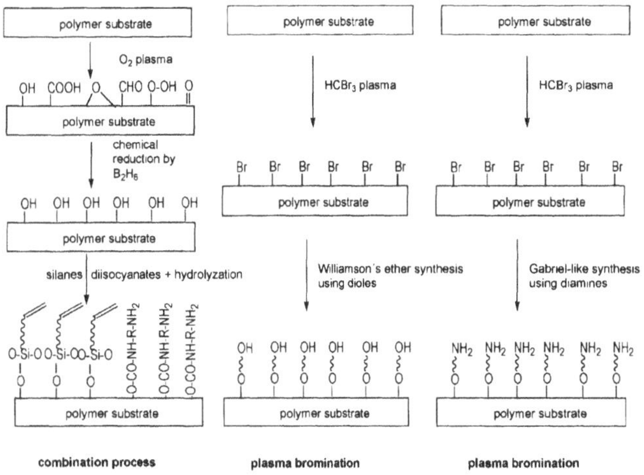

Fig. 5 Scheme of synthesis at brominated or hydroxylated polymer surfaces

Williamson's Ether Synthesis. The C – Br groups of brominated PP surfaces reacted with sodium ethylene glycolates with $n = 2, 3, 4, 5$ and 22 (PEG 1000) glycol units yielding in spacer-bonded OH groups. The concentration of these spacer-bonded terminal OH groups was in the same order as that of the directly to the surface bonded OH groups. It was also possible to attach a C_6 spacer onto C – Br groups with an amino end group using sodium hexamethylene diamine (Fig. 5).

Both end groups at the spacer molecule, OH and especially NH_2, are important for further biochemical synthesis. As follows from XPS measurements, in all cases only 2% Br was not consumed and remained at the PP surface. It was assumed that either steric problems played a role or these bromine groups were not situated directly at the surface. For all glycolate spacers a "functionalization density" of about 5 spacer-bonded OH groups per 100 C was measured. Additionally, two directly at the surface bonded (without spacer) OH groups per 100 C atoms were also detected. 4–5 NH_2 groups/100 C were found using the hexamethylene diamine spacer. In each case the starting concentration of bromine groups was about 10 C – Br groups, so that about 50% of these groups were consumed by attachment of spacer molecules.

Silane Spacer. Undecenyltrichlorosilane was reacted with OH groups produced by oxygen plasma modification and hydroboration (B_2H_6/H_2O_2) (cf. Fig. 5). The yield of this reaction (attachment of C_{11} spacers) amounts to about 7. This means a selectivity of about 70% starting with 10 OH groups per 100 C. Additionally 2–3 unreacted and directly bonded OH groups per 100 C atoms remained at the polymer surface. Subsequently, the undecenyl spacer was modified and now bearing bromo, cyano or amino groups in comparable yields and selectivities.

Azomethine Group Containing Spacers and Coupling of Amino Acids. Another approach to covalently fix spacer molecules onto a hydroxylated polymer surface consisted in the attachment of γ-aminopropyltrimethoxysilane. This results in a polymer surface coated with primary amino groups. These groups reacted with glutaraldehyde acting as a spacer. The spacer can facilitate the attachment of bioactive molecules such as phospholipids or amino acids and can be also used as a starting-point for biochemical synthesis. Therefore, simple amino acids, alanine and cysteine, were attached to the spacer (cf. Fig. 6). 4 to 6 carboxylic end groups were formed by this reaction pathway starting from 10 OH/100 C produced by oxygen plasma treatment and diborane reduction.

Isocyanate Grafting. Weakly O_2 plasma-treated and diborane reduced polypropylene foils with about 7 OH/100 C were reacted with diisocyanates (hexamethylene diisocyanate (HDI), methylene-di-p-phenylene isocyanate (MDI) and toluylene-2,4-diisocyanate (TDI)) as schematically shown in Fig. 7. Subsequently, different fluorescence

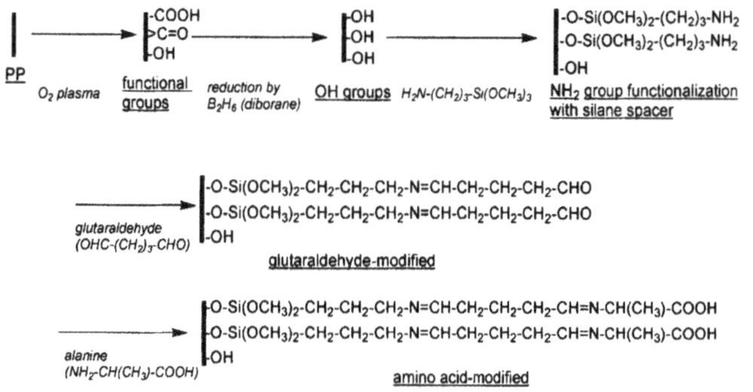

Fig. 6 Reaction scheme of covalent alanine grafting onto polypropylene surfaces

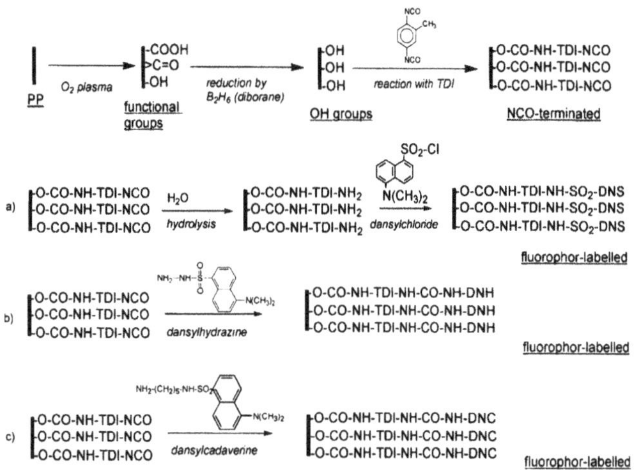

Fig. 7 Schematics of the reactions of hydroxylated PP surfaces with diisocyanates and fluorophores

labels as dansylchloride (DNS), dansylhydrazine (DNH) and dansylcadaverine (DNC) were grafted.

The reaction with HDI was too slow and the hydrolysis (Fig. 7a) not clearly reproducible. Using MDI and TDI the reactions 7b and 7c result in the attachment of ca. 1 fluorophore/100 C and in satisfying absorbance-fluorescence spectra.

Stability of Surface Functionalization. Precondition for any chemical processing of functional groups at polymer surfaces is their stability during storage and chemical consumption. According to XPS data, long-time storage of O_2 plasma treated and diborane reduced polymer samples resulted in a loss of functional groups. 30 to 50% of all OH groups disappeared after storage of 6 month. This decrease in concentration of functional groups during exposure to the air could be mostly prevented by storage at −20 °C. However, the reactivity of these long-time stored samples was high enough for further biochemical processing. One possible explanation for the diminution of functional groups was given by H. Yasuda who had observed diffusion of functionalized macromolecule segments away from the surface to the bulk of polymers [21].

Moreover, wet-chemical processing, especially the diborane reduction, has quenched all trapped C radical sites in polymers (cf. [22]), inevitably formed by each plasma processing of polymers. Thus, auto-oxidation processes during storage are nearly eliminated as XPS results let suggest. The formation of low molecular weight species in the sub-surface during plasma treatment and then their subsequent diffusion to the surface was generally avoided by using short plasma treatment times (O_2 plasma $t \leq 2$ s, HCBr plasma $t \leq 5$ s. More recently, it could be demonstrated that such short exposure to the oxygen plasma neither change the supermolecular structure of polymers nor produce polymer degradation [23].

Deposition of Functional Groups-carrying Plasma Deposited Homopolymers as Anchoring Points for Grafting

Pulsed Plasma Polymerization. Allyl alcohol, allylamine and acrylic acid were polymerized in pulsed plasma to retain a maximum in functionalities in the resulting plasma polymer. The retention of functional groups during the plasma process, introduced by functional-group carrying monomers and followed by deposition to polymer layers, was primarily measured by XPS including the chemical derivatization of these groups as described in Experimental. These layers were also checked for side-products by respective IR spectra. The results are summarized in Table 1.

Table 1 Absolute and relative yields in functional groups at the surface of deposited pulsed plasma polymer layers measured with XPS after derivatization (cf. Experimental, 100 W)

Group	Monomer	Theoretical stoichiometry	Measured stoichiometry	Yield in functional groups (per 100 C)	Characteristic side-products
OH	Allyl alcohol	100 C/ 33 O	100 C/ 33 O	31 (95%)	$>C=O$
NH_2	Allyl-amine	100 C/ 33 N	100 C/ 36 N/4 O	21 (65%)	$C-N-C$, $C\equiv N$, $C-O$
COOH	Acrylic acid	100 C/ 67 O	100 C/ 60 O	26 (79%)	$>C=O$
-	Ethy-lene	100 C	100 C/ 3 O	-	$>C=O$, $C-O$, $C=C$

Homopolymerization. OH and COOH groups containing homopolymer layers could be reproducibly deposited without significant amounts of by-products and without ageing tendency if exposed to the air. Therefore, these polymer layers were best qualified as models for monotype functionalized substrates with variable concentrations of functional groups should be proved. To vary the density of functional groups the plasma-initiated copolymerization of acrylic acid or allyl alcohol with ethylene or 1,3-butadiene was performed. The results are shown in terms of measured COOH- and OH-concentration as a function of the composition of the respective comonomer mixture (Fig. 8). Depending on the comonomer reactivity, a more linear correlation between precursor composition and produced COOH groups (butadiene) or a parabolic behaviour with subsequent saturation (ethylene) was observed. For each type and concentration of functional groups its concentration was determined by chemical derivatization followed by XPS analysis as described in Experimental.

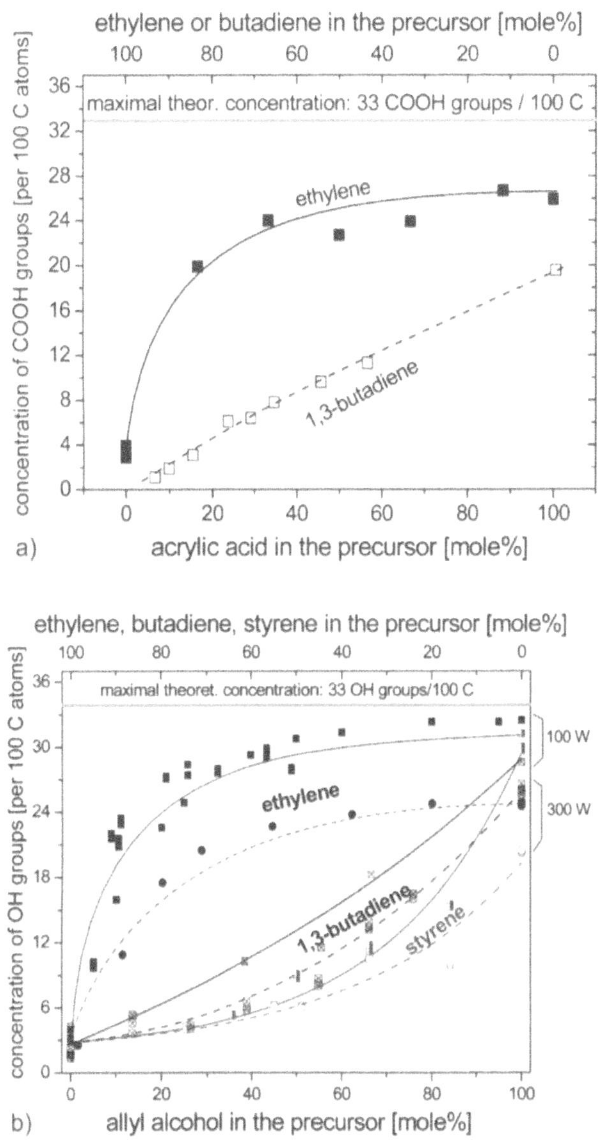

Fig. 8 Yield in COOH groups (**a**) and OH groups (**b**) in dependence on the composition of the precursor mixture and the types of comonomers (a – acrylic acid + ethylene or butadiene; b – allylalcohol + ethylene or butadiene or styrene)

As shown in Fig. 8, similar tendencies were detected considering the allyl alcohol copolymerization with ethylene, butadiene or styrene. Here, different curve progressions are observed depending on the types of comonomers, their ratio in the precursor mixture, and the plasma parameters. These deviations were attributed to the different tendencies to undergo a copolymerization, for example expressed in copolymerization parameters, which are also valid for the plasma-initiated gasphase copolymerization.

Taking these curves as reference or calibration it was possible to adjust a defined concentration of functional groups in the deposited layer by choosing the appropriate comonomer ratio (Fig. 8).

In the same way allylamine-copolymer and pure polyethylene-like layers were deposited. moreover, the nature and the quantity of side-products of these polymers were investigated. Generally, the number of side-products was minimum for low power values (cf. Fig. 8 and Table 2).

It should be remembered that the aim of this work was to produce defined layers with monosort functional groups, which can be used for grafting. Now, in contrast to the irregularly structured continuous-wave plasma polymers, the structure of pulsed plasma polymers was so much improved that partial or complete solubility was observed. Therefore, the further chemical processing in solvents and water led to dissolving the layer. Here, also chemically crosslinking copolymers as butadiene, divinylbenzene and trivinylcyclohexane were necessary as schematically shown in Fig. 9.

Introduction of Fluorescence Labels onto the Surface of Plasmapolymer Layers. Fluoresceinisothiocyanate should

Table 2

Group	Monomer	Maximum concentration (X per 100 C)	Suited comonomers	Range of concentration (X per 100 C)
OH	Allyl alcohol	31	C_2H_4, C_4H_6, C_8H_8	0–31
NH_2	Allylamine	18	C_2H_4, C_4H_6, C_8H_8	0–18
	Diaminocyclohexane	12	C_2H_4	0–12
COOH	Acrylic acid	25	C_2H_4, C_4H_6, C_8H_8	0–25
Epoxy	Glycidylmethacrylat	10	–	–

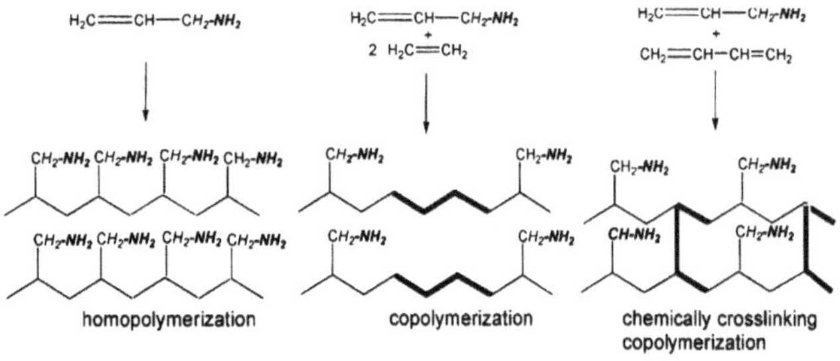

Fig. 9 Schematics of chemically crosslinking by copolymerization

be covalently bonded to polymer surfaces via NH_2 groups instead of OH groups. For this purpose, allylamine (and allylamine-butadiene copolymers – in process) polymer layers were deposited onto polypropylene foils (cf. Fig. 10). These layers were relatively stable at processing in aqueous solution or in thf. by means of derivatization of NH_2 groups with pentafluorobenzaldehyde or *p*-trifluoromethylbenzaldehyde their concentration was calculated to about 12 NH_2/100 C. Two reaction pathways follow, the consumption with the thioisocyanate (a) and with glutaraldehyde and dansylhydrazine or dansylcadaverine (Fig. 10). Immediate transfer to wet-chemical processing avoided problems with the auto-oxidation of the amino group containing plasmapolymers, when exposed to air.

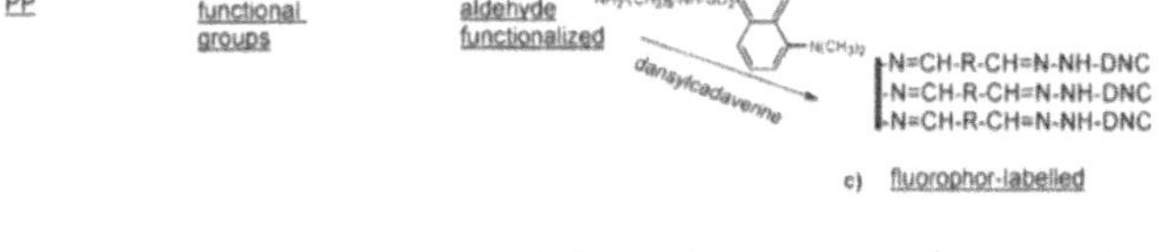

Fig. 10 Covalent bonding of fluorophores onto primary amino groups of allylamine pulsed plasma polymer layers (**a** – fluorescein bonding; **b** – dansylhydrazine bonding; **c** – dansylcadaverine bonding)

Application of Plasmapolymers as Adhesion-promoting Layers in Metal-polymer Composites. Functional group carrying plasma polymers are also well suited as adhesion-promoting interlayers in metal-polymer composites [24]. Monotype-functional groups at polypropylene surfaces and tuning of their density are prerequisites for the investigation of the individual contribution of aluminium-functional group interactions to the respective bond strength of Al-evaporated (200 nm) – polypropylene composites. A linear dependence of aluminium adhesive bond strength on the concentration of functional groups at the interface was found for peel strengths from 0 to at least 300 N/m (Fig. 11). This linearity is limited by changing from pure interface failures to cohesive failures in the metal-polymer systems. The slope of these linear curve sections correlates well with the tendencies of functional groups to undergo a reaction with aluminium atoms and with the strength of formed covalent Al-polymer bonds. An important message from these experiments was that chemical interactions of metal-functional groups play a dominant role as demonstrated by the high strength of composites with functional groups, which can form chemical interactions or bonds to Al.

It should be remarked that the plasma surface functionalization by oxygen plasma with and without reduction to monotype OH functional groups yields a lower peel strength compared to the use of the monotype functional groups bearing adhesion-promoting plasmapolymer layers (cf. Fig. 11).

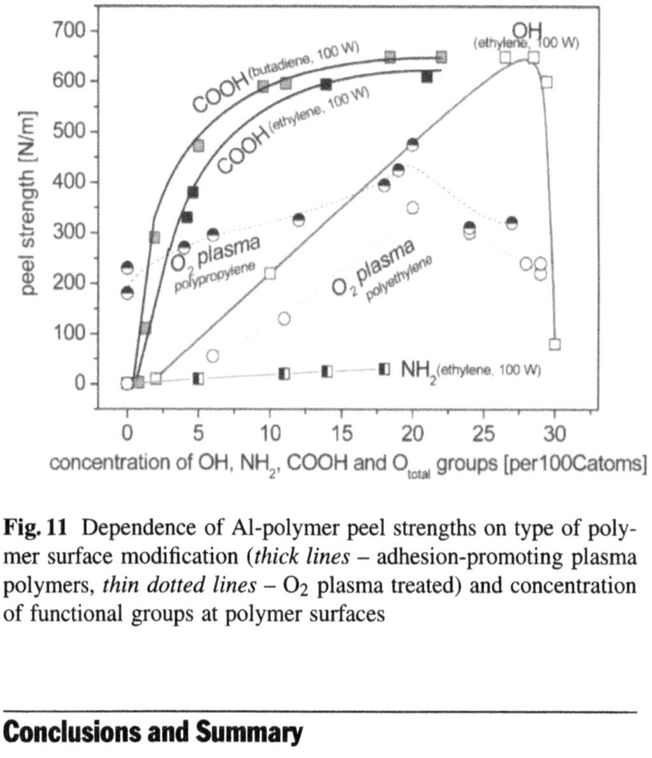

Fig. 11 Dependence of Al-polymer peel strengths on type of polymer surface modification (*thick lines* – adhesion-promoting plasma polymers, *thin dotted lines* – O_2 plasma treated) and concentration of functional groups at polymer surfaces

Conclusions and Summary

A few synthetic routes were presented for preparation of inert polymer surfaces with chemically reactive functional groups as Br, OH, COOH, NH_2 etc. These functional groups are accessible for grafting of other chemical species as organic molecules with specific functions and prepolymers. Another application of these functional groups is their reactivity towards metal atoms for producing well-adhering metal-polymer composites. For the first time, these functional groups are provided as more or less monotype functional groups. This monosort functionalization on polymer surfaces is a pre-requisite for both the starting of chemical graft reactions of biomolecules as well as the investigation of the specific contribution of each type of (mono-) functional groups to the metal-polymer interactions in composites.

Monotype functional groups (OH, NH_2, COOH, Br) could be formed by a combination process (plasma oxidation + chemical reduction, process **i**), by bromination (process **ii**) and by coating with monotype functional groups carrying polymer layers (process **iii**). the combination process leads to moderate yield in OH group formation (10–14 OH/100 C) and shows moderate selectivity (60% of all O-functionalities appear as OH groups). On the other hand, this process allows to covalently bind the OH groups directly to the polymer molecules. The OH modification of polymer surfaces was stable if stored and exposed to air over longer periods of time (> 1 year). The desired adjustment of the OH group density at the polymer surface could not be achieved on this way.

The tuning of the density of functional groups was reached applying the bromination. This functionalization was very selective and high concentrations of C – Br groups could be introduced to the polymer surface. The functional groups were also directly bonded to polymer backbones. Nearly all Br-sites were situated at the surface and, therefore, open for chemical graft reactions. The concentration of C – Br groups was maximum compared with the other functionalization methods. Up to 40 C – Br/100 C could by produced with only negligible amounts of side-products. The concentration of C – Br groups could be slightly tuned by variation of the bromination time.

Plasmapolymers from functional groups-carrying monomers allowed a broad range of variation of both the type of monosort functional groups and the concentration. The concentration could be infinitely varied by copolymerization from 0 to 31 groups per 100 C atoms. In such a way OH and COOH groups carrying layers were produced using allyl alcohol respectively acrylic acid. These polymer layers were resistant to long-time exposure to air. In contrast, NH_2 groups containing layers possess generally a moderate yield in NH_2 retention, indications for reactions at the amino groups and for a pronounced tendency of post-plasma oxidation. This oxidation during exposure to air may be ascribed to the auto-oxidation with oxygen from air initiated by trapped C radical sites in the layer.

Applications of these layers as supports for sensor devices and biocompatible materials as well as for adhesion promotion have illustrated the ability of polymer surface chemistry.

References

1. Hollahan JR, Stafford BB, Falb RD, Payne ST (1969) J Appl Polym Sci 13:807
2. Friedrich JF, Gähde J, Frommelt H, Wittrich H (1976) Faserforsch Textiltechn/Z Polymerenforsch 27:604
3. Meyer-Plath A (2003) PhD-thesis, Ernst-Moritz-Arndt University Greifswald
4. Rossmann K (1956) J Polym Sci 19:141
5. Nuzzo RG, Smolinsky G (1984) Macromolecules 17:1013
6. Kühn G, Weidner S, Decker R, Ghode A, Friedrich JF (1999) Surf Coatings Technol 116–119:796
7. Kühn G, Retzko I, Lippitz A, Unger WES, Friedrich JF (2001) Surf Coat Technol 142–144:494
8. Friedrich JF, Unger WES, Lippitz A, Koprinarov I, Ghode A, Geng Sh, Kühn G (2003) Composite Interface 10:139
9. Friedrich JF, Gähde J, Frommelt H, Wittrich H (1976) Faserforsch Textiltechn/Z Polymerenforsch 27:517
10. Yasuda HK (1979) ACS Symp Ser 108:37
11. Friedrich JF, Kühn G, Mix R, Unger WES (2004) Polymer Processing and Plasmas 1:28
12. Friedrich JF, Kühn G, Mix R, Retzko I, Gerstung V, Weidner S, Schulze R-D, Unger WES (2003) Plasma Polymer Adhesion Promoters to be used for Metal-Polymer Composites. In Mittal KL (ed) Polyimides and other High Temperature Polymers: Synthesis, Characterization and Applications. VSP, Utrecht, p 359–388
13. Friedrich JF, Retzko I, Kühn G, Unger W, Lippitz A (2001) Metal doped plasma polymer films. In: Mittal KL (ed) Metallized Plastics, vol 7. VSP, Utrecht p 117–142
14. Retzko I, Friedrich JF, Lippitz A, Unger WES (2001) J Electr Spectr Rel Phenom 121:111
15. Chilkoti A, Ratner BD (1996) Derivatization of OH groups. In: Sabbattini L, Zambonin PG (eds) Surface Characterization of Advanced Polymers. VCH Publishers, Weinheim, p 221–259
16. Geng Sh (1996) PhD thesis, University of Potsdam
17. Everhart DE, Reilley CN (1981) Anal Chem 53:665
18. Nakayama Y, Takayuka T, Soeda F, Hatada K, Nagaoka K, Suzuki J, Ishitani A (1988) J Polym Sci: Part A 26:559
19. Alexander MR, Wright PV, Ratner BD (1996) Surf Interf Analysis 24:217
20. Sagiv J (1980) J Amer Chem Soc 102:92
21. Yasuda T, Okuno T, Yoshida K, Yasuda H (1988) J Polym Sci B: Polym Phys 26:1781
22. Garbassi F, Morra M, Occhiello E (1998) Polymer Surfaces. Wiley, Chichester, p 65, p 318–349
23. Friedrich JF, Unger WES, Lippitz A, Koprinarov I, Kühn G, Weidner S, Vogel H (1999) Surf Coat Technol 116 TO 119:772
24. Friedrich JF, Loeschcke I, Gähde J (1986) Acta Polymerica 37:687

Progr Colloid Polym Sci (2006) 132: 72–81
DOI 10.1007/2882_037

Published online: 8 March 2006

Alla Synytska
Leonid Ionov
Victoria Dutschk
Sergiy Minko
Klaus-Jochen Eichhorn
Manfred Stamm
Karina Grundke

Regular Patterned Surfaces from Core-Shell Particles. Preparation and Characterization

Alla Synytska · Victoria Dutschk · Klaus-Jochen Eichhorn · Manfred Stamm · Karina Grundke (✉)
Leibniz Institute of Polymer Research Dresden e.V., Hohe Str. 6, 01069 Dresden, Germany
e-mail: grundke@ipfdd.de

Leonid Ionov
Max-Planck-Institute of Molecular Cell Biology and Genetics, Pfotenhauerstr. 108, 01307 Dresden, Germany

Sergiy Minko
Chemistry Department, Clarkson University, Potsdam, NY13699-5810, USA

Abstract A simple route for fabrication of regularly patterned surfaces with specifically designed surface roughness and chemistry is reported using colloidal particles. The surface was built up from self-assembled submicrometer- and micrometer-sized monodisperse core-shell particles of different radius (0.1–10 μm) forming ordered arrays. In this way, an increase in the vertical roughness is achieved with increasing particle radius, but without changing the Wenzel roughness factor. The morphology of the ordered particle arrays was characterized using an optical imaging method (MicroGlider), scanning force (SFM) and scanning electron (SEM) microscopy. The organic shell was either prepared by covalent grafting of polymer brushes or by chemisorption of a silane with a long fluoroalkyl tail. From FTIR-ATR, diffuse reflection IR spectroscopy, and capillary penetration experiments, it was concluded that the grafted polymer completely covers the surface of the silica particles. The solid surface tension of the organic shell obtained from contact angle measurements on smooth surfaces decreased in the following order: polystyrene brush-PS ($\gamma_{sv} = 28.9\,\mathrm{mJ/m^2}$) > copolymer of polystyrene and 2,3,4,5,6-pentafluoropolystyrene brush-FPS ($\gamma_{sv} = 24.3\,\mathrm{mJ/m^2}$) > chemisorbed (tridecafluoro-1,1,2,2-tetrahydrooctyl) dimethylchlorosilane-FSI ($\gamma_{sv} = 18.3\,\mathrm{mJ/m^2}$). Water contact angle measurements revealed an influence of the surface height roughness and the shell chemistry on the wettability. For all surfaces investigated, the contact angle hysteresis increased on the rough model surfaces compared to the smooth surfaces due to the increase of the advancing contact angle and the decrease of the receding angle. The lower the surface free energy of the shell chemistry, the smaller is the contact angle hysteresis on the closely packed surface arrays. Further the contact angles varied with increasing height roughness. A possible explanation for this behaviour is that the vertical roughness influences the curvature radius of the liquid in trapped air pockets at the solid-liquid interface as was already assumed in the literature for nanostructured metal surfaces and paraffin-coated steel balls.

Keywords Core-shell particles · Modification · Patterned surfaces · Roughness · Wettability

Introduction

It is well known that the surface properties of polymers are strongly influenced by the topography of the surface. In particular, wetting processes can be controlled by the surface roughness [1–21]. The understanding of wetting on rough surfaces is therefore of great importance for fundamental research, but also for numerous industrial applications such as coatings, paintings, adhesives, microfluidic technology, microelectronics, etc. Rough polymer surfaces possessing a low surface free energy are very promising as super water-repellent synthetic materials with reduced adhesion properties [1–21]. In the centre of interest is here the question of how does a microtexture reinforce the natural hydrophobicity of a polymer surface. In most cases, wetting is studied on rough substrates prepared by plasma treatment [22], corona discharge [23] or initially porous substrates [24]. Non-regularity and non-uniformity of such surfaces hamper the interpretation of the contact angles and comparison with theoretical considerations. Therefore, regular structured surfaces appear to be suitable objects for modelling of the liquid wetting behaviour. They can be prepared by different techniques including moulding [25], photolithography [8, 26] or soft lithography [27], microsphere sedimentation [28], parallel plate fluidic confinement [29], microfluidic [30] and electrochemical deposition [31].

In this work, we report a new, versatile and experimentally simple approach for designing regularly patterned surfaces for thorough study of wetting phenomenon. This approach is based on use of core-shell colloidal particles. The surface is built up from self-assembled submicrometer- and micrometer-sized monodisperse core-shell particles forming ordered arrays. The preparation of structured surfaces from core-shell particles has several advantages: (i) immobilization of various polymers and low molecular weight compounds on the surface of silica particles allows control of wetting properties of the particle layer; (ii) spheres of any size can be coated; (iii) prepared surfaces are resistant to mechanical, chemical and thermal treatment; (iv) easily controlled surface roughness parameters (most important in this work) are available. Such surfaces can potentially be both model surfaces for fundamental research and templates for technologically relevant micro scale structures with unique geometric and topographic properties. In this work we report about the preparation, characterization and application of these surfaces for wetting experiments aimed to a better understanding of wetting on hydrophobic rough surfaces.

Experimental Section

Materials

Highly polished single-crystal silicon wafers with native silicon oxide layers (Semiconductor Processing, Germany) were used as substrates. They were pre-cleaned by rinsing 3 times in dichloromethane in an ultrasonic bath for 5 min followed by washing in a mixture of deionized water, ammonia solution (25%) and hydrogen peroxide (30%) in the 1 : 1 : 1 volume ratio at 60 °C for 1 h. After washing, the substrates were thoroughly rinsed with deionized reagent-grade water and then dried with nitrogen flux.

Carboxy-terminated polystyrene PS-COOH (Mn = 45 900 g/mol, PDI = 1.05, purity 99.9%) synthesized by anionic polymerisation was purchased from Polymer Source (Germany) (Fig. 1). Random carboxy-terminated copolymer of polystyrene (Aldrich) and 2,3,4,5,6-pentafluoropolystyrene (Aldrich), FPS-COOH (Mn = 40 000 g/mol, PDI = 2.4) was prepared by free radical polymerisation in toluene with the ratio 80 : 20 (Fig. 1). Toluene (Aldrich) was distilled after drying over sodium. (Tridecafluoro-1,1,2,2-tetrahydrooctyl) dimethylchlorosilane (FSI) from Gelest (Germany) and 3-glycidoxypropyl trimethoxysilane (GPS) from ABCR (Germany) were used as received (Fig. 1). Silica particles of different radius varying from 0.1 to 10 μm were purchased in dry state from Geltech (Germany), Microparticles (Germany) and Duke Scientific Corporation (USA).

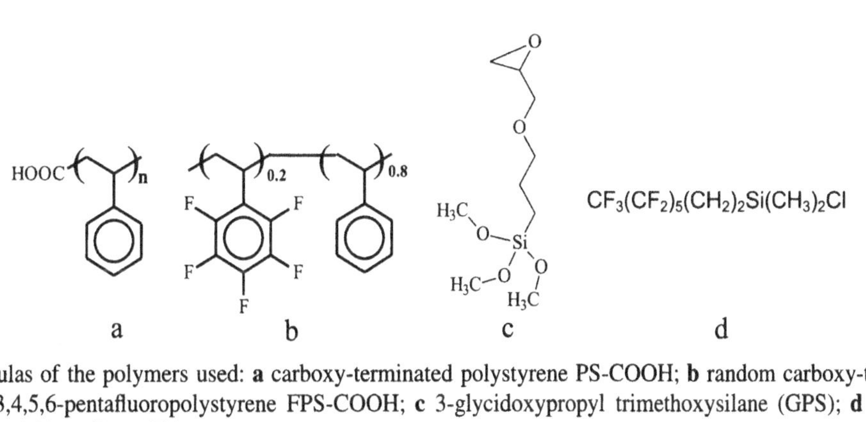

Fig. 1 Chemical formulas of the polymers used: **a** carboxy-terminated polystyrene PS-COOH; **b** random carboxy-terminated copolymer of polystyrene and 2,3,4,5,6-pentafluoropolystyrene FPS-COOH; **c** 3-glycidoxypropyl trimethoxysilane (GPS); **d** tridecafluoro-1,1,2,2-tetrahydrooctyl) dimethylchlorosilane FSI

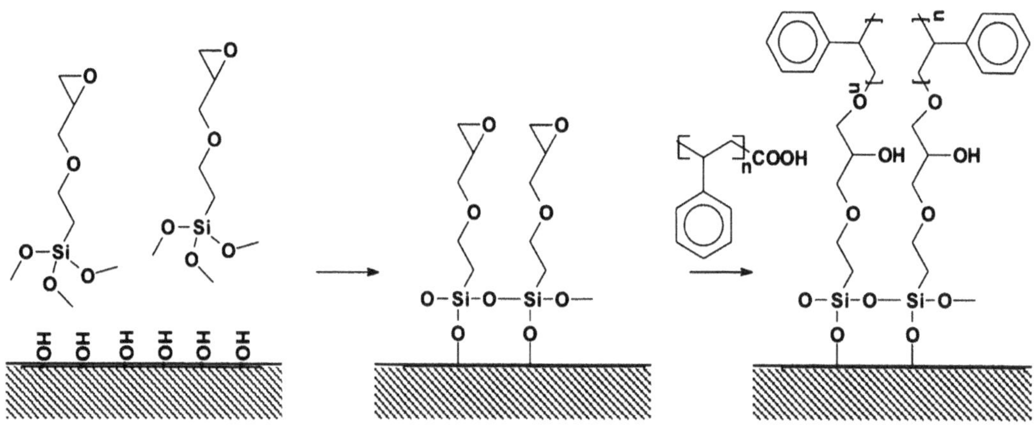

Scheme 1 Scheme of "grafting to" approach of carboxy-terminated polymer onto anchoring layer of GPS

Surface modification

The "grafting to" method was used to anchor polymer chains onto the surface of silica particles and silicon wafers (Scheme 1). The synthetic procedure starts with covalent grafting of GPS to the surface. A self-assembled monolayer of GPS on silicon wafer surfaces was prepared according to the procedure suggested by Luzinov [32]. For this, Si wafers were immersed in a 1% GPS toluene solution for 15 h under dry Ar atmosphere (< 1 ppm H_2O). After treatment, GPS modified wafers were washed 3 times in dry toluene under dry Ar atmosphere to avoid polymerisation of non-grafted GPS and precipitation of particles. Afterwards, the silanized wafers were washed 2 times with ethanol in ultrasonic bath for 5 min followed by drying with nitrogen flux. The thickness of the GPS layer was determined by null ellipsometry.

The similar procedure was used to adsorb GPS onto the surfaces of silica particles. Before silanization, particles were washed 3 times in dichloromethane followed by drying in a vacuum oven at 110 °C. GPS was chemisorbed from a 2% toluene solution. The particles concentration in the solution was 5 wt %. To remove non-adsorbed GPS, the particles were washed and centrifuged 6 times in toluene.

In the next step, polymer films (thickness: ~ 200 nm) of PS-COOH or FPS-COOH were spin coated on the top of the GPS layer from 2% toluene solutions. After casting, the films were annealed in a vacuum oven at 150 °C for 15 h. Non-grafted polymer was removed by Soxhlet extraction in toluene for 6 h. The thickness of PS and FPS grafted layers was determined using null ellipsometry.

For grafting of polymer onto the particle surface, GPS modified particles were mixed with a 3% toluene solution of desired polymer at room temperature during 24 h and then dried. The particle concentration in the solution was 3 wt %. Grafting was performed at 150 °C during 15 h producing a dense polymer layer. Several circles of washing in toluene and centrifuging were needed to remove non-grafted polymer. FSI was chemisorbed from its vapour onto both Si wafer and layers of ordered particles at 100 °C during 4 h. The quality of FSI layers on Si wafer was controlled by water contact angle experiments and null ellipsometry. It was found that the contact angle became constant, 113°, after 2 h of deposition. Surfaces of Si wafers modified with grafted polymers or FSI were smooth. Root mean square roughness (rms) for all non-structured surfaces was less than 1 nm.

Deposition of particles

Silica particles were deposited onto silica wafers using a vertical deposition technique [33]. For this, a 1-wt % suspension of polymer-modified particles in toluene was prepared. A suspension of non-modified particles was prepared in ethanol. Particles with grafted polymer were deposited on the silicon wafer modified with the same polymer. Bare particles were deposited onto bare Si wafers. For this purpose, the wafer was vertically immersed into a colloid suspension that was evaporated at 70 °C during 4 h. Particles of a radius larger than 1.2 μm were deposited via sedimentation on a slightly inclined silica wafer modified with a thin layer of the grafted polymer. The quality of the layers was controlled by optical imaging instrument (MicroGlider), scanning force (SFM) and scanning electron (SEM) microscopy. In all cases substrates were covered by multilayers of particles.

Sample characterization

Ellipsometry. Layer thickness of GPS, PS, FPS and FSI on the flat silica wafers was evaluated at $\lambda = 633$ nm and an angle of incidence of 70° with a SENTECH SE-402 microfocus null ellipsometer (lateral resolution is defined by the beam spot of about 20 μm). The measurements were performed for each sample after each step of modification to use the data of the previous step as a reference for the simulation of ellipsometric data. Initially, the thickness of the native SiO_2 layer (usually 1.4 ± 0.2 nm) is calculated at refractive indices $n = 3.858 - i0.018$ and $n = 1.4598$

for the Si wafer and the SiO_2 layer, respectively. The thickness of the GPS layer (typically 0.8 ± 0.1 nm) was evaluated using the two-layer model $Si/SiO_2/GPS$ with $n = 1.429$. The thickness of PS, FPS was evaluated with the three-layer model $Si/SiO_2/(GPS)/(PS$ or FPS) with $n = 1.59$ [34]. The thickness of FSI layers on the surface of silicon wafers was obtained using the two-layer model $Si/SiO_2/FSI$ with $n = 1.47$.

FTIR-ATR spectra were taken with an IFS 55 (Bruker, Germany) spectrometer for both the chemisorbed GPS, FSI, and grafted PS-COOH and FPS-COOH layers using the Si prism with a native SiO_2 layer as a substrate for the grafting.

Diffuse reflection IR spectroscopy was used for determination of grafted amount of polymer. It was evaluated using calibration curves plotted from the absorbency spectra of the samples, made from a mixture of the polymer and particles (50 mg). The characteristic band at $1493\,cm^{-1}$ was chosen for this estimation.

Capillary penetration experiments. Using a series of nonpolar and polar probe liquids, the solid surface tension of unmodified and modified particles was evaluated by capillary penetration experiments [35, 36]. The strategy is based on a modified Washburn equation:

$$K\gamma_{lv}\cos\theta = \left(\eta/\rho^2\right)\left(m^2/t\right) , \quad (1)$$

where K is a factor representing the geometry of the capillary system, (η/ρ^2) reflects properties of the test liquid (viscosity η and density ρ) and (m^2/t) is determined in the experiment. The quantity (m^2/t) can be determined by measuring the weight m of a penetrating liquid into a particle packing as a function of time t using a microbalance. More experimental details about the set-up for capillary penetration measurements including the preparation of powder packings can be found in references [34–36]. The solid surface tension, γ_{sv}, of the porous particle system was estimated directly from plots of $K\gamma_{lv}\cos\theta$ versus γ_{lv} [35]. The accuracy of this method depends on the differences between the surface tension values of the liquids used for the characterization.

Surface topography. The surface topography was investigated using scanning force microscopy (SFM), scanning electron microscopy (SEM) and optical analysis (MicroGlider, Germany).

Scanning Force Microscopy (SFM) studies were performed on a Dimension IV (Digital Instruments, USA) microscope. The tapping mode was used to map the surface morphology at ambient conditions. Standard silicon tips with radius of about 10–30 nm, apex angle 65° and frequency of about 300 kHz as well as ultra-sharp ones with tip radius < 10 nm and apex angle 12° for the highest resolution imaging were used.

An imaging measuring instrument for the optical analysis of roughness (MicroGlider, Germany) was used as a complementary method because of its large-scale vertical resolution from 10 nm to 300 mm and the possibility to investigate samples with a maximal size area of $100 \times 100\,mm^2$. The lateral resolution is, however, determined by the size of the reflected light on the sample, 1–2 μm.

The root mean square roughness R_q [38], the roughness factor r_s, which is the ratio of the actual area of the rough surface to the projected area, and the power spectral density (PSD) of the images were evaluated using SFM and FTR Mark III software, respectively. Images of the surfaces were obtained at least at five different locations. Roughness parameters were determined for a particular surface sample based on these images from five different locations.

A Scanning Electron Microscope (SEM), (DSM 982 Gemini, ZEISS, Germany) was used for the evaluation of surface topography as well.

Contact angle measurements. Advancing and receding contact angles were measured by the sessile drop method using axisymmetric drop shape analysis (ADSA) and a conventional goniometer technique (Krüss DSA 10, Hamburg, Germany). Details of the procedure and the experimental setup for low-rate dynamic contact angle measurements by ADSA are given elsewhere [35]. Deionized reagent grade water was used for contact angle measurements. In the case of the goniometer technique, at least 5 droplets were put on each sample. Three different samples were used to calculate a mean contact angle value. The error of the mean contact angle values, calculated as the standard deviation, did not exceed 2 and 3° for advancing and receding contact angles, respectively. All contact angle measurements were carried out at 24 ± 0.5 °C and relative humidity of $40 \pm 3\%$ which were kept constant.

The advancing contact angles, measured on smooth silicon surfaces coated with the polymer brushes or the silane, were used to calculate the solid surface tension γ_{sv} according to the equation of state approach (EQS) [39]:

$$1 + \cos\theta_Y = 2\sqrt{\frac{\gamma_{sv}}{\gamma_{lv}}}\,e^{-\beta(\gamma_{lv}-\gamma_{sv})^2} \quad (2)$$

where β is a constant which was experimentally found to be $0.0001247\,(mJ/m^2)^{-2}$, γ_{sv}, γ_{lv} denote the interfacial tensions of the solid-vapour and liquid-vapour interfaces, respectively; θ_Y is the Young contact angle.

Results and Discussion

Grafting of polymers

Scheme 1 illustrates the "grafting to" procedure of the polymers onto a solid surface. It starts with covalent anchoring of GPS to the surface of a Si wafer and silica particles as well. The thickness of the GPS layer immobilized onto a silica wafer was found to be 0.8–1.2 nm that corresponds to 1–1.5 theoretical layers. SFM images show that the GPS layer was homogenous and smooth, the

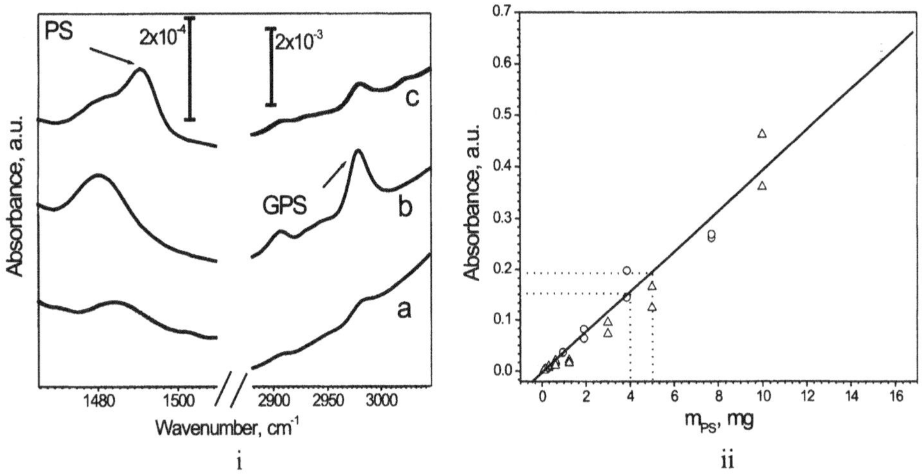

Fig. 2 **i** FTIR spectra of 0.1 μm silica particles in radius: (a) bare; (b) with chemisorbed GPS; (c) with grafted PS; **ii** Calibration plot for PS grafted layer in the mixture of PS powder with silica particles: o 0.1 μm; and △ 1.2 μm in radius in terms of the FTIR spectrum integral absorbance vs. mass fraction of the polymer

rms roughness was less than 1 nm. The immobilization of GPS onto the particle surfaces was monitored by FTIR-ATR measurements (Fig. 2ib). The characteristic bands of epoxy groups (736–854, 863–950 and $1260\,cm^{-1}$) could not be detected in a very thin layer of grafted GPS. Only the hydrocarbon bands (C – H stretching vibrations) at $2800–3100\,cm^{-1}$ were identified. Recently it was found, that the GPS chemisorption involves not only hydrolysis and reaction of trimethoxysilane groups with surface silanol groups (Scheme 1) but also opening of the epoxy ring of GPS [32, 34]. Consequently, the chemisorbed GPS forms a layer with different functional groups and at least one of them (epoxy groups) can be used for grafting of the carboxyl-terminated polymers.

The next step of the synthetic procedure consists in grafting of the carboxy-terminated polymers (PS-COOH or FPS-COOH) onto the surface of the GPS modified particles. Reaction of esterefication between terminating carboxyl groups and epoxy groups on the surface leads to chemical bond formation between polymer chains and the surface. FTIR-ATR spectra, shown in Fig. 2ic for 0.1 μm particles in radius, provide further evidence for grafting of PS-COOH onto the surface. Very pronounced bands at 1601, 2923, and $3027\,cm^{-1}$ appear after grafting of PS-COOH. The grafted amount of PS-COOH was determined by diffuse reflection IR spectroscopy from the calibration curves (Fig. 2ii). The results show that about $4–5\,mg/m^2$ of polymer was grafted onto the Si particle surface, which corresponds to a total thickness of 4–5 nm. Previous findings show that such a thick polymer layer completely covers the surface [32, 34]. In the case of larger particles, the polymer layer thickness could not be determined due to the very low ratio of the PS band to the noise level.

Additionally, the changes in the surface-energetic properties of the silica particles due to the grafting of polymers

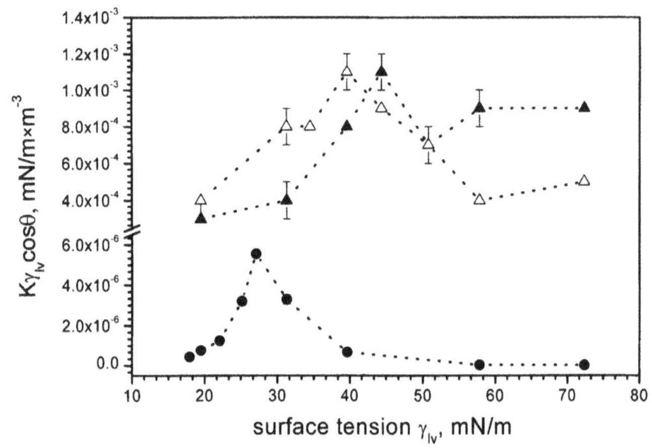

Fig. 3 Plot of $K\gamma_{lv}\cos\theta$ vs. γ_{lv} for △ cleaned and ▲ non-cleaned bare silica particles; • polystyrene modified particles (0.5 μm in radius)

were investigated. By plotting $K\gamma_{lv}\cos\theta$ versus the liquid surface tension, γ_{lv}, the solid surface tension of the particles, γ_{sv}, could be estimated. Figure 3 shows the results for bare silica particles (non-cleaned and cleaned) and modified particles where the shell consists of polystyrene brushes. Each curve shows a maximum that corresponds to the solid surface tension of the particles (Fig. 3). In the case of the PS modified particles, this maximum is shifted to distinctly lower values compared to the bare silica particles. It is in reasonable agreement with the γ_{sv} value ($28.9\,mJ/m^2$) obtained from contact angles measured on polystyrene brushes that were grafted to flat and smooth solid surfaces (Table 1). Moreover, γ_{sv} the solid surface tension of the polystyrene modified silica particles is also in good agreement with the solid surface tension of a polystyrene powder [36]. Hence, the capillary penetration experiments confirm that the surface of the silica

Table 1 Advancing and receding water contact angles, contact angle hysteresis and the surface energy values for smooth surfaces covered by PS and FPS and chemisorbed FSI

Polymer	θ_A, deg	θ_R, deg	$\Delta\theta = \theta_A - \theta_R$, deg	γ_{sv} [1)], mJ/m^2	r_s
PS	89.5±0.2	82.0±1.0	7.5	28.9	1.001
FPS	97.0±0.3	93.0±1.5	4.0	24.3	1.001
FSI	113.0±0.2	107.0±1.2	6.0	18.3	1.001

[1)] Values of surface solid energy obtained form the advancing contact angles using EQS considering the measured value of water surface tension of 71.8 mN/m

particles is completely covered by the grafted polymer layer.

Surface topography and roughness

Figure 4 shows representative SEM, SFM, and optical microscopy images (MicroGlider) for patterned surfaces from core-shell particles of different radius (0.1 µm, ... 2.5 µm). All techniques revealed densely packed and well-ordered particle layers. SFM and optical microscopy imaging provide quantitative roughness parameters. Surfaces composed from particles smaller than 2.5 µm in radius were characterized by SFM, whereas MicroGlider was applied to the patterned surfaces consisting of particles larger than 2.5 µm. The PSD images (Fig. 4a–d, insets) show an almost perfect hexagonal order of the particle arrays. Six sharp first order peaks and the presence of higher order reflections can be clearly seen. For particle arrays consisting of particles with 0.1 and 1.2 µm in radius, the sectional analyses of the height profiles along the [100] axis in the 2D hexagonal lattice is presented in Fig. 5. These results show the regular embossments caused by the particles, but no additional roughness on the nm scale indicating that the surface of the modified particle is very smooth.

In Fig. 6, experimentally determined roughness parameters are compared with the calculated roughness. It can be clearly seen that the height roughness (root mean

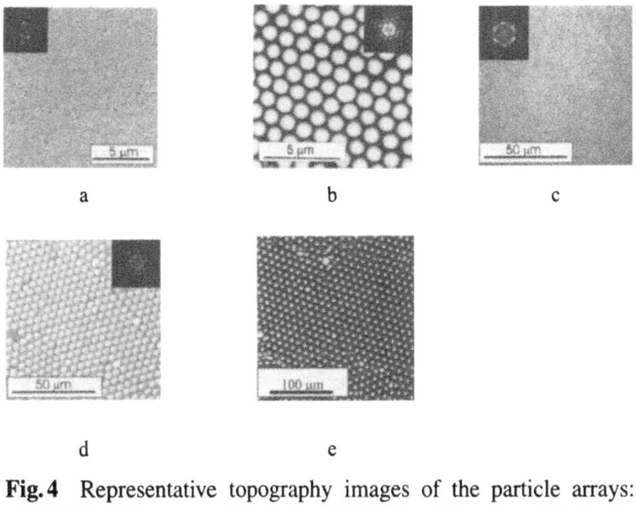

Fig. 4 Representative topography images of the particle arrays: **a** SEM for 0.1 µm particle layer; **b** SFM for 0.5 µm particle layer; **c** optical microscopy (MicroGlider) for 1.2 µm particles; **d** optical microscopy (MicroGlider) for 2.5 µm particle layer; **e** optical microscopy (MicroGlider) for 2.5 µm particle layer (larger scale)

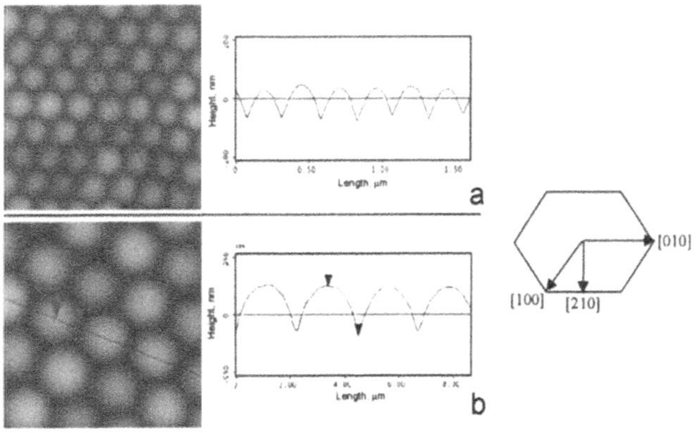

Fig. 5 Representative SFM images and corresponding height profiles along the [100] direction for 0.1 µm (**a**) and 1.2 µm particles in radius (**b**)

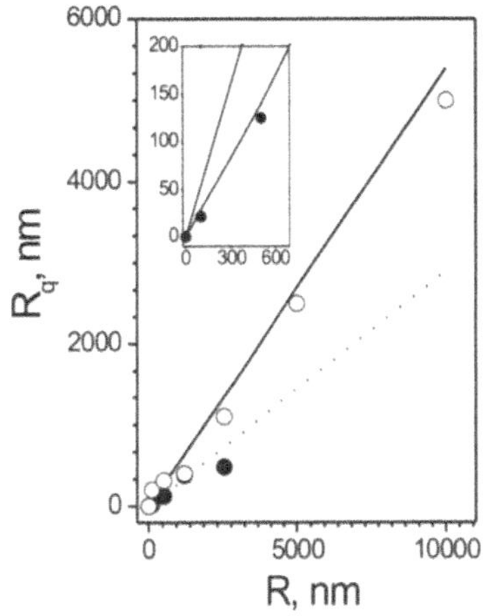

Fig. 6 Experimental values (symbols) of Rq: • SFM; ○ MicroGlider and calculated (*solid line*: Model 1, *dotted line*: Model 2)

square roughness values R_q) determined from the SFM and MicroGlider images increases with increasing particle radius. However, the R_q value of the patterned surface with 2.5 μm large particles obtained by MicroGlider is higher than that obtained by SFM. We suppose that this result originates from different resolution of SFM and MicroGlider. The resolution of SFM is strongly limited by the tip shape geometry (radius and apex angle) and sample topography to a large extend. On the other hand, the lateral resolution of the MicroGlider is limited by the width of the light beam (1–2 μm). Therefore, the absolute values of the height roughness values may be tampered by the limitations of each technique. Nevertheless, both techniques reflect clearly the steady increase in the height roughness with increasing particle radius.

Simple mathematical calculations of the surface composed from hexagonally packed spheres were undertaken. For this, we considered two possible geometries of the particle array, taking into account different probing depth of various experimental techniques (SFM, MicroGlider). Since the lower half of the particles is not visible and one observes only the pore between them, the following models are suggested. Both models (Fig. 7a,b) consider the surface consisting of the hexagonally packed hemispheres with a radius R. In the second model, the hexagonally packed hemispheres are situated on the cylinders with a height R (Fig. 7b). The elementary cell of such structures is a triangle connecting the points where spheres touch each other (Fig. 7c). The roughness characteristics calculated for this elementary cell are representative for the whole ordered surface. These two models are used for calculation of R_q and roughness factor r_s for structured

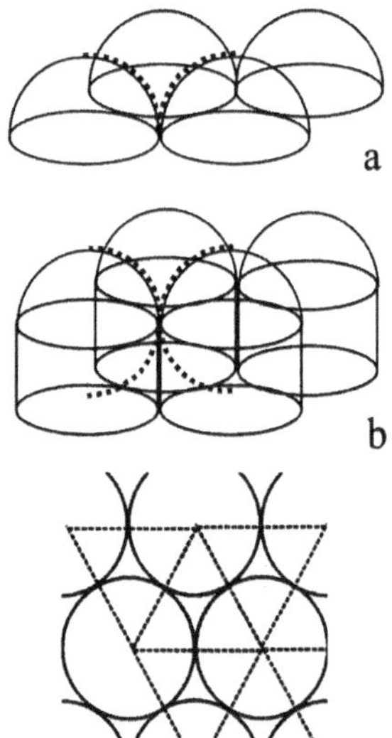

Fig. 7 Possible geometries of the particle layer: **a** Model 1 and **b** Model 2 and **c** elementary cell as a triangle connecting the touched points of adjacent spheres within the hexagonally packed particle layer

particle layers. The root mean square roughness is defined by

$$R_q = \sqrt{\frac{1}{S}\int_S (Z - Z_0)^2 \, dS}\,, \tag{3}$$

where Z is the height in some particular location, Z_0 is averaged height, S is the effective surface area.

It was found, that R_q predicted from the first and second mathematical model is in reasonable agreement with those values experimentally obtained by SFM and MicroGlider (Fig. 6). From the first model, R_q is equal to $0.32R$ and from the second to $0.54R$. In fact, the SFM tip is unable to penetrate into pores between particles and reflects a smoother surface compared to the real rough surface. The first model considers this fact. On the other hand, the light beam of MicroGlider reaches the bottom surface of particle layers resulting in R_q values, which are similar to the calculated ones according to the second model.

Since the Wenzel roughness factor is an important parameter in many theoretical considerations of wetting phenomena [11–21], it was also calculated as the ratio between the actual surface area to the projected one (Eq. 4). A surface of hexagonally packed hemispheres is considered. In this case, the roughness factor is independent of the particle radii and approximately equal to 1.91 [31, 40, 41]:

$$r_s = \frac{S_{\text{actual}}}{S_{\text{geometric}}} = \frac{S_{\text{sphere segment}} + S_{\text{pore}}}{S_{\text{triangle}}} = \frac{\pi R^2 + R^2\sqrt{3} - \frac{\pi R^2}{2}}{R^2\sqrt{3}} \approx 1.9069\,. \tag{4}$$

r_s values determined by SFM were again lower due to the tip shape geometry and do not reflect the real roughness for this type of surface.

Wettability

There are two general states in which a drop can reside on a given rough surface [16–21]. The liquid drop either sits on the asperities of the rough surface ("heterogeneous wetting regime") or wets their grooves ("homogeneous wetting regime") [11, 20, 21]. For the latter case, Wenzel derived that roughness increases the liquid contact angle on a hydrophobic surface ($\theta_Y > \pi/2$). If air is entrapped inside the grooves, underneath the liquid ("heterogeneous wetting regime"), the drop is sitting on a "composite" surface consisting of the asperities and air. This behaviour can be described by the Cassie-Baxter equation [12, 20]. Recently, Marmur [20] has shown that the Wenzel and the Cassie-Baxter equations are necessary but not sufficient conditions for describing wetting equilibrium on rough surfaces. From theoretical considerations he deduced, that different wetting regimes could occur depending on the detailed solid surface topography (semicircular grooves, semicircular protrusions, or saw-toothed profiles). In this work, the fabricated ordered arrays can be compared with the case of "semicircular protrusions".

However, the difficulty that results from the experimental contact angle measurements is contact angle hysteresis. Instead of an equilibrium contact angle, a number of mechanically stable contact angles exists. Therefore, a maximum (θ_a) and minimum (θ_r) contact angle should be measured for the advancing and the receding contact line. The difference, $\Delta\theta = \theta_a - \theta_r$, is called contact angle hysteresis. In the "homogeneous wetting regime" large values of contact angle hysteresis are observed due to an increase in the advancing and a decrease in the receding contact angle (liquid fill up the roughness grooves). Drop pinning is favoured. In the "heterogeneous wetting regime" the contact angle hysteresis is small or extremely small as in the case of ultrahydrophobicity. Johnson and Dettre predicted for a sinusoidal surface a transition from a nonporous (Wenzel or "homogeneous wetting regime") to a porous surface (Cassie-Baxter or "heterogeneous wetting regime") with increasing surface roughness r_s [13, 14]. According to these results hysteresis increases until this critical roughness is reached and then decreases. We could verify this transition by experimental results for anodically oxidized aluminium surfaces and oxygen plasma etched polytetrafluoroethylene surfaces [7] and [42]. However, contact angle hysteresis was strongly dependent on the geometrical nature of the roughness features, and the type of roughness rather than its absolute size was the determining factor to obtain a transition to an ultrahydrophobic surface [7].

In this work, ordered arrays of core-shell particles were used as model surfaces to study the water wetting behaviour of these surfaces. Two factors were varied in the wetting experiments: (i) the shell chemistry and hence the solid surface tension of the organic shell, and (ii) the height roughness from sub-μm up to μm roughness values whereas the Wenzel roughness factor was kept constant.

Figure 8 summarizes the influence of the height roughness (particle radius) and the chemistry of the shell polymer on the wetting behaviour of the hexagonally packed particle arrays. For each surface, the maximum (advancing angle θ_a) and the minimum (receding angle θ_r) contact angles were determined. In general, contact angle hysteresis was higher on the rough patterned surfaces in comparison to the smooth surfaces of the same shell chemistry. In Fig. 8b, the results of two different contact angle techniques, a conventional goniometer technique and ADSA contact angle measurements, are shown for FPS modified surfaces. The measured advancing contact angles are in very good agreement for both techniques. In the case of receding angles, ADSA yields sometimes no receding angle (drop pinning occurs) or lower angles compared to the goniometer technique. However, the general trend for the receding angles with increasing height roughness is the same for both contact angle techniques.

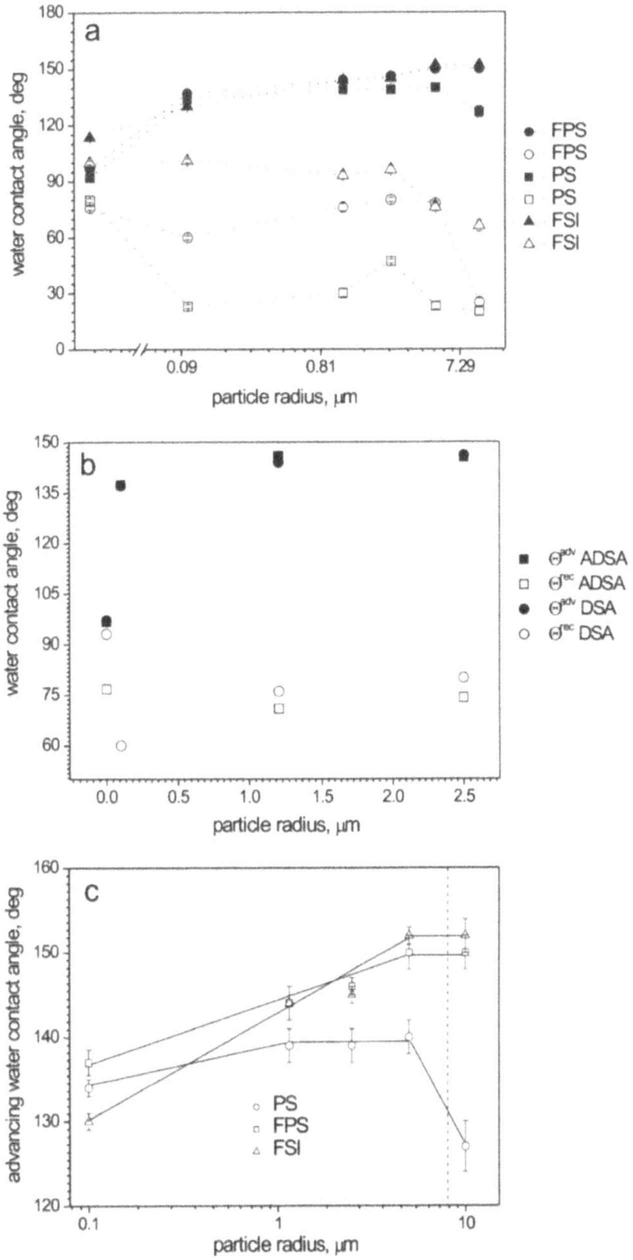

Fig. 8 **a** Water advancing and receding contact angle vs. particle radius; **b** Water advancing and receding contact angle measured by ADSA and DSA vs. particle radius; **c** Advancing water contact angles vs. particle radius on a logarithmic scale for different "shell" chemistry

On PS particle modified arrays, which were less hydrophobic (solid surface tension of $\gamma_{sv} = 28.9\,\mathrm{mJ/m^2}$), the largest increase in contact angle hysteresis was observed due an increase in the advancing and a decrease in the receding angle (Fig. 8a). This behaviour is typical for a "homogeneous wetting regime" when the drop wets the grooves of the rough surface. If the surface chemistry was changed to more hydrophobic surfaces and hence lower solid surface tensions (PS ($\gamma_{sv} = 28.9\,mJ/m^2$) > FPS ($\gamma_{sv} = 24.3\,mJ/m^2$) > FSI ($\gamma_{sv} = 18.3\,mJ/m^2$) (Table 1) then the contact angle hysteresis decreased in the same order. Obviously, the grooves or pores were less wetted by water when the solid surface tension becomes lower resulting in higher receding angles. The lowest hysteresis was observed for the surfaces modified by chemisorbed FSI. Here, the advancing contact angle was increased with increasing height roughness while the receding contact angle stays nearly constant (above 90°). However, with increasing the height roughness (particle radius > 2.5 μm), the contact angle hysteresis becomes also higher on patterned FSI modified surfaces. Of course, this behaviour cannot be explained by the Wenzel equation because this equation does not predict contact angle hysteresis. Moreover, for the present case of rough surfaces, the Wenzel roughness parameter is 1.91, independent of the particle radius. For this reason Wenzel equation would predict a constant contact angle of 89° with increasing height roughness for the PS modified particle surfaces with an intrinsic contact angle of 89.5°.

In the literature, the increase of the advancing contact angle with increasing vertical roughness but a constant Wenzel roughness parameter was explained by a change in the radius of curvature of the liquid in trapped air pockets at the solid/liquid interface [31,40]. Figure 8c indicates that the advancing water contact angle on FPS and FSI shows a behaviour that is roughly consistent with this predicted behavior. However, more studies, both experimental and theoretical, are needed to thoroughly understand the wetting behaviour on this special type of rough surfaces characterized by hydrophobic hemispherical close-packed protrusions.

Conclusions

A simple and effective approach for designing regularly patterned surfaces with specifically designed surface roughness and chemistry using core-shell colloidal particles was demonstrated. The chemistry was varied by covalent grafting of polymer brushes onto silica particles or by chemisorption of fluorosilane. The modified colloidal spheres were organized into closely packed hemispherical hexagonal arrays either by a vertical deposition technique (0.1–1.12 μm large particles in radius) or by sedimentation on slightly inclined coated silicon wafers (2.5, 5.0, and 10.0 μm large particles in radius). In this way, an increase in the vertical roughness was achieved with increasing particle radius, but without changing the Wenzel roughness factor of 1.91. It was found that contact angles varied with increasing height roughness on this type of surfaces. The wetting results show an enhance of the surface hydrophobicity with increasing the "shell" hydrophobicity in the sequence PS–FPS–FSI. With increasing the particle radius, the water contact angle hysteresis remained large even for the most hydrophobic FSI "shell". Hence, ultrahydrophobic properties could not be obtained on this type of regularly patterned surfaces from core-shell particles.

Acknowledgement Authors thank Dr. M. Motornov for providing assistance in grafting of particles, Ms. G. Adam, Ms. G. Zedler and Mr. U. Streller for FTIR-ATR, capillary penetration and SEM investigations, respectively.

References

1. Barthlott W, Neinhuis C (1997) Planta 202:1
2. Tsujii K, Yamamoto T, Onda T, Shibuichi S (1997) Angew Chem 109:1042
3. Nakajima A, Fujishima A, Hashimoto K, Watanabe T (1999) Adv Mater 11:1365
4. Youngblood J, McCarthy TJ (1999) Macromolecules 32:6800
5. Chen W, Fadeev AY, Hsieh MC, Öner D, Youngblood J, McCarthy T (1999) Langmuir 15:3395
6. Minko S, Müller M, Motornov M, Nitschke M, Grundke K, Stamm M (2003) J Am Chem Soc 125:3896
7. Grundke K, Nitschke M, Minko S, Stamm M, Froeck C, Simon F, Uhlmann S, Pöschel K, Motornov M (2003) In: Mittal KL (ed) Contact Angle, Wettability and Adhesion, vol 3, p 1–25
8. Öner D, McCarthy TJ (2000) Langmuir 16:7777
9. Onda T, Shibuichi S, Satoh N, Tsujii K (1996) Langmuir 12:2125
10. Shibuichi S, Onda T, Satoh N, Tsujii K (1996) J Phys Chem 100:19512
11. Wenzel RN (1936) Ind Eng Chem Res 28:988
12. Cassie A (1948) Discuss Faraday Soc 3:11
13. Johnson RE, Dettre RH (1964) In: Contact angle, Wettability and Adhesion. Adv Chem Ser 43:112
14. Dettre RH, Johnson RE (1964) In: Contact angle, Wettability and Adhesion. Adv Chem Ser 43:136
15. Johnson RE, Dettre RH (1968) J Phys Chem 64:1744
16. Bico J, Tordeux C, Quere D (2001) Europhys Lett 55:214
17. Quere D (2003) Nanotechnology 14:1109
18. Patankar NA (2003) Langmuir 19:1249
19. Patankar NA (2004) Langmuir 20:7097
20. Marmur A, Krasovitski B (2003) Langmuir 19:8343
21. Zhang J, Kwok DY (2005) J Colloid Interface Sci 282:434
22. Coulson SR, Woodward I, Badyal JPS, Brewer SA, Willis C (2000) J Phys Chem B 104:8836
23. Shirtcliffe N, Thiemann P, Stratmann M, Grundmeier G (2001) Surf Coat Technol 142:1121
24. Dae-Hwan J, Park J, Choi YK, Lee SB, Park HS, Rühe J (2002) Langmuir 18:6133
25. Bartell FE, Shepard JW (1953) J Phys Chem 57:211

26. Krupenkin TN, Taylor JA, Schneider TM, Yang S (2004) Langmuir 20:3824
27. Yang SM, Migues H, Ozin GA (2002) Adv Func Mater 12:425
28. van Blaaderen A, Ruel R, Wiltzius PP (1997) Nature 385:321
29. Park SH, Qin D, Xia Y (1998) Adv Mater 10:1028
30. Yang SM, Ozin GA (2000) Chem Commun 24:2507
31. Lee W, Jin MK, Yoo WC, Jang ES, Choy JH, Kim JH, Char K, Lee JK (2004) Langmuir 20:287
32. Luzinov I, Julthongpiput D, Malz H, Pionteck J, Tsukruk VV (2000) Macromolecules 33:1043
33. Goldenberg LM, Wagner J, Stumpe J, Paulke BR, Görnitz E (2002) Langmuir 18:3319
34. Minko S, Patil S, Datsyuk V, Simon F, Eichhorn KJ, Motornov M, Usov D, Tokarev I, Stamm M (2002) Langmuir 18:289
35. Kwok DY, Gietzelt T, Grundke K, Jacobasch HJ, Neumann AW (1997) Langmuir 13:2880
36. Grundke K, Augsburg A (2000) J Adhesion Sci Technol 14:765
37. Grundke K (2001) In: Holmberg K (ed) Handbook of Applied Surface and Colloid Chemistry: Wetting, Spreading and Penetration, chapter 7, vol 2, p 119–140
38. Scanning Probe Microscopy (2000) Training Notebook, Digital Instruments, Veeco Metrology Group, Santa Barbara, CA, p 40
39. Kwok DY, Neumann AW (1999) Adv Colloid Interface Sci 81:167
40. Nakae H, Inui R, Hirata Y, Saito H (1998) Acta Mater 46:2313
41. Synytska A, Ionov L, Minko S, Motornov M, Eichhorn KJ, Stamm M, Grundke K (2004) Polym Mater Sci Eng 90:624
42. Hennig A, Grundke K, Frenzel R, Stamm M (2002) Tenside Surfactants Detergents 39:243

Progr Colloid Polym Sci (2006) 132: 82–94
DOI 10.1007/2882_029

Published online: 9 March 2006

A. Drechsler
N. Petong
C. Bellmann
P. Busch
M. Stamm
K. Grundke
O. Wunnicke
J. Reichelt

The Effect of Adsorbed Cationic Surfactant on the Pattern Collapse of Photoresist Lines in Photolithographic Processes

A. Drechsler (✉) · N. Petong ·
C. Bellmann · P. Busch · M. Stamm ·
K. Grundke · J. Reichelt
Leibniz Institute of Polymer Research
Dresden e.V., Hohe Strasse 6,
01069 Dresden, Germany
e-mail: drechsler@ipfdd.de

O. Wunnicke · J. Reichelt
Infineon Technologies SC300,
Koenigsbruecker Str. 180, 01099 Dresden,
Germany

Abstract A crucial problem in the manufacturing of high aspect ratio structures in the microchip production is the collapse of photoresist patterns caused by unbalanced capillary forces. A new concept to reduce the pattern collapse bases on the reduction of the capillary forces by adsorption of a cationic surfactant. The application of a cationic surfactant rinse step in the photolithographic process leads to a reduction of the pattern collapse. Physico-chemical investigations elucidate the mechanism of surfactant adsorption and its effect on the surface properties of the photoresist. It is shown that the best pattern collapse reduction is obtained at a surfactant concentration referring to monolayer coverage. In this concentration range the capillary forces are minimized due to a hydrophobizing of the processed photoresist by adsorbed cationic surfactant.

Keywords Adsorption · Capillary forces · Cationic surfactant · Hydrophobizing · Pattern collapse

Introduction

The microelectronics industry is continuously reducing the feature size of integrated circuits. In 2006, a DRAM half-pitch, i.e. line widths of 70 nm will be put into practice, until 2010 a reduction to 45 nm is laid down in the International Technology Roadmap for Semiconductors [1]. An integrated circuit consists of a series of patterned functional layers (insulators, metal wires). The structure of each layer is transferred from a mask via a photolithographic process followed by etching or ion implantation. These manufacturing processes must be able to produce the required feature sizes.

In the photolithographic process, the functional layer is covered by a photoresist film. State-of-the-art circuits are fabricated with chemically amplified photoresists consisting of a polymer with an acid-labile pendant protection group, photoacid generator molecules (PAG), and additional additives [2]. Upon exposure to UV radiation through a patterned mask, the PAG is decomposed generating a low concentration of acid. In a post-exposure bake the acid diffuses and catalyzes a deprotection reaction that cleaves the pendant group of the insoluble polymer resulting in a polymer soluble in the developer solution. This deprotection reaction can be generally written as shown in Scheme 1.

In a wet-development using a 2.38 wt % TMAH (tetramethyl ammonium hydroxide solution), the exposed resist is removed [3]. A pattern of unexposed photoresist lines remains. To fulfill the demands for minimizing the feature size, the width of the photoresist structures must shrink adequately. Their height, however, cannot be reduced in the same way since the etch resistance must be retained. This leads to a drastically increasing ratio of line height to line width, the aspect ratio. With increasing aspect ratio, the mechanical strength of the photoresist lines decreases leading to a collapse of the structures during the development process. An example for collapsed and non-collapsed structures is shown in Fig. 1. This so-called pattern collapse has been found to be caused by unbalanced capillary forces acting between the lines during the drying step after the development [4, 5].

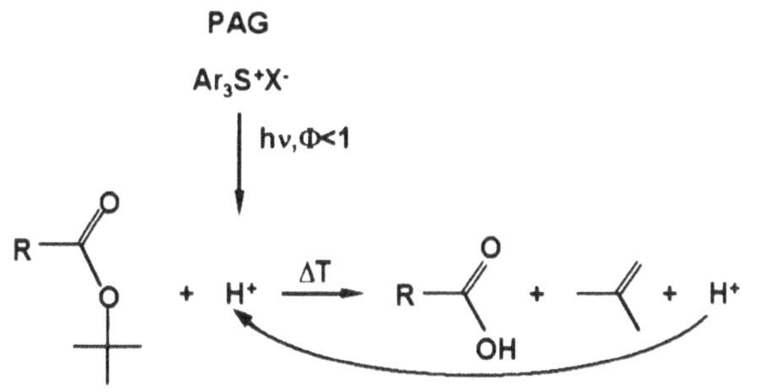

Scheme 1 Acid-catalyzed deprotection reaction. In the post-exposure bake, the acid generated by the photoacid generator molecules (PAG) in a photochemical reaction catalyzes a deprotection reaction that cleaves the pendant group of the insoluble polymer, resulting in a polymer that is soluble in the developer

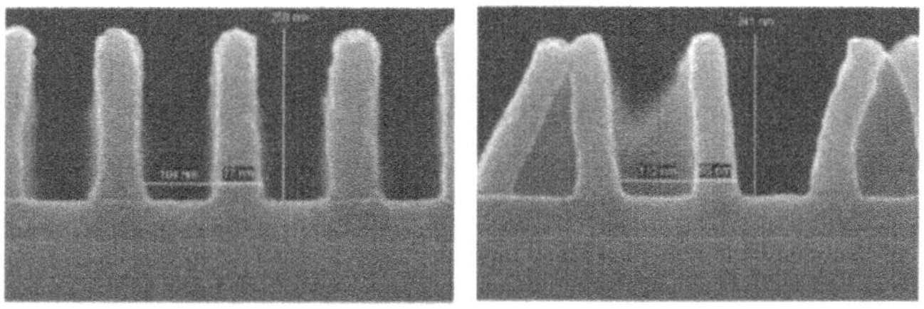

Fig. 1 Cross-section SEM images of free standing photoresist lines (*left*) and photoresist structures after pattern collapse (*right*).

To maintain the reduction of the feature size, it is of essential importance to prevent the pattern collapse. Several approaches have been proposed:

1. The mechanical stability of the photoresist has been improved in several ways: by changing its polymer backbone, by using nanocomposites [6] or by increasing the cross-linking density [7].
2. The adhesion between photoresist and the sub material is improved by special sub-coatings or by increasing the contact area [8].
3. The capillary forces are reduced by using hydrophobic photoresists, by rinse additives, by freeze-drying or by drying with supercritical CO_2 [9].

Approaches 1 and 2 require the design of new resists and coatings and are rather long-term items. Freeze-drying and drying with supercritical fluids is very effective but complicated and expensive. The most promising way is therefore the reduction of capillary forces by adding suitable substances to the rinsing liquid [5].

The pressure difference between both sides of a photoresist line caused by capillary forces can be quantified by the Laplace equation assuming the meniscus is shaped as a section of a cylinder between two infinite lines [10]:

$$\Delta P = 2\gamma_{\mathrm{lv}} \cos \Theta \left(\frac{1}{S_1} - \frac{1}{S_2} \right) , \qquad (1)$$

where γ_{lv} is the surface tension of the rinse liquid, Θ the contact angle of the solution on the photoresist and S_1 and S_2 the spacings besides the structure. Since the spacings are given, the pressure difference can be reduced by decreasing the wetting tension $\gamma \cos \Theta$, i.e. by lowering the surface tension of the rinse liquid or by raising the contact angle of the photoresist.

Several studies deal with the reduction of pattern collapse by adding surfactants to the rinse liquid. A wide variety of surfactants has been applied. At concentrations near the critical micelle concentration (cmc), the surface tension is lowered significantly. Therefore it is assumed that the capillary forces are reduced, too. It was shown that surfactant rinse indeed is able to shift the pattern collapse to higher aspect ratios [4]. The results of various groups are, however, very contrary. Hien et al. [11] found a positive effect only with an amphoteric but not with nonionic surfactants. Tanaka et al. [7] discovered a maximum of the pattern collapse reduction for a nonionic surfactant at a concentration above the cmc. Several authors [10, 12–14] determined the wetting tension $\gamma \cos \Theta$ but only Junarsa et al. [10] could find a direct correlation between minimum wetting tension and maximum pattern collapse reduction using an amphoteric fluorosurfactant.

The highest efficiency to reduce pattern collapse was found for concentrated surfactant solutions. This may, however, have serious disadvantages. Watanabe et al. [15] noticed that high surfactant concentrations melt the photoresist and amplify the pattern collapse. In contrary, Miyahara et al. [16] found that some nonionic surfactants stabilize the photoresist surfaces and prevent melting. In addition, concentrated surfactant solutions tend to foam which also might cause structure defects. Zhang et al. [17] therefore proposed a less foaming diol-type surfactant for the reduction of pattern collapse.

It can be summarized that the addition of surfactants to the rinse liquid in many cases leads to reduced pattern collapse. There is, however, no systematic study of the performance of surfactants; neither of the surfactant concentration nor of the surfactant type or chain length. None of the authors investigated the adsorption of the surfactant at the interface photoresist – fluid in more detail. Therefore the effect of surfactants to reduce pattern collapse is not really understood by now.

The aim of the present study is to elucidate the mechanisms of pattern collapse reduction by surfactant rinse for a special class of surfactants, the cationic surfactants. The basic idea was to use a rinse additive that at the same time reduces the surface tension of the rinse liquid and increases the contact angle of the liquid on the photoresist. It is known that ionic surfactants adsorb with their polar headgroup onto oppositely charged surfaces rendering them hydrophobic [18]. In the case of photoresists being negatively charged in the rinse liquid, cationic surfactants fulfill this requirement. For negatively charged silica and quartz, a maximum value of the contact angle of about 60° is reported at cationic dodecylpyridinium chloride concentrations far below the cmc [19, 20].

On one hand it is investigated how the addition of cationic surfactants affects the pattern collapse of 193 nm photoresist lines. On the other hand, the adsorption of the surfactant on model photoresist surfaces is explored by a variety of surface chemical methods. Of special interest is how the surfactant changes the surface properties of the photoresist as surface potential and wettability. For an optimum modelling of the properties of real photoresist structures, both unexposed photoresists and photoresists that have been UV exposed, baked and developed are studied.

Experimental

Substances. A commercially available 193 nm resist based on a methacrylate platform was used. It was chosen due to its collapse behaviour at 75 nm dense lines. The surfactant under study is a commercially available cationic surfactant of high purity (> 99%). Surfactant solutions of various concentrations have been prepared by dilution with purified water. Since all concrete data of the surfactant are the subject of concealment, the surfactant concentration will be given in this paper on a relative scale. A relative concentration of 1 refers to the cmc which was determined from the surface tension of the solutions.

Investigation of pattern collapse. For the investigation of pattern collapse, a photoresist layer with a thickness of 230 nm was used to reach an aspect ratio of 3. Lines of 55 to 80 nm width were produced by deep ultraviolet light (DUV) exposure (193 nm), baking and wet development. The line width was varied by changing the exposure dose. The surfactant rinse step was inserted prior to spin-drying the wafers. The surfactant solution is added as a puddle standing on the wafer between 1 and 900 s. Top-down analysis of the photoresist structures was carried out using scanning electron microscopy (SEM). For each exposure dose, a 5 μm^2 area of interest is defined. If one or more lines are collapsed, the field is labelled as "collapsed". For a more global view, defect density measurements were performed. More details of the procedure are given in [21].

As a measure of pattern collapse, the maximum usable exposure latitude (MUEL) was defined:

$$\mathrm{MUEL} = \frac{E_{\mathrm{collapse}} - E_{\mathrm{size}}}{E_{\mathrm{size}}}\,, \tag{2}$$

where E_{collapse} is the exposure dosis at which the patterns collapse and E_{size} is the dose needed to produce the target line width. The larger the MUEL, the larger are the capabilities to reduce pattern collapse [21].

Physico-chemical characterization. For the physico-chemical characterization of the photoresist, flat model surfaces are needed that are accessible for the methods described below. They were prepared by spin-coating the photoresist directly onto a silicon wafer with a native oxide layer. Then, they were exposed to DUV radiation with various doses by open frame exposure. The exposed wafers were baked, developed and rinsed without surfactant. Photoresist layers that have undergone these preparation steps will be named "processed" in the following. For characterization experiments the wafers were broken into small uniformly exposed pieces of the required size. Details of this procedure are described elsewhere [22].

Scanning Force Microscopy. The sample topography was investigated in air with a scanning force microscope (SFM) DI-3100 (Digital Instruments, Santa Barbara, USA). The SFM was used in tapping mode to avoid damage of the photoresist surface. The cantilevers were operated at their resonance frequency of about 275 kHz. Different scan ranges ranging from 1 μm × 1 μm, to achieve a high lateral resolution, up to 10 μm × 10 μm, to detect possible large agglomerates, were applied. From the data, the root mean square (RMS) roughness over the complete scan area was calculated.

Zeta potential measurements. Electrokinetic measurements were performed using the streaming potential method with an ElectroKinetic Analyser EKA (Anton Paar GmbH, Graz, Austria). A home-built rectangular cell for small flat pieces was applied. Two small pieces of the wafer ($10 \times 20\,\mathrm{mm}^2$) were arranged face to face to build a streaming canal inside this cell. The measuring fluid (electrolyte solutions with different pH or containing surfactant) was streaming through this canal. We used Ag/AgCl-electrodes for determining the streaming potential (U). The zeta potential (ζ) was calculated according to Smoluchowski [23]:

$$\frac{\Delta U}{\Delta p} = -\frac{\varepsilon_{\mathrm{r}} \cdot \varepsilon_0 \cdot \zeta}{\eta \cdot \kappa}\,. \tag{3}$$

U is the streaming potential, p is the pressure, ε_{r} and ε_0 are the dielectric constant and the electric constant, respectively. η is the viscosity and κ is the conductivity of the solution. Since the instrument requires a minimum conductivity of the solution, measurements were done in a background electrolyte (KCl) concentration of 10^{-5} to 10^{-3} M.

Ellipsometry. The thickness of the photoresist layers was measured in air using a null ellipsometer in a polarizer-compensator-sample-analyzer arrangement (Multiskop, Optrel Berlin) with a He-Ne laser light source of 632.8 nm wavelength. The angle of incidence was 70°. A multilayer model (silicon/silicon oxide/photoresist) was used to calculate the thickness of the photoresist layer from the ellipsometric angles [24]. The adsorption of the surfactant was monitored in situ in the same device using a special liquid cell with an angle of incidence of 68°. Details of this procedure are described elsewhere [21].

Contact Angle Measurements. Axisymmetric drop shape analysis – profile (ADSA-P): The hydrophobicity/hydrophilicity of a solid surface is usually expressed in terms of wettability, which can be quantified by contact angle measurements. ADSA-P is a technique to determine liquid-fluid interfacial tensions and contact angles from the shape of sessile and pendant drops or captive bubbles [25, 26]. Assuming that the experimental drop is Laplacian and axisymmetric, ADSA-P finds a theoretical profile that best matches the drop profile extracted from an image of a real drop, from which the contact angle, liquid surface tension, drop volume, surface area and three-phase contact radius can be computed simultaneously. Advancing and receding contact angles can be obtained with high accuracy (±0.3°) by increasing and decreasing the volume of the drop using a motorized syringe. In this study, sessile water droplets and captive air bubbles were used to characterize the resist surfaces.

The "captive bubble method" was applied to quantify the wettability of the resist in contact with water, with surfactant solutions of different concentration and with water after contact with the surfactant solution. The wafer piece is mounted with the photoresist layer down in a cuvette filled with the solution of interest. Through a small hole in the wafer an air bubble is placed under the photoresist surface. The shape of the drop is analyzed while its volume is slowly increased and decreased and the contact angle of the bubble is computed. It has to be converted into the water contact angle by subtracting its value from 180°.

As a complementary method, the wetting behavior of dry photoresist surfaces that have been pretreated with surfactant solutions was investigated. Pieces of the photoresist-covered wafers were inserted in surfactant solutions of various concentrations for 3 minutes. Afterwards, the solution was removed and the surface dried with a nitrogen flow. After several hours, the contact angle of water droplets was measured with the Dynamic Contact Angle Tester FIBRO DAT 1100 (FIBRO Systems, Sweden). A water droplet of 8 μl volume was placed onto the surface. Its contact angle was measured in time intervals of 20 ms. From the slope of the contact angle curve as a function of time, the contact angle at first contact ($t = 20$ ms) was extrapolated. This value gives a measure of the contact angle on the dry surfactant layer adsorbed on the photoresist before the desorption of the surfactant starts.

Surface Tension Measurement. The surface tension of the surfactant solution was determined by means of the Dynamic Contact Angle Tester FIBRO DAT 1100 (FIBRO Systems, Sweden) using the pendant drop method. It was also an output of the ADSA captive bubble contact angle measurements with surfactant solutions.

Results and Discussion

Influence of the Cationic Surfactant Rinse on the Pattern Collapse

To investigate whether the rinse with cationic surfactants does affect the pattern collapse, the photolithographic process as described in Experimental was performed without surfactant rinse and with an additional rinse step with four different concentrations of the surfactant. The results of these experiments have been discussed in detail by Wunnicke et al. [21]. Here only a short summary is given.

In Fig. 2, top-down SEM images of photoresist structures of variable line width are shown after cationic surfactant rinse (upper row) and after conventional rinse without surfactants (lower row). The improvement is evident: While without surfactant rinse the lines start to collapse at a width between 65 and 70 nm; they collapse after surfactant rinse below 60 nm. Defect density measurements, too, showed a significant reduction of defects due to pattern collapse by surfactant rinse [21].

For the rinse process with various surfactant concentrations, the maximum usable exposure latitude (MUEL) has

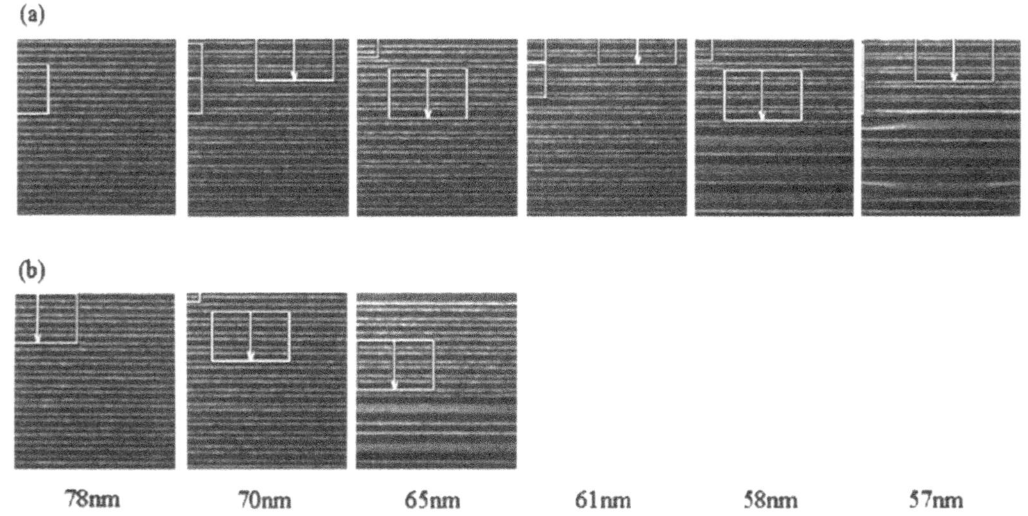

Fig. 2 Top-down SEM images (860 × 860 nm^2) of photoresist structures with variable line width after surfactant rinse (**a**) and after conventional rinse (**b**). The *numbers* give the average line width. *Small lines* stand upright, *broad lines* represent collapsed patterns

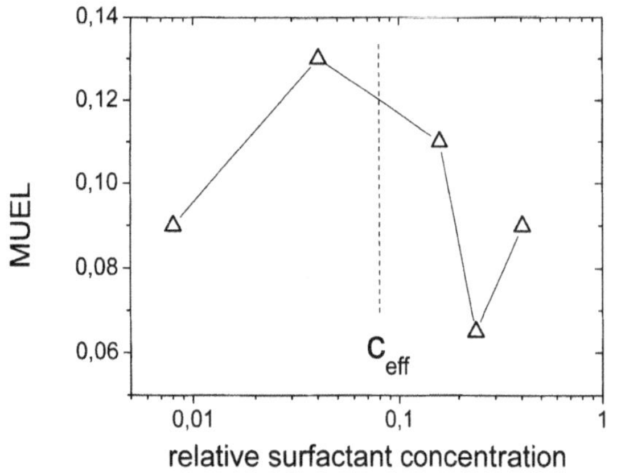

Fig. 3 Maximum usable exposure latitude (MUEL) as a function of the concentration of the surfactant solution used in the surfactant rinse. The *dashed line* gives the concentration c_{eff} assumed to have the highest efficiency to reduce pattern collapse

been calculated according to Eq. 2. It is shown in Fig. 3 as a function of the surfactant concentration. Increase of the MUEL represents a shift of the pattern collapse onset towards higher exposure doses, i.e. lower line width. It can be seen that all surfactant concentrations yield a positive MUEL, i.e. an improvement against the conventional rinse. A maximum of MUEL is expected at the concentration c_{eff} between the relative surfactant concentrations 0.04 and 0.2. At higher concentrations the MUEL decreases again. The concentration c_{eff} will be used in all physico-chemical investigations to mark the concentration of maximum pattern collapse reduction.

Further investigations have been carried out to elucidate the influence of the puddle time of the surfactant rinse on the pattern collapse. It was varied between 1 and 900 s. It has been found that the performance of the surfactant rinse did not depend on the time [21]. This fact is very important since in manufacturing processes short process times are essential.

Characterization of the Model Photoresist Surface

Dry photoresist layers. The capillary forces causing pattern collapse act on the vertical sidewall of the photoresist structures. These surfaces are formed during the development at the interface between soluble and insoluble photoresist. It is generally accepted that a photoresist with a deprotection level above 80% is soluble in the developer [3]. Therefore it can be concluded that the structure sidewalls consist of a blend of more than 20% protected and less than 80% deprotected photoresist. Because of their small dimensions, the sidewalls are not accessible for most surface chemical methods. To allow an investigation, they have to be modelled by flat photoresist surfaces. To simulate the properties of the real sidewalls, these surfaces have to be processed, i.e. exposed, baked and developed. One goal of this study is to find out which of the flat surfaces processed with different exposure doses is the optimum model for the real photoresist sidewalls.

Photoresists used in microelectronic manufacturing are designed that way that they are not changed significantly by processing if the exposure dose is lower than a threshold dose (also named "dose-to-zero"). If the threshold dose is exceeded, the deprotection exceeds 80% and the photoresist is dissolved nearly completely in the developer solution. Figure 4 illustrates the change of some surface properties of processed photoresist as a function of the exposure dose which is given in % of the threshold dose.

Figure 4a shows the thickness of the photoresist layer. Below the threshold dose, nearly no thickness change is observed. When the threshold dose is reached, the layer thickness goes to near zero. Above the threshold dose, a small residual layer was detected that disappears at about 120% of the threshold dose. This behaviour leads to the conclusion that the sidewalls of processed photoresist structures are in a state comparable to that of a pro-

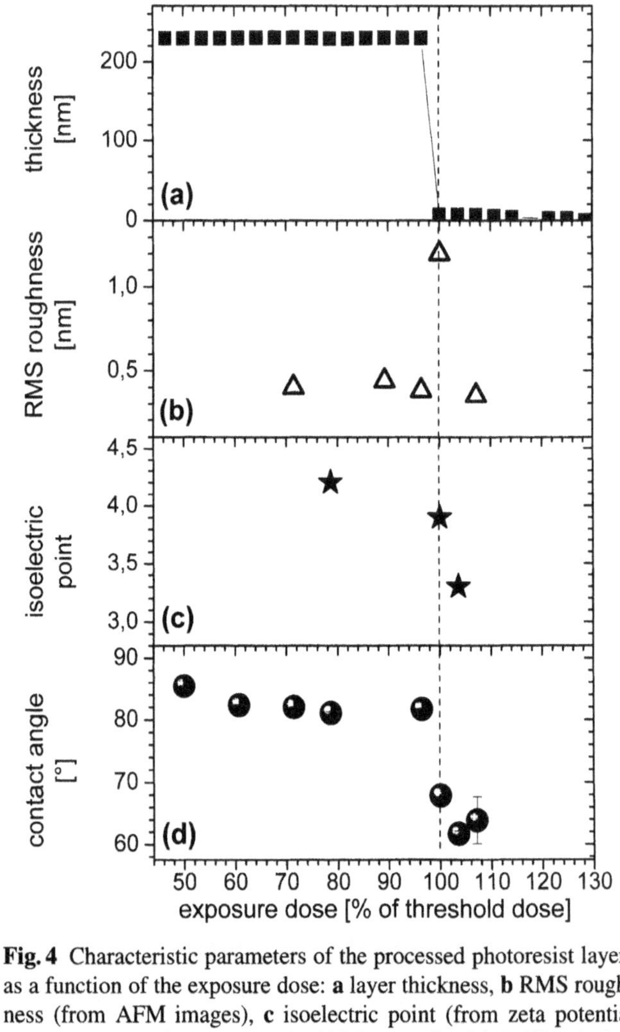

Fig. 4 Characteristic parameters of the processed photoresist layers as a function of the exposure dose: **a** layer thickness, **b** RMS roughness (from AFM images), **c** isoelectric point (from zeta potential measurements), **d** contact angle (ADSA sessile drop). The *dashed vertical line* denotes the threshold dose

cessed flat photoresist layer in the vicinity of the threshold dose [22]. To gain insight in the change of the properties of the photoresist around the threshold dose, both unexposed and processed photoresists treated with variable exposure doses have been characterized by a variety of methods.

SFM images of unexposed photoresists and photoresists processed well below the threshold dose show a homogeneous, flat surface with a RMS roughness of 0.3 to 0.4 nm (Fig. 5 left). When the threshold dose is reached, the surface becomes structured with a RMS roughness of about 1 to 1.4 nm (Fig. 5 right). This structure supports the hypothesis that the remaining photoresist layer is a blend of unchanged and deprotected photoresist where the latter is partly dissolved during the development [22]. With an exposure dose of 107% of the threshold dose, a low roughness is observed again. The residual layer seems to be homogeneous. Figure 4b compares the RMS roughness of photoresist layers processed at various exposure doses.

To investigate the chemical functionality or the order of deprotection of the photoresist surface, the zeta potential of some chosen photoresist samples was measured in electrolyte solutions as a function of the pH value. It is shown in Fig. 6. In consequence of dissociation and adsorption processes at the surface the zeta potential decreases monotonically. The pH value at which the zeta potential function passes zero is called the isoelectric point (iep). Figure 4c shows the iep as a function of the exposure dose. It is shifted to smaller values if the exposure dose approaches the threshold value. This is caused by the generation of acid groups at the surface during processing (cf. Scheme 1). In the case of unexposed photoresist no functional groups exist at the surface. Ion adsorption processes determine the electrokinetic properties, the iep is 4.2. By processing the photoresist more and more acid functional groups are generated at the surface. The electrokinetic properties are governed by dissociation processes of the functional groups. In the case of acid groups the iep decreases. For samples processed above the threshold dose, an iep of 3.4 was measured which is comparable to the iep of bare silicon wafers.

Of special importance for the capillary forces responsible for the pattern collapse is the wettability of the photore-

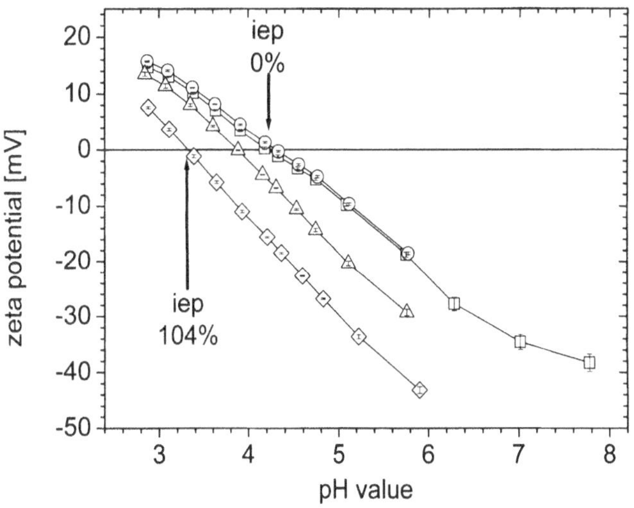

Fig. 6 Zeta potential of flat photoresist layers processed with various exposure doses: 0% (*squares*), 79% (*circles*), 100% (*triangles*), 104% (*diamonds*) of the threshold dose as a function of the pH value, measured in 10^{-3} M KCl. The arrows mark the isoelectric point (iep) for the lowest and highest exposure dose

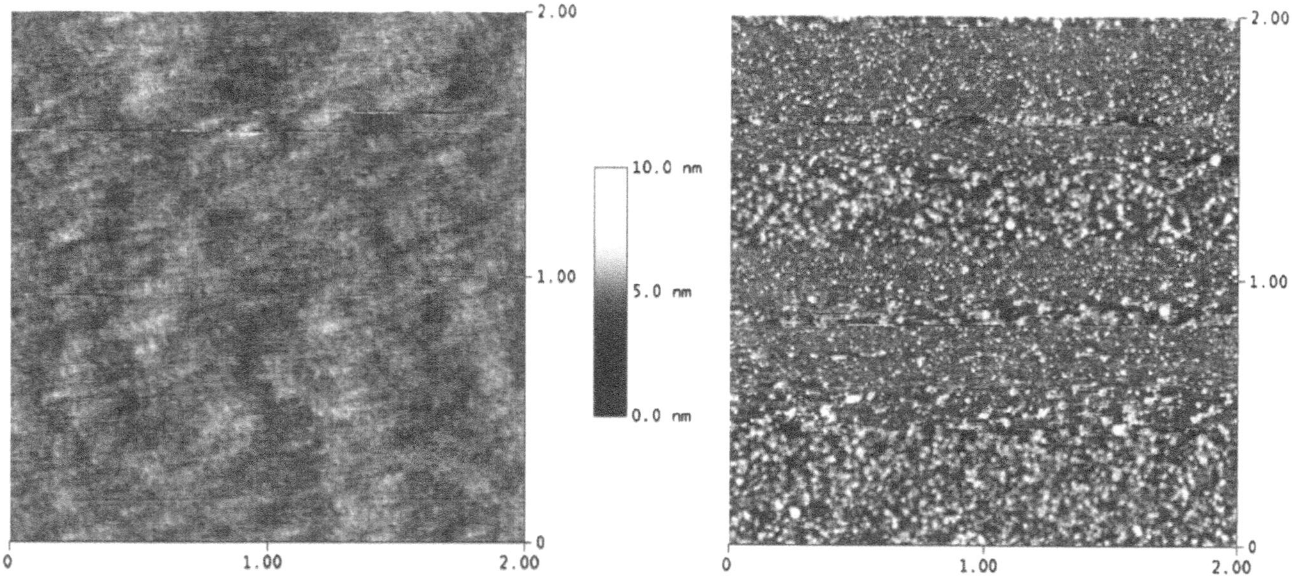

Fig. 5 SFM images of photoresist layers processed with 80% of the threshold dose (*left*) and with the threshold dose (*right*). Image size: $2 \times 2\,\mu m^2$, z-scale 10 nm

sist surface. To reduce the pattern collapse, hydrophobic polymers are used as photoresists. Our investigations show that the contact angle of sessile water drops on the unexposed photoresist surface is $86 \pm 2°$. Processing the photoresist leads to lower water contact angles and, hence, to a more hydrophilic surface as shown in Fig. 4d. Already at 75% of the threshold dose the contact angle is reduced by about 5°. This reflects the slightly increasing content of deprotected polymer in the photoresist. If the threshold dose is applied, the contact angle sinks by about 15° due to the deprotection of nearly 80% of the polymer [3]. Above the threshold dose, contact angles of about 60° are detected. This value is distinctly higher than on a bare silicon wafer surface suggesting that still a residual photoresist layer is present.

It can be summarized that the processing of the photoresist layers with increasing exposure doses leads to a decrease of thickness, isoelectric point and contact angle (Fig. 4). These properties reflect the creation of the deprotected polymer possessing an increasing amount of acidic groups with increasing exposure doses [3, 22]. A significant effect is, however, noticed only in the range of the threshold dose when the degree of deprotection reaches 80% making the polymer soluble in the developer. SFM images illustrate the heterogeneity of the surface in this region. At doses above 110% of the threshold value, all measurements yield the properties of a bare silicon wafer surface.

Behaviour of the photoresist in long-time measurements in water. The photolithographic development is a very short process. It usually takes not more than a few minutes. Thus effects as swelling and slow dissolution play a minor role. In contrary, most of the physico-chemical surface characterization methods used in this study extend over more than ten minutes; if equilibrium is to be waited for, over hours. Therefore, changes of the properties of photoresist layers during immersion in water is of great interest.

The swelling of photoresist layers that were unexposed or processed with exposure doses below the threshold dose was investigated using null-ellipsometry. Whereas the unexposed layer swells over about one hour increasing its thickness by about 5%, the processed wafers swell continuously over a long time. Even after several days no equilibrium state was reached. In inverse contact angle measurements, a gel-like layer was observed on processed wafers leading to non-reproducible contact angles.

Photoresist layers processed at the threshold dose have a thickness of about 5 nm which is too low to show a swelling effect detectable by ellipsometry. However, long-time contact angle measurements using the ADSA captive bubble method revealed a decrease of the contact angle with time as shown in Fig. 7. At the beginning, the water contact angle was 50°. After 50 h a value of only 39° was obtained. It is assumed that the decrease in contact angle is caused by swelling. A partial dissolution of the photoresist is unlikely as the surface tension of water did not change during the long-time experiment. To eliminate the effect of time in investigations of the surfactant adsorption, the curve shown in Fig. 7 was used for baseline correction in later contact angle measurements.

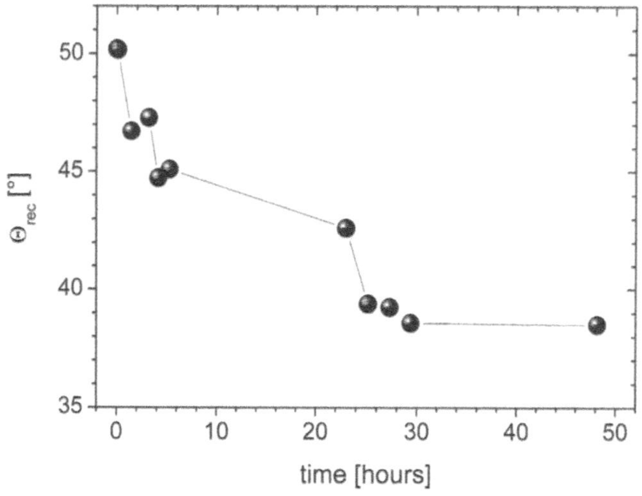

Fig. 7 Receding contact angles of a photoresist layer processed with the threshold dose in a long-time experiment in water (method: inverse ADSA). The *lines* are guides to the eye

In-situ Monitoring of Surfactant Adsorption on the Photoresist Surface

To understand the mechanisms that contribute to the reduction of the pattern collapse by surfactant rinse it was necessary to study the adsorption behaviour of the surfactant molecules on the photoresist surface directly in the surfactant solution.

The thickness change of the photoresist layer plus adsorbed surfactant in surfactant solution was monitored directly by null-ellipsometry. Most of these measurements were done with unexposed photoresist. After an initial swelling by about 5% of its thickness, the ellipsometric angles remained constant and were defined as start values referring to an adsorbed layer of thickness zero. Then a certain amount of surfactant solution was added. The change of the ellipsometric angles was interpreted basing on a multilayer model assuming the formation of a distinct surfactant adsorption layer on top of the photoresist layer. The results of the ellipsometric measurements have been discussed in detail elsewhere [21]. Here only a brief summary is given.

Investigations of the adsorption kinetics showed that a surfactant adsorption layer is formed. In the beginning, the adsorption ran very fast. After one minute, already 40% of the equilibrium amount was adsorbed. Then the adsorption became slower until after 10 to 30 min the adsorption equilibrium is reached. The fast adsorption gives an explanation of the fact that the puddle time, i.e. the time the surfactant solution is applied in the photolithographic process, has no significant influence on the reduction of the pattern collapse. Already after a time of 20 sec as usual in the process, an appropriate amount of surfactant is adsorbed. This justifies that most of our investigations are done in equilibrium although the photolithographic process is far from reaching equilibrium.

Both the adsorption velocity and the adsorbed amount in equilibrium rise with increasing concentration. The adsorption isotherm (adsorbed amount as a function of the concentration) shows a plateau in the vicinity of c_{eff}. That was interpreted basing on a two-step model [27] as formation of a surfactant monolayer. From the adsorbed amount of about $1.2 \pm 0.3\ \mu\text{mol/m}^2$ at c_{eff} an area per molecule of $1.3 \pm 0.3\ \text{nm}^2$ was calculated which is typical for a loose monolayer of the used type of surfactant.

Ellipsometric measurements with processed photoresists showed a long-time swelling over several days. Attempts to investigate the adsorption after addition of surfactant delivered only an continued swelling. It could not be distinguished between the effects of swelling and of the surfactant adsorption.

It can be summarized that ellipsometric measurements proved the formation of a surfactant adsorption layer on the photoresist surface. At c_{eff} it is assumed to form a monolayer. To get more information about the adsorption layer and its influence on the surface properties of the photoresist, an electrokinetic characterization of unexposed and processed photoresist in solutions of the cationic surfactant was carried out. The zeta potential of the photoresist layers is given in Fig. 8 as a function of the surfactant concentration. The measurement was performed at pH = 6 in a background electrolyte (KCl) concentration of 10^{-5} M to ensure the minimum conductivity of the solution necessary for the measurement.

In water and low concentrations of the cationic surfactant, the photoresist is negatively charged. The photoresist processed at or above the threshold dose has a higher negative potential due to its acid groups. With increasing concentration, more and more positively charged surfactant ions adsorb onto the surface. The amount of the zeta potential decreases. At the point of zero charge (pzc), the sign of the zeta potential changes. Up to this concentration the adsorption is governed mainly by an electrostatic attraction between the negatively charged surface and the positively charged surfactant headgroups [28]. The pzc refers to a complete compensation of the surface charge by adsorbed surfactant and counter-ions. As a first approximation, at the pzc the surfactant is assumed to form a monolayer on the photoresist surface.

The pzc is marked in Fig. 8 for photoresist layers processed with < 80% and ≥ 100% of the threshold dose. For the latter, the pzc is about one order of magnitude higher than for unexposed photoresist, i.e. a higher surfactant concentration is necessary to neutralize the acid groups

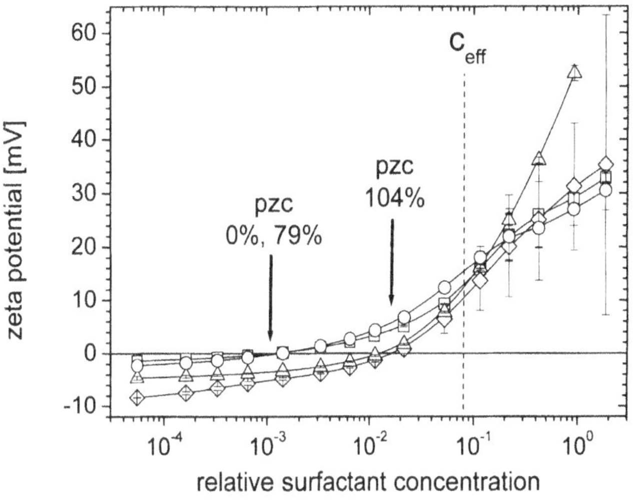

Fig. 8 Zeta potential of flat photoresist layers processed with various exposure doses: 0% (*squares*), 79% (*circles*), 100% (*triangles*), 104% (*diamonds*) of the threshold dose as a function of the surfactant concentration measured in 10^{-5} M KCl.The arrows mark the points of zero charge (pzc) for various exposure doses, the dashed line the concentration c_{eff}

on the highly deprotected photoresist. Therefore a higher adsorbed amount would be expected on these processed layers in the vicinity of the pzc. Unfortunately this could not be proved by ellipsometry due to the strong swelling as discussed above.

Raising the surfactant concentration above the pzc, the zeta potential increases steadily in the positive range. At c_{eff}, the most efficient concentration for the reduction of the pattern collapse, both unexposed and processed photoresists have a positive zeta potential between 10 and 16 mV. This is surprising since the maximum pattern collapse reduction is assumed (and will be shown in the next section) to be correlated to maximum hydrophobizing by a surfactant monolayer in the vicinity of the pzc [29]. In contrary, a positive zeta potential is interpreted as an excess of positively charged surfactant molecules forming a bilayer or admicelles. One reason for this difference might be the influence of the background electrolyte needed for the zeta potential measurement. It is known that in the presence of electrolytes adsorption and aggregation processes of ionic surfactants are shifted to lower concentrations [30]. The effect of a 10^{-5} M KCl solution is, however, expected to be negligible. Another possible explanation is that the hydrophobic tails of the surfactant molecules aggregate around c_{eff} to form positively charged hemi-micelles on the surface [30]. At the interface photoresist-solution this may be energetically more favorable. If the solution front recedes during evaporation or in the captive bubble experiment, the hemi-micelles might transform into a monolayer which is more favorable in contact with air. This hypothesis as well as the influence of electrolytes will be the subject of further investigations.

The Influence of Surfactant on the Capillary Forces

It was outlined that the surfactant rinse is expected to reduce the pattern collapse by lowering the capillary forces between the photoresist structures. For a given geometry, the capillary forces are proportional to the wetting tension $\gamma_{lv} \cos \Theta$, the product of surface tension and the cosine of the contact angle (cf. Eq. 1). In the case of an evaporation process, the receding contact angle of the evaporating solution has to be taken into account.

The inverse Axisymmetric Drop Shape Analysis (ADSA-P) allows to measure simultaneously the surface tension and the contact angle of a receding solution front by increasing the volume of an air bubble that is in contact with the photoresist layer. Therefore it is suited very well to simulate the conditions during the evaporation of the solution in the photolithographic process as is illustrated in Scheme 2. There is, however, one significant difference between the evaporation process and the captive bubble method: The evaporation takes place within seconds; the contact angle measurement can be started at least than 10 minutes after insertion of the sample in the liquid and may exceed over hours. To eliminate this influence of swelling on the contact angle, for each contact angle measured by the captive bubble method the corresponding change of the contact angle in time was estimated from Fig. 7. This correction value was added to the measured contact angle.

Several series of contact angle measurements were done on unexposed photoresist and photoresist processed below the threshold dose. All these measurements showed a continuous decrease of the contact angle with increasing surfactant concentration even if the contribution of swelling was subtracted. No hydrophobizing due to the adsorption of surfactant was observed in surfactant solutions.

To find out whether a hydrophobizing effect can be obtained by surfactant adsorption, photoresist layers processed with exposure doses between 50% and 120% of the threshold dose have been investigated by the captive bubble method. Their receding contact angle was first measured in water. Then the wet photoresist layers were inserted for 15 min in a surfactant solution of concentration c_{eff}. When a sufficient amount of surfactant was assumed to be adsorbed, the wafer pieces were shortly rinsed with water and transferred into water to measure the receding contact angle again under the same conditions. Figure 9 shows the receding contact angles before and after surfactant adsorption. With this method, for all exposure doses an increase of the contact angle (2° to 8°) after the surfactant adsorption is recorded. At the threshold dose, the increase of the contact angle is remarkably higher, about 15°.

The experiment described above does surely not yield the contact angles that determine the capillary forces since the surfactant layer is desorbed partially in water. It proves, however, that a noticeable hydrophobizing effect after surfactant adsorption can be found for the rather hydrophilic photoresist surfaces processed with the threshold dose.

Therefore, in another captive bubble experiment, the receding contact angle of the surfactant solutions on photoresist processed with the threshold dose was measured when the photoresist was directly in contact with the solution. The photoresist sample was inserted in water and the surfactant concentration was stepwise increased. Contact angles were determined after each concentration change when equilibrium was reached and corrected by the time effect. Figure 10a compares the receding angles of the surfactant solution on the photoresist surface with the simultanously measured surface tension of the solution. The contact angles were obtained in two independent measurement series.

In the investigated concentration range, the surface tension of the surfactant solution is always above 70°, i.e. it is

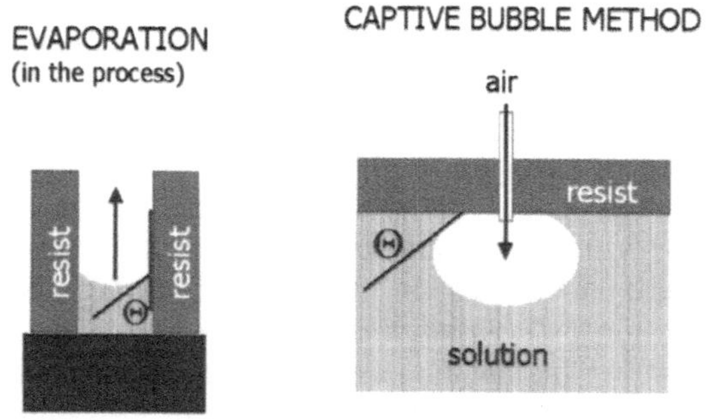

Scheme 2 Comparison of the geometries of the receding contact angle responsible for the capillary forces between photoresist lines in the photolithographic process (*left image*) and the receding contact angle measured by the captive bubble method (*right image*)

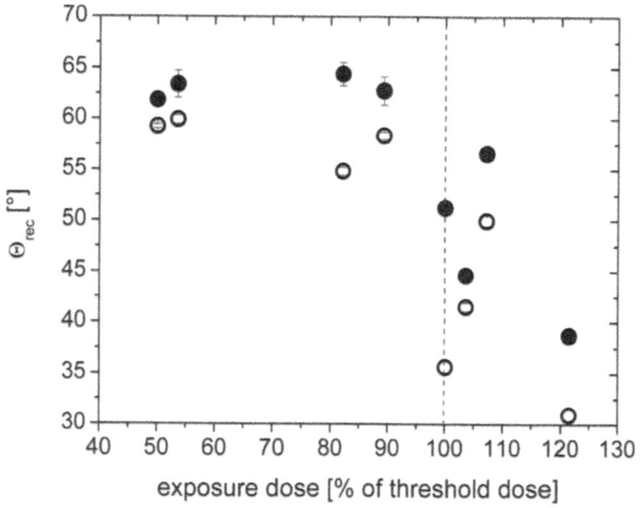

Fig. 9 Receding contact angles (inverse ADSA) of photoresist layers processed with various exposure doses in water (*open circles*) and in water after 15 min contact with surfactant solution of concentration c_{eff} (*full circles*) as a function of the exposure dose. The *dashed line* marks the threshold dose.

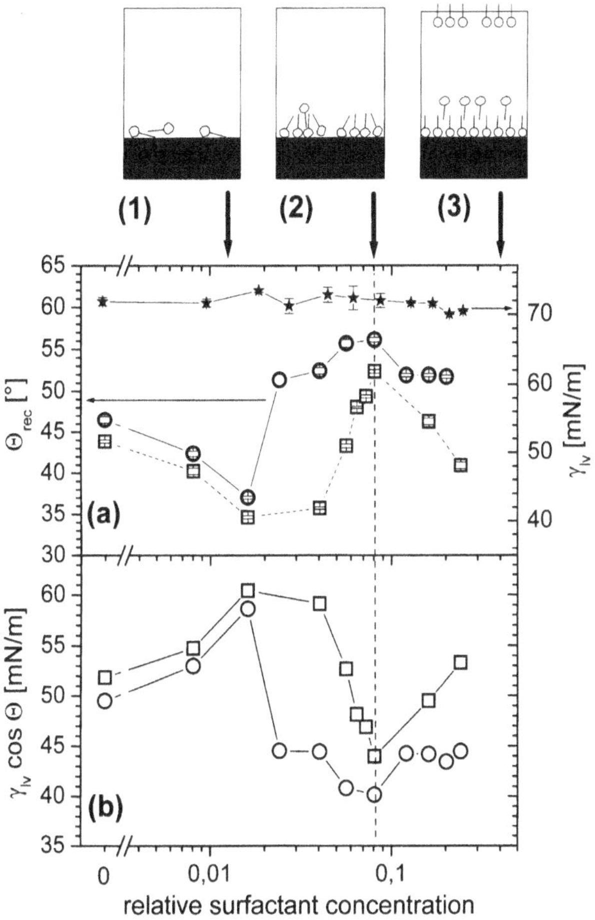

Fig. 10 **a** Surface tension of the surfactant solution (*stars, right axis*) and receding contact angle of the solutions (*squares, circles, left axis*) on a photoresist layer processed at the threshold dose corrected by the effect of swelling. **b** Wetting tension $\gamma_{lv}\cos\Theta$ calculated from the values given in (**a**). The dashed line marks the concentration c_{eff}. Above **a**, the assumed distribution of the surfactant molecules at various concentrations is drawn schematically

not significantly lowered compared to the surface tension of water. On the other hand, the receding contact angle changes remarkably. Starting from a value of 45° in water, it decreases at very low concentrations ($c_{rel} < 0.02$). Approaching c_{eff}, the contact angle increases and reaches a maximum with a contact angle about 10° higher than in water. At concentrations above c_{eff} the receding contact angle of the solution decreases. Independent investigation of surfactant solutions without contact with the photoresist surface using the pendant drop method showed a significant decrease of the surface tension at relative concentrations > 0.2.

The behaviour of the surface tension and the contact angle allows conclusions concerning the adsorption mechanism. They are illustrated schematically in Fig. 10. At concentrations < c_{eff}, the surfactant adsorption is driven mainly by the electrostatic attraction between the negatively charged photoresist and positively charged surfactant ions. Nearly all surfactant ions adsorb randomly on the photoresist surface (1). Around c_{eff}, the photoresist surface is densely covered with surfactant ions. They are assumed to form a monolayer in the vicinity of the three-phase line between photoresist, surfactant solution and air. This monolayer hydrophobizes the photoresist surface. If no air is present, it might be energetically more favorable to form hemi-micelles as was postulated from zeta potential measurements. The amount of surfactant molecules adsorbed on the solution surface is negligible (2). At higher concentrations, hydrophobic interactions between the hydrocarbon chains of the surfactant dominate. This leads to the formation of a bilayer or surface micelles and an aggregation of the surfactant at the interface solution – air reducing both surface tension and contact angle (3).

To estimate the influence of the surfactant adsorption on the capillary forces, the wetting tension $\gamma_{lv}\cos\Theta$ was calculated from the values given in Fig. 10a. The results drawn in Fig. 10b show for both measurement series a minimum of the capillary forces exactly at the concentration c_{eff}. The capillary forces are reduced by about 20% compared to water. This confirms the hypothesis that the reduction of the pattern collapse is caused by a hydrophobizing of photoresist processed with the threshold dose by cationic surfactant adsorption. Unfortunately the inverse ADSA method could not be applied at relative surfactant concentrations > 0.2 since the bubbles became unstable due to the lower surface tension. Thus it cannot be estimated how the wetting tension evolves at higher concentrations.

It must be kept in mind that the data given in Fig. 10 may contain an error due to the correction by the effect of swelling. This correction can only be an approximation since the photoresist might swell in surfactant solutions in another way than in water. It cannot be excluded, too, that surfactant molecules penetrate into the photoresist.

To prove the hydrophobizing avoiding swelling, a complementary method was used that characterizes dry photoresist layers after short contact with the surfactant solution. Unexposed photoresist samples and photoresist processed with 70–96% and 100–110% of the threshold dose were inserted for 3 minutes in surfactant solutions of different concentrations. Then the solutions were removed quickly by a nitrogen flow. The contact angle of sessile water drops on the dried layers was measured using a fast dynamic method.

In some cases the initial contact angle of photoresist treated with surfactant solutions was significantly higher than that of the untreated reference samples but decreased fast due to the desorption of surfactant molecules. Since the drop is not mechanically stable in the first ca. 100 milliseconds, the contact angles were extrapolated to time $t = 20$ ms, i.e. to the first contact. The values obtained for a series of 8–10 drops were averaged. The results are shown in Fig. 11 as a function of the concentration of the surfactant solution used for pretreatment.

Processed photoresists that have not been treated with surfactant solutions (concentration = 0), show – depending on the exposure dose used in the processing – very different contact angles ranging from 46° to 74° (see also Fig. 4d). After treatment with surfactant solutions, however, the distribution of the contact angles is narrower (58° to 70°). It is assumed that as a result of the pretreatment similar surfactant adsorption layers with similar contact angle are formed on all photoresist samples. This coincides with the results of Versluijs [14] who found the best pattern collapse reduction using surfactant solutions that minimized the contact angle differences between unexposed and exposed resist. The maximum of this contact angle is in the range of 60° to 65°, i.e. it is lower than that of the photoresist processed below the threshold dose. It remains nearly constant in a broad relative concentration range (0.01–0.3) and decreases at higher concentrations.

This observation confirms that a hydrophobizing effect is indeed found mainly for photoresists processed with the threshold dose having an initial contact angle < 60°. For photoresists exposed with lower doses, the contact angle remains constant or decreases as has been found in inverse ADSA measurements in surfactant solutions.

Furthermore we concluded that the photoresist layers processed with 100–110% of the threshold dose reflect quite well the properties of the structure sidewalls in the photolithographic process. The surfactant treatment of these model photoresist layers should be comparable with the surfactant rinse in the photolithographic process. Therefore the hydrophobizing of these layers can serve as a proof of the hydrophobizing of the photoresist structures at surfactant concentrations around c_{eff}. In our experiment, a hydrophobizing effect similar to that at the concentration c_{eff} is obtained already with a ten times lower concentration. This may be explained by the fact that – in contrary to the surfactant rinse in the photolithographic process – the surfactant solution was not diluted by other liquids and the adsorption time was about 20 times longer resulting in a higher adsorbed amount.

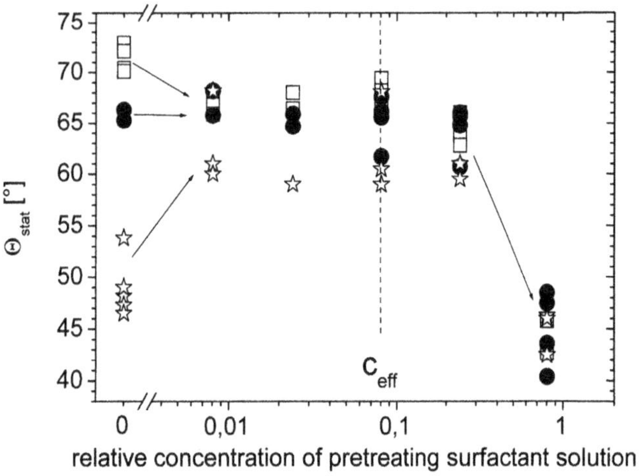

Fig. 11 Initial contact angles of sessile water drops on dry photoresist layers pretreated with surfactant solutions. The photoresist layers have been processed with various exposure doses: unexposed (*squares*), 70–96% (*full circles*), 100–110% of threshold dose (*stars*). The horizontal axis gives the concentration of the solution used for pretreatment

Generally, it must be kept in mind that the physicochemical characterization was carried out under conditions that are different from those of the surfactant rinse in the photolithographic process. In the process, the solution is added for about 20 s to a solution containing developer and dissolved photoresist. The physico-chemical characterization has been done under idealized conditions, i.e. mostly in equilibrium and in pure surfactant solution, but long-time effects as swelling and dissolution have to be taken into account.

Therefore, a surfactant monolayer might be formed during drying already at concentrations below c_{eff} as indicated by the pzc (cf. Fig. 8) or the contact angle of the dried photoresist layers.

Nevertheless a good coincidence between the reduction of the capillary forces under laboratory conditions with the reduction of pattern collapse in the photolithographic process was found. This supports our hypothesis that cationic surfactants are able to reduce pattern collapse by hydrophobizing the photoresist surface. It shows also that the hydrophobizing must occur in a similar way under real conditions. Thus, flat photoresist layers processed at the threshold dose are appropriate model surfaces for the sidewalls of photoresist structures.

The cationic surfactant reduces the capillary forces already at concentrations far below the cmc. This may be a great advantage since problems occurring at concentrations around the cmc like melting of the photoresist structures [15] or defect creation by foaming [17] can be avoided.

Conclusions

Goal of the present study was to investigate how solutions of cationic surfactants affect the pattern collapse in the photolithographic sub-100 nm structuring. On one hand, the solutions were applied directly in the photolithographic process to investigate their ability to reduce the pattern collapse. On the other hand, the adsorption of the surfactant on flat model photoresist layers was studied using a variety of physicochemical characterization methods and its influence on the capillary forces was determined.

The investigation of the pattern collapse in the photolithographic process revealed that a short rinse step with cationic surfactant solutions yields a maximum pattern collapse reduction at a surfactant concentration far below the cmc. The rinsing time had no influence on the pattern collapse.

The physico-chemical characterization of processed flat photoresist layers showed that the properties of the photoresist change significantly if it is processed with the threshold exposure dose that makes the photoresist soluble in the developer. Caused by a deprotection reaction, the photoresist becomes acidic and more hydrophilic. It was postulated that the thin residual photoresist layers processed with the threshold dose are appropriate model surfaces for the sidewalls of the photoresist structures in the photolithographic process.

By ellipsometry it was proved that the surfactant adsorbs on the photoresist surface. In the first minute, the adsorption runs very fast. In or slightly below the concentration range that accounts for highest pattern collapse reduction the major part of the surfactant ions adsorbs onto the photoresist layer. It is assumed that they form a monolayer or hemimicelles that transform into a monolayer during contact with air. A remarkable hydrophobizing effect was found for photoresist layers that had been processed with the threshold dose. The contact angle of these layers was increased from 45° to 55°. Although the surface tension was not lowered significantly, the hydrophobizing causes a reduction of the wetting tension by about 20%.

The minimum of the capillary forces in the model system correlates with the maximum of the pattern collapse reduction in the real photolithographic process. It is concluded that the sidewalls of the photoresist structures are hydrophobized in a similar way as the model surfaces by the cationic surfactant rinse even although the conditions in the process differ from those of the characterization experiments. The resulting reduction of the capillary forces during the drying of the structures is the essential factor in the reduction of pattern collapse.

A great advantage of the cationic surfactant rinse is that the minimum of the capillary forces is obtained already at concentrations far below the cmc. Low amounts of surfactant are needed and problems as melting of the structures or foaming occurring at high surfactant concentration are avoided.

Acknowledgement Financial support for this work provided by the German Federal Ministry of Education and Research (Project No. 01M3154D) and by Infineon Technologies SC300 is gratefully acknowledged. The authors also would like to thank A. Caspari for the zeta potential measurements, B. Pinter, I. Mäge, T. Pearce, M. Voigt for the defect density measurements and the unit process team – resist of Infineon Technologies and J. Kobisch for wafer processing.

References

1. International Technology Roadmap for Semiconductors (2004) http://public.itrs.net
2. Wallraff GM, Hinsberg WD (1999) Chem Rev 99:1801
3. Lin EK, Soles CL, Goldfarb DL, Trinque BC, Burns SD, Jones RL, Lenhart JL, Angelopoulos M, Wilson CG, Satija SK, Wu W (2002) Science 297:372
4. Cao HB, Nealey P, Domke WD (2000) J Vac Sci Technol B 18:3303
5. Domke WD, Graffenberg VL, Patel S, Rich GK, Cao HB, Nealey PF (2000) Proc SPIE 3999:139
6. Shibata T, Ishii T, Nozawa H, Tamamura T (1997) Jap J Appl Phys 36:7642
7. Tanaka K, Naito R, Kitada T, Kiba Y, Yamada Y, Kobayashi M, Ichikawa H (2003) Proc SPIE 5039:1366
8. Lee SK, Jung JC, Lee MS, Lee SK, Kim SY, Hwang YS, Bok CK, Moon SC, Shin KS, Kim SJ (2003) Proc SPIE 5039:166
9. Namatsu H, Yamazaki K, Kurihara K (1999) Microelectronic Engineering 46:129
10. Junarsa I, Stoykovich MP, Yoshimoto K, Nealey PF (2004) Proc SPIE 5376:842
11. Hien S, Rich GK, Molina G, Cao HB, Nealey PF (2002) Proc SPIE 4690:254
12. Lee G, Lee SK, Hwang YS, Jung JC, Bok CK, Moon SC, Shin KS (2003) Proc SPIE 5039:1416
13. Jung MH, Lee SH, Kim HW, Woo SG, Cho HK, Han WS (2003) Proc SPIE 5039:1298
14. Versluijs JJ, Misat SI, Maenhoudt M, Grozev G, Furusho R (2003) Proc Interface 2003 Symposium, Arch Chemicals
15. Watanabe M, Tomo Y, Yamabe M, Kiba Y, Tanaka K, Naito R (2003) Proc SPIE 5037:925
16. Miyahara O, Tanaka K, Wakamizu S, Kitano J, Yamada Y (2004) Proc SPIE 5376:830
17. Zhang P, King DM, Karwacki EJ (2002) MICROMagazine June 2002, http://www.micromagazine.com/archive/02/06/zhang.html
18. Stache H (1981) Tensid-Taschenbuch, 2nd edition. Carl Hanser, München Wien
19. Koopal LK, Goloub T, de Keizer A, Sidorova MP (1999) Colloids Surf A 151:5
20. Fuerstenau DW, Jia R (2004) Colloids Surf A 250:223
21. Wunnicke O, Reichelt J, Busch P, Petong N, Bellmann C, Pinter B, Mäge I, Pearce T, Voigt M, Hermsdorf N, Drechsler A, Stamm M, Grundke K (2004) High aspect ratio patterning by utilizing surfactant rinse process, Proc Interface 2004 Symposium. Arch Chemicals
22. Wunnicke O, Hennig A, Grundke K, Stamm M, Czech G (2002) Proc SPIE 4690:332
23. Smoluchowski M (1921) Handbuch der Elektrizität und des Magnetismus, Band II. Barth-Verlag, Leipzig
24. Azzam RMA, Bashara NM (1999) Ellipsometry and Polarized Light. North Holland, Amsterdam
25. Lahooti S, del Rio OI, Cheng P, Neumann AW (1996) In: Neumann AW, Spelt K (eds) Applied Surface Thermodynamics, Marcel Dekker, p 441
26. Grundke K, Bogumil T, Werner C, Janke A, Pöschel K, Jacobasch HJ (1996) Colloids Surf A 116:79

27. Atkin R, Craig VSJ, Wanless EJ, Biggs S (2003) Adv Colloid Interface Sci 103:219
28. Parfitt GD, Rochester CH (1983) In: Parfitt GD, Rochester CH (eds) Adsorption from Solution at the Solid/Liquid Interface. Academic Press London, p 247
29. Churaev NV, Sergeeva IP, Sobolev VD, Jacobasch HJ, Weidenhammer P, Schmitt FJ (2000) Colloids Surf A 164:121
30. Goloub TB, Koopal LK (1997) Langmuir 13:673
31. Fuerstenau DW (1970) Pure Appl Chem 24:981

Progr Colloid Polym Sci (2006) 132: 95–101
DOI 10.1007/2882_043

Published online: 8 March 2006

Tatiana B. Ermakova
Inessa P. Sergeeva
Anna D. Anuchkina
Vladimir D. Sobolev
Nikolay V. Churaev

Electrokinetic Study of Layer-by-Layer Polyelectrolyte and Surfactant Adsorbed Layers

Tatiana B. Ermakova (✉) · Inessa P. Sergeeva · Anna D. Anuchkina · Vladimir D. Sobolev · Nikolay V. Churaev
Laboratory of Thin Liquid Films, Institute of Physical Chemistry and Electrochemistry by A.N. Frumkin of Russian Academy of Science, Leninsky prospect, 31, 119991 Moscow, Russia
e-mail: etb@mail.ru

Abstract Investigations of alternate adsorption regularities of cationic polyelectrolytes a) copolymer of styrene and dimethylaminopropylmaleimide (CSDAPM) and b) poly(diallyldimethylammonium chloride) (PDADMAC) and anionic surfactant – sodium dodecyl sulfate (SDS) on fused quartz surface were carried out by capillary electrokinetic method. The adsorption/desorption kinetics, structure and properties of adsorbed layers for both polyelectrolytes and also for the second adsorbed layer were studied in dependence on different conditions: molecular weight of polyelectrolyte, surfactant and polyelectrolyte concentration, the solution flow rate through the capillary during the adsorption, adsorbed layer formation time. For PDADMAC adsorbed layers evolution of the structure and rheological properties was observed. Adsorption of anionic surfactant leads to the overcharge of preliminarily modified quartz surface by cationic polyelectrolyte and depends on the SDS concentration and prehistory of the first adsorbed layer. The desorption study showed that in all cases the partial desorption of SDS took place, the value of this desorption depended on the structure of the first adsorbed layer and SDS concentration.

Keywords Adsorbed layer deformation · Adsorption · Aging · Polyelectrolytes · Streaming potential method

Introduction

In recent years the investigation of polymer-surfactant interactions is a rapidly growing field of interest of modern colloid science [1–4]. The mixtures and multilayer structures of polyelectrolytes and surfactants are widely used for industrial application to govern the wetting, adhesion, flotation processes and so on.

The formation of multilayer structure can be carried out by several ways: 1) adsorption on liquid/liquid interface – Langmuir-Blodgett films [5,6]; 2) adsorption on solid/liquid interface – alternate adsorption of oppositely charged polyelectrolytes (PE) and surfactants on flat surfaces or spherical particles [7,8]. To control the process of multilayer systems formation, it is necessary to understand the different factors influence on the structure and properties of layers formed, such as charge distribution density on the solid surface, the state of polyelectrolyte in a bulk solution (i.e. its charge and conformation), molecular weight of polyelectrolyte and salt concentration.

It is important to concentrate special attention on the first step of the process, viz on the study of polyelectrolyte adsorption on charged surface. The time dependence of layer growing and structural reconstruction can give the information of formation mechanism of the multilayer.

The initial step of any multilayer formation model is the surface overcharge as a result of one polyelectrolyte layer adsorption. Only in this case the adsorption of the second oppositely charged polyion/surfactant ion layer is possible [9]. In our work the negatively charged surface of fused quartz was used as a substrate, that's why the first layer is the layer of cationic polyelectrolyte. The main driving force in the adsorption process in this case is the electrostatics. So the multilayer thickness depends on the charge value of the substrate (it increases with the charge increase) and on the salt concentration. The ionic strength of the solution influences on the conformation of polyelectrolyte molecules and, consequently, on the adsorbed amount and the structure of the layer [10–12]. Our experiments were carried out at fixed salt concentration and pH to limit the number of factors influenced on the multilayer system.

When the multilayer is formed the question of adsorption reversibility of polycation is very important. It was determined experimentally that the polymer desorption is a very slow process and only very small amount can be desorbed. There are only structural changes if the adsorbed amount is fixed [13, 14].

In this paper we investigate the process of alternate adsorption of cationic polyelectrolyte and anionic surfactant, structure and properties of adsorbed layers depending on different factors (molecular weight of PE, concentration of polyelectrolyte and surfactant, adsorbed layer formation time, the flow rate of the solution) by measuring potential and streaming current using the capillary electrokinetic method.

Materials and Methods

Materials. As a ***substrate***, the surface of thin quartz capillaries, $r = 5$–$8\,\mu$m, freshly drawn from thin cylindrical quartz tubes of high purity ($SiO_2 > 99.99\%$) was used. Capillary is cylindrical (its conicity not exceeded 10^{-6} rad), the surface is molecular smooth. For the experiments the capillaries with ζ potential value about 100–110 mV in KCl 10^{-4} M solution at pH 6.5 were chosen. **Cationic polyelectrolytes:** 1) copolymer of styrene with dimethylaminopropylmaleimide (CSDAPM), synthesized in Institute of Polymer Research, Dresden, Germany, $M_w = 20\,000\,\mathrm{g\,mol^{-1}}$ (see Fig. 1A) (The positive charge is due to protonation of tertiary nitrogen atoms, one molecule contains 72 repeating units, approximately 60% of which are ionized in the neutral pH region); 2) poly(diallyldimethylmammonium chloride) (PDADMAC), $M_w = 400\,000$–$500\,000\,\mathrm{g\,mol^{-1}}$, Aldrich (see Fig. 1B) ($PDADMA^+$ carries a constant charge of one elementary charge per segment and the average distance between the charges is 5.2 Å [15]).

As ***anionic surfactant*** sodium dodecyl sulfate (SDS), $M_w = 288.4$, Merck was used (cmc 8.3×10^{-3} M).

All solutions were prepared using background solution KCl 10^{-4} M and three-distilled water with electrical conductivity $10^{-6}\,\Omega^{-1}\,\mathrm{cm}^{-1}$ and pH 6.5 to avoid problems connected with conformational and surface charge changes with increasing of salt concentration and pH.

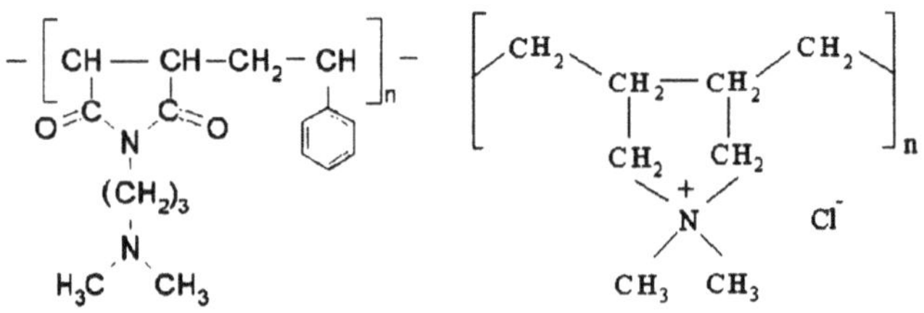

Fig. 1 Structural formula of copolymer of styrene and dimethylaminopropylmaleimide (CSDAPM) and poly(diallyldimethylammonium chloride) (PDADMAC)

Electrokinetic measurements. In this study the capillary electrokinetic method was applied [16].

A solution of a certain composition was forced through a capillary under pressure of nitrogen supplied from a compressed gas cylinder. This method allows to measure the voltage-current characteristics $E(I)$ of the solution in the capillary. Electrical resistance R of the solution was determined from the slope of the linear graph $R = \Delta E/\Delta I$. Since in all cases the radius of the capillaries was much greater than the Debye radius (≤ 30 nm), the electrokinetic potential could be determined with Helmholtz–Smoluchowski equation: $E_0/P = \varepsilon\zeta/4\pi\eta K$, where ε is dielectric constant, η is solution viscosity and K is the conductivity of the solution in the capillary, $K = l/\pi r^2 R_0$, where r and l are the radius and the length of the capillary, respectively. The value of K determined in this way includes the effect of surface conductivity.

It is usually considered that ζ potential value is independent on the pressure value that testifies for the constant position of slipping plane. This is fairly for solid non-deformed surfaces. However, in some cases the position of slipping plane depends on the surface structure, which can be affected by tangential shear stress [17]. So in the case of deformed surfaces ("soft layers") the calculated ζ potential would depend on shear stress value, i.e. pressure value. This should be taken into account when experimental data are interpreted.

Besides there is uncertainty in hydrodynamics inside adsorbed layer, which can be (or not) permeable for flux. Hydrodynamic profile can deviate from parabolic Poiseuille profile, that's why in this case the Helmholtz–Smoluchowski equation does not correctly transform the streaming potential into the zeta-potential values. So we indicate calculated ζ potential as apparent zeta-potential (ζ^*).

Experimental procedure. The experiment was initiated by pumping of background electrolyte solution KCl 10^{-4} M through the capillary. The initial surface potential was measured. Than a polyelectrolyte solution, as the first adsorbed layer, and anionic surfactant solution, as the second one, were pumped through the same capillary. Between these two stages the capillary was rinsed with KCl

10^{-4} M solution. It should be emphasized that desorption and deformation studies were also carried out in solution free of polyelectrolyte and surfactant molecules.

Results and Discussion

Kinetics of Layer-by-Layer Adsorption of Polyelectrolyte and Surfactant

Since the negative charged surface of fused quartz was used as a substrate, the first adsorbed layer was the layer of cationic polyelectrolyte. In our previous works [18–20] the adsorption kinetics of cationic polyelectrolyte CSDAPM was studied in detail. The estimation of adsorption was carried out by changing of ζ^* potential of charged quartz surface during the cationic polyelectrolyte adsorption.

Here we represent the comparative results of adsorption on the quartz surface of two polyelectrolytes with tertiary ammonium (CSDAPM) and quaternary ammonium groups (PDADMAC), which are weak and strong polyelectrolytes, respectively (Figs. 2 and 3).

All adsorbed layers were formed at different pressures (from 2 atm up to 10 atm). But measurements were done at $\Delta P = 10$ atm (as our data will show hereinafter, there is no any deformation of adsorbed layers above $\Delta P = 10$ atm). The constant ζ^* potential value establishment doesn't depend on the flow conditions at each concentration.

As can be seen from the Figs. 2 and 3, the character of dependencies doesn't qualitatively differ for polyelectrolytes of different nature. The quantitative characteristics such as the time of establishment of surface potential magnitude and its values are different.

For both polyelectrolytes the adsorption of polymer molecules occurs mostly due to the forces of electrostatic attraction at concentrations below 10^{-4} g/l. At higher PE

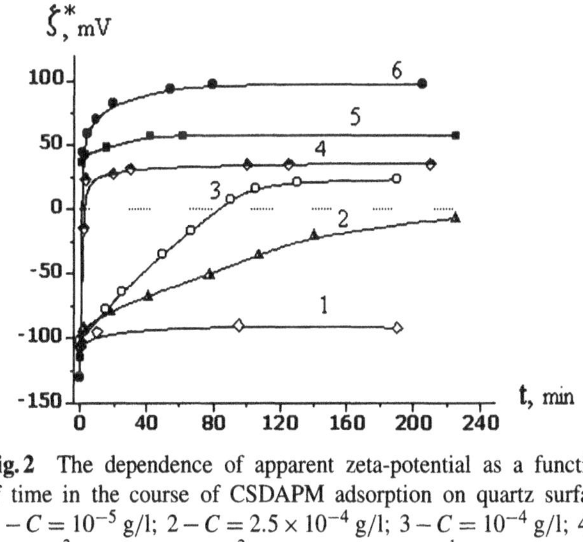

Fig. 2 The dependence of apparent zeta-potential as a function of time in the course of CSDAPM adsorption on quartz surface ($1 - C = 10^{-5}$ g/l; $2 - C = 2.5 \times 10^{-4}$ g/l; $3 - C = 10^{-4}$ g/l; $4 - C = 10^{-3}$ g/l; $5 - C = 10^{-2}$ g/l; $6 - C = 5 \times 10^{-1}$ g/l)

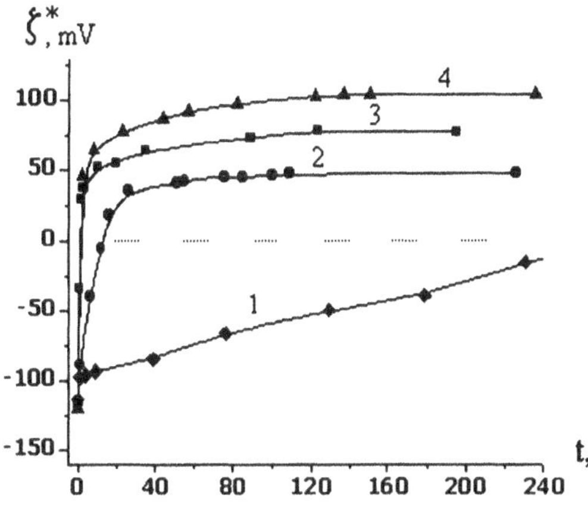

Fig. 3 The dependence of apparent zeta-potential as a function of time in the course of PDADMAC adsorption on quartz surface ($1 - C = 10^{-5}$ g/l; $2 - C = 10^{-4}$ g/l; $3 - C = 10^{-3}$ g/l; $4 - C = 10^{-2}$ g/l)

concentration ($C \geq 10^{-4}$ g/l) the rapid decrease in the negative values of ζ^* potential is observed, following by the charge reversal of the surface, which occurs the earlier the higher the polymer concentration is. The constant positive values of ζ^* potential correspond to the adsorption saturation of the quartz surface.

The observed charge reversal can prove the presence of two types of the PE adsorption sites on the capillary surface. At low concentration, the electrostatic adsorption of positively charged PE molecules predominantly occurs on the negatively charged sites of quartz surface. Thereafter (or simultaneously), on the surface of a capillary covered with a polymer adsorbed layer, the adsorption of the PE molecules can occur due to the forces of molecular attraction and attraction between hydrophobic sites of polyelectrolyte and surface (e.g. siloxane groups). Their competition with the electrostatic repulsion forces that increase in the course of further adsorption of PE molecules determines the completion of the adsorption and the formation of equilibrium (with the solution) adsorbed layer.

After the establishment of constant PE adsorption value the capillary was rinsed with background KCl 10^{-4} M solution and then the adsorption of the second layer was carried out.

The second adsorbed layer was formed due to the pumping through the capillary with preadsorbed polyelectrolyte layer of anionic surfactant solution of different concentrations below cmc. To clear out the influence of the first adsorbed PE layer on the formation of the second anionic surfactant layer, we studied the adsorption of SDS on PE layers of different structures (see Deformation of Adsorbed Layers): when the PE molecules adsorbed in flat conformation (CSDAPM at $C = 10^{-4}$ g/l) and when the extended layer with loops and tails was formed (CSDAPM at $C = 10^{-2}$ g/l and PDADMAC at $C = 10^{-2}$, 10^{-3} and 10^{-4} g/l).

A) CSDAPM + SDS. Figures 4 and 5 represent the kinetics of ζ^* potential change during the SDS adsorption on the modified quartz surface. As can be seen from Fig. 4, when the SDS adsorption takes place from solutions with concentrations 10^{-4} M (curve 1) and 10^{-3} M (curve 2) on the cationic polyelectrolyte layer with concentration 10^{-4} g/l, the strong overcharge of the surface occurs. This fact may testify that there is another adsorption mechanism except electrostatic one possibly due to the hydrophobic interaction of non-polar tails of surfactant molecules with hydrophobic sites of cationic polyelectrolyte adsorbed layer. The overcharge and consequently the adsorption increase with the increasing of surfactant concentration, although the surface charge caused by PE adsorption is equal in both cases.

The similar dependencies are observed when the preadsorbed layer was formed from 10^{-2} g/l CSDAPM solution (Fig. 5).

The comparison of the findings shows that the ζ^* potential and the SDS adsorption values depend on the PE solution concentration from which the first adsorbed layer is formed and also on the SDS concentration: at $C_{PE} = 10^{-4}$ g/l and $C_{SDS} = 10^{-4}$ M (Fig. 4, curve 1) and 10^{-3} M (Fig. 4, curve 2) the adsorption values estimated by ζ^* potentials difference ($\zeta^*_{PE} - \zeta^*_{SDS}$) are 74 mV and 123 mV, respectively. And at $C_{PE} = 10^{-2}$ g/l and $C_{SDS} = 10^{-4}$ M (Fig. 5, curve 1) and 10^{-3} M (Fig. 5, curve 2) $\zeta^*_{PE} - \zeta^*_{SDS}$ are 105 mV and 142 mV, respectively.

B) PDADMAC + SDS. The results of adsorption of SDS with concentrations of 10^{-3} and 10^{-4} M on the quartz surface preliminarily modified by PDADMAC of concentration range from 10^{-4} to 10^{-2} g/l are shown in the Table 1.

The comparison of results of the SDS adsorption on the CSDAPM and PDADMAC preadsorbed layers shows that in the second case the adsorption of anionic surfactant also leads to the overcharge of the surface and increases with the increasing of surfactant concentration. Whereas the time of SDS adsorption on different polyelectrolytes is various. For example, when SDS 10^{-4} M adsorbs on

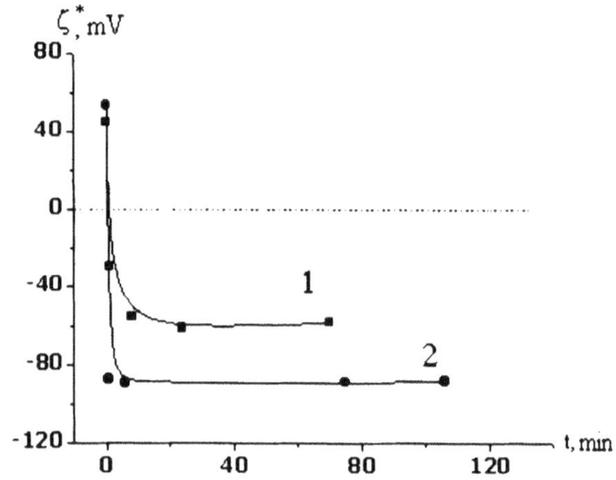

Fig. 5 The adsorption kinetics of SDS with concentrations $C = 10^{-4}$ M (*curve 1*) and $C = 10^{-3}$ M (*curve 2*) on the quartz surface preliminary adsorbed by cationic polyelectrolyte CSDAPM $C = 10^{-2}$ g/l

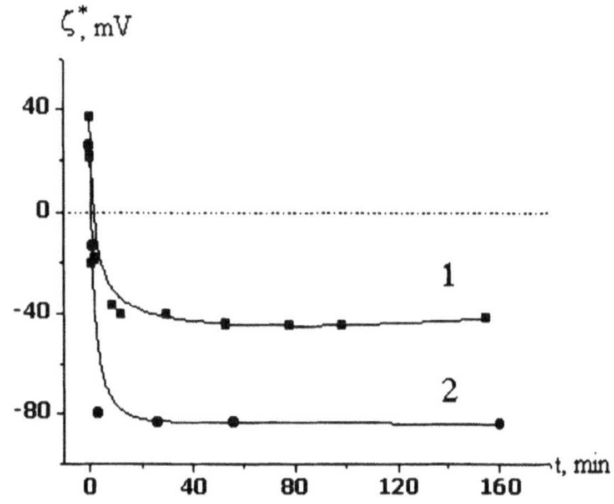

Fig. 4 The adsorption kinetics of SDS with concentrations $C = 10^{-4}$ M (*curve 1*) and $C = 10^{-3}$ M (*curve 2*) on the quartz surface preliminary adsorbed by cationic polyelectrolyte CSDAPM $C = 10^{-4}$ g/l

Table 1

C_{PE}, g/l	ζ^*_{PE}, mV	$\Delta\zeta^*_{PE}$	C_{SDS}, M	ζ^*_{SDS}, mV	$\zeta^*_{PE} - \zeta^*_{SDS}$	$\Delta\zeta^*_{SDS}$	ζ^*, mV, rinsing at $\Delta P = 10$ atm
10^{-4}	+65	non-deformed in 6 days	10^{-4}	−20	85	40	+13
10^{-4}	+53	14	10^{-3}	−76	129	57	−43
10^{-3}	+75	30	10^{-4}	−37	113	80	+22
10^{-3}	+73	10	10^{-3}	−80	153	60	−50
10^{-2}	+103	20	10^{-4}	−30	133	74	+30
10^{-2}	+111	10	10^{-3}	−70	181	70	−35

CSDAPM 10^{-4} g/l ζ^* potential values reach the equilibrium in about 50–60 minutes while at the same SDS concentration on PDADMAC 10^{-4} g/l – in 120 minutes. Such difference can be connected with the nature of polyelectrolytes, molecular weight and different deformation of adsorbed layers on which the SDS adsorption takes place.

Deformation of Adsorbed Layers

The capillary electrokinetic method allows to get the information about the structure of the adsorbed layers by measuring the dependence of ζ^* potential on the pressure enclosed (ΔP) (Figs. 6 and 7).

In the case of CSDAPM (Fig. 6) the measured values of zeta-potentials are independent of the solution flow rate at the concentrations of PE $C < 10^{-2}$ g/l (curve 1). At higher solution concentration $C \geq 10^{-2}$ g/l the considerable dependence shows up.

According to these data, one may draw a conclusion that at low concentrations PE molecules adsorb in flat conformation and at high concentrations more extended layer with loops and tails is formed. These data about conformation of polyelectrolyte molecules are in a good agreement with other experimental and theoretical works [21–23]. **Note** that the curves are reversible and ζ^* potential values establish immediately at each pressure value after pressure rising and decreasing. This is the argument that the deformation of adsorbed layers but not desorption of macromolecules takes place on experimental time scale, since our measurements are carried out in polyelectrolyte-free solution.

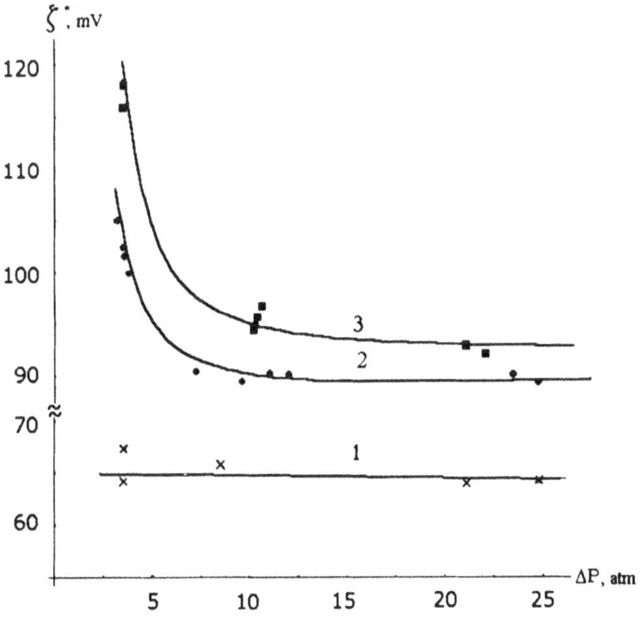

Fig. 6 The dependence of ζ^* potential on the pressure enclosed for CSDAPM solution for concentrations of 10^{-4} g/l (*curve 1*), 10^{-3} g/l (*curve 2*), 10^{-2} g/l (*curve 3*)

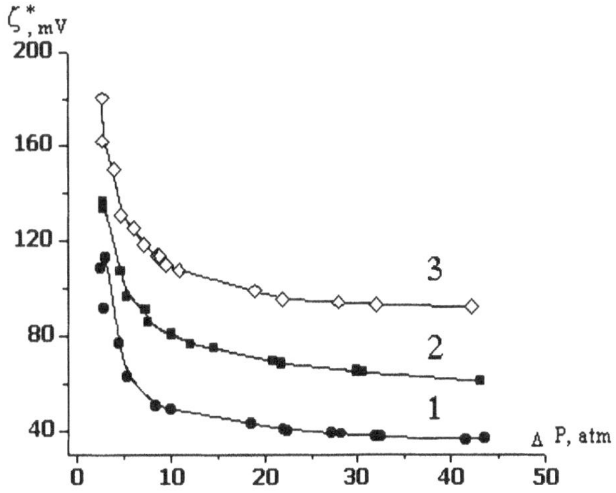

Fig. 7 The dependence of ζ^* potential on the pressure enclosed for PDADMAC solution for concentrations of 10^{-4} g/l (*curve 1*), 10^{-3} g/l (*curve 2*), 10^{-2} g/l (*curve 3*)

In the case of PDADMAC the dependence of ζ^* potential on the pressure was observed at the whole concentration range investigated (Fig. 7). Even at low PE concentration the conformation of macromolecules differs from the flat one. It can be a consequence of the high molecular weight of PDADMAC.

One more interesting fact noted was the time evolution of PDADMAC structure and rheological properties, namely, there were the change of deformation value (here, the change of the value of $\Delta\zeta^* = \zeta^*_{\Delta P=2.5\text{ atm}} - \zeta^*_{\Delta P=45\text{ atm}}$) of PDADMAC adsorbed layer with time. The dependence of the ζ^* potential on the pressure was measured immediately after the reaching of equilibrium values of ζ^* potential (so-called "freshly" formed layer), in one day and in some cases in 6 days. For example, in Fig. 8 the dependence of ζ^* potential on the pressure for the PDADMAC solution with concentration of $C = 10^{-4}$ g/l is represented. Here, ζ^* changed from +113 to +36 mV for "freshly" formed layer (ΔP = 3–45 atm) (curve 1), for the adsorbed layer in one day ζ^* changed from +74 to +51 mV (curve 2) and in 6 days there were no ζ^* potential dependence on the pressure ($\zeta^* = 64$ mV) (curve 3).

The absence of CSDAPM adsorbed layer aging can be the sequence of this PE nature: this polyelectrolyte is weak and has low molecular weight in comparison with PDADMAC. Hence, the reconstruction inside the adsorbed layer may occur during less time, in our case during the adsorption time scale.

Now it is rather well established that ionic surfactants associate with cationic polyelectrolytes both in the bulk solution and at the interface. In our study it was found that the deformation increases after the SDS adsorption in the whole concentration range investigated. As can be seen from the Table 1, the SDS adsorption on practically non-deformed first adsorbed layer leads to the increase in ζ^* potential change value with pressure.

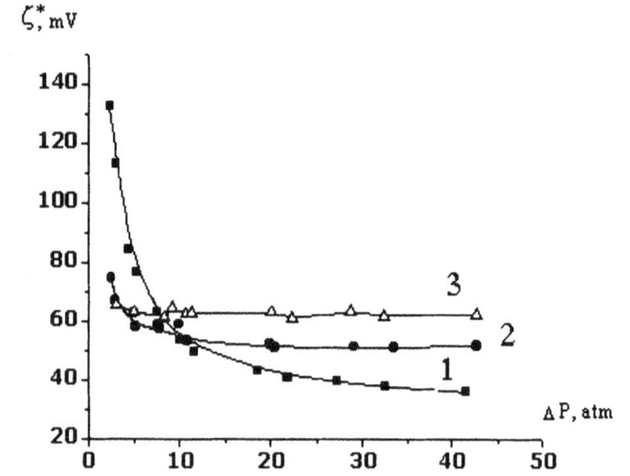

Fig. 8 The dependence of ζ^* potential on the pressure for the PDADMAC solution with concentration of $C = 10^{-4}$ g/l for "freshly" formed adsorbed layer (*curve 1*) and in one day (*curve 2*)

This fact can't be explained only by the thickness growth due to the adsorption of not large SDS molecules, but it can also be connected with conformational changes in PE layer followed by reduction of affinity between PE chain and negatively charged surface. This comes in a good agreement with AFM study performed by V. Bergeron and P.M. Claesson [24]. They showed that before addition of SDS the polyelectrolyte coated the surface homogeneously, but after SDS addition the association and material redistribution took place and large features were observed. So we may say that the anionic surfactant causes the significant "swelling" of the preadsorbed layer. Such swelling tendency was also observed in several works [3, 25, 26].

Strictly speaking the dependencies of ζ^* potential on pressure may be caused not only by the thickness changes due to the deformation but also by the changes of flow conditions inside the layer [27]. That's why values calculated here are evaluative to a greater extent. However, the dependencies obtained describe the processes taking place in adsorbed layers in reality.

Desorption Study

The desorption study of polyelectrolytes (CSDAPM and PDADMAC) by pumping through the capillary of the background electrolyte solution under high pressure ($\Delta P \approx 30$ atm) during one day showed that the adsorption is not reversible. One of the reason may be that the polyelectrolytes are firmly bound to the quartz surface via great number of attachments, which is a consequence of their large size [3, 28–30].

A) Surface/CSDAPM/SDS system. The desorption was carried out by pumping of the background electrolyte solution through the capillary under different pressure values. Electrokinetic potential values of SDS adsorbed layer formed on non-deformed PE layer (10^{-4} g/l) vary from −30 to −40 mV. If the rinsing was carried out at $\Delta P = 5$ atm, ζ^* reached the value of −3 mV in 15 minutes and +30 mV in 3 hours. The pressure increasing up to 10 atm didn't change the ζ^* potential value, i.e. the SDS layer came off almost completely. These results testify that the ζ^* potential change is not connected with the PE desorption after adsorption of anionic surfactant since the potential values of the first layer become former as a result of rinsing.

If SDS adsorption occurred on deformed PE (10^{-2} g/l), the rinsing of the capillary during long time didn't lead to the full desorption of SDS molecules; its surface stayed negatively charged. This may be the result of adsorption of SDS molecules inside the more extended PE layer. The desorption of these molecules wasn't observed even if the pressure was increased up to 20–30 atm during one day.

B) Surface/PDADMAC/SDS system. The desorption study was carried out for the SDS adsorbed layers formed during 1 day (see Table 1). As the bilayer structure became more compact, the value of desorption was determined by SDS concentration. It was shown that bilayer structure with the second SDS adsorbed layer of $C = 10^{-3}$ M is more stable in comparison with SDS $C = 10^{-4}$ M: in the first case the negatively charged surface stays negative and when the SDS concentration equals to 10^{-4} M the overcharge of the surface takes place.

Partial desorption for both systems can be explained by the fact that the part of surfactant molecules could induce into preadsorbed cationic PE layer and form "internal micelles" that was reported in some works [31–33]. And in this case only surfactant molecules that are not in such "internal micelles" structures would be desorbed.

Conclusions

In this paper we have investigated the regularities of bilayer adsorption of cationic polyelectrolytes a) copolymer of styrene and dimethylaminopropylmaleimide (CSDAPM) and b) poly(diallyldimethylammonium chloride) (PDADMAC) and anionic surfactant (SDS) on fused quartz surface by capillary electrokinetic method. It was found that the adsorption of the first layer depended on molecular weight and concentration of PE. The adsorption increased with PE concentration and the time of the constant adsorption value establishment was less for PE with higher molecular weight (PDADMAC).

The adsorption in all cases was irreversible. No desorption was observed even at high velocity of the solution flow through the capillary.

Our study revealed the time effect on the structure of adsorbed layer. It was noted that the deformation ability of PDADMAC layer decreased in one day after its formation. This means that the conformation rearrangement time is more than the adsorption time.

Adsorption of oppositely charged SDS on preadsorbed PE layer resulted in the surface overcharge. The bilayer deformation study showed the increase in deformation values as a result of SDS adsorption, hence the anionic surfactant caused "swelling" of the preadsorbed layer.

The desorption study indicated that the SDS adsorption did not result in PE desorption, whereas desorption of SDS took place for all bilayer structures. The desorption value depends on preadsorbed layer structure and SDS concentration.

Acknowledgement We thank Dr. S. Zschoche (IPF, Dresden, Germany) for providing us with polyelectrolyte (CSDAPM). This work was supported by the Russian Foundation for Basic Research, project no. 04-03-32455.

References

1. Kwak JCT (1998) Polymer-Sufactant Systems. Marcel Dekker, New York
2. Goddard ED, Ananthapadmanabhan KP (1993) Interaction of Surfactants with Polymers and Proteins. CRC Press, Boca Raton FL
3. Kjellin URM, Claesson PM, Audebert R (1997) J Colloid and Int Sci 190:476
4. Geffroy C, Lebeau MP, Wong K, Cabane B, Cohen Stuart MA (2000) Colloids and Surf A: Phys and Eng Aspects 172:47
5. Poloucek P, Pietch U, Geue T, Symietz CH, Brezesinski G (2001) J Phys D: Appl Phys 34:450
6. Lvov Y, Decher G, Sukhorukov G (1993) Macromolecules 26:5396
7. Schwarz S, Eichorn KJ, Wischerhoff E, Laschewsky A (1999) Coll Surf A 159:491
8. Hoogeveen NG, Cohen Stuart MA, Fleer GJ (1996) Langmuir 12:3675
9. Caselnovo M, Joanny J-F (2000) Langmuir 16:7524
10. Van der Schee HA, Lyklema J (1984) J Phys Chem 88:6661
11. Shubin V, Linse P (1997) Macromolecules 30:5944
12. Steitz R, Leiner V, Siebecht R, v Klitzing R (2000) Colloids and Surf A: Physicochem Eng Aspects 163:63
13. Plech A, Salditt T, Münster C, Peisl JJ (2000) Colloid Interface Sci 223:74
14. Sukhishvili SA, Dhinojwala A, Granick S (1999) Langmuir 15:8471
15. McAloney RA, Sinyor M, Dudnik V, Goh MC (2001) Langmuir 17:6655
16. Churaev NV, Sergeeva IP, Sobolev VD, Zorin ZM, Gasanov EK (1993) Colloids Surf A 76:23
17. Churaev NV, Sergeeva IP, Sobolev VD (1995) J Colloid and Int Sci 169:300
18. Sergeeva IP, Ermakova TB, Sobolev VD, Churaev NV (2004) Colloid J 66:378
19. Sergeeva IP, Ermakova TB, Sobolev VD, Churaev NV (2003) Colloid J 65:606
20. Sergeeva IP, Ermakova TB, Churaev N, Grundke K, Yudin O (2002) Colloid J 64:478
21. Marra J, Hair ML (1988) J Phys Chem 92:6044
22. Biggs S, Proud AD (1997) Langmuir 13:7202
23. Bremmell KE, Jameson GJ, Biggs S (1998) Colloid and Surf A: Phys and Eng Aspects 139:199
24. Bergeron V, Claesson PM (2002) Adv in Colloid and Int Sci 96:1
25. Shubin V, Petrov P, Lindman B (1994) Colloid Polym Sci 272:1590
26. Plunkett MA, Claesson PM, Rutland MW (2002) Langmuir 18:1274
27. Cohen Stuart MA, Waajen FHWH, Cosgrove T, Vincent B, Crowley TL (1984) Macromolecules 17:1825
28. Parfit GD, Rochester CH (1983) Adsorption from Solution at the Solid/Liquid Interface. Academic Press, London,NY
29. Shubin V, Linse P (1988) J Phys Chem 92:6044
30. Lowack K, Helm CA (1998) Macromolecules 31:823
31. Lindman B, Thalberg K (1993) In: Goddard ED, Ananthapadmanabhan KP (eds) Interactions of surfactants with polymers and proteins. CRC Press, Boca Raton FL
32. Voisin D, Vincent B (2003) Adv in Colloid and Int Sci 106:1
33. Claesson PM, Dedinaite A, Blomberg E, Sergeyev VG (1996) Ber Bunsenges Phys Chem 100:1008

Progr Colloid Polym Sci (2006) 132: 102–109
DOI 10.1007/2882_039

Published online: 10 March 2006

S. Schwarz
J. Nagel
A. Janke
W. Jaeger
S. Bratskaya

Adsorption of Polyelectrolytes with Hydrophobic Parts

S. Schwarz (✉) · J. Nagel · A. Janke
Leibnitz-Institut für Polymerforschung Dresden e.V., Hohe Str. 6, 01069 Dresden, Germany
e-mail: simsch@ipfdd.de

W. Jaeger
Fraunhofer Institut für Angewandte Polymerforschung, Geiselbergstr. 69, 4476 Golm, Germany

S. Bratskaya
Institute of Chemistry, Far East Department of Russian Academy of Sciences, 159, Prosp. 100-letiya Vladivostoka, 690022 Vladivostok, Russia

Abstract Polyelectrolytes offer numerous possibilities for modification of particles, planar inorganic or polymer surfaces, and control of colloidal stability. Significant enhancement of flocculation properties can be achieved through introduction of hydrophobic functionalities into polyelectrolyte backbone that increases surface activity and allow formation of polyelectrolyte aggregates. Here the flocculation performance of a high molecular weight cationic polyelectrolyte with hydrophobic functionalities, PMBQ (poly(methacryloyloxyethyl dimethylbenzyl-ammonium chloride)) was investigated in stable kaolin dispersions using turbidity measurements for efficiency estimation. Polyelectrolyte aggregates of PMBQ obtained at high ionic strength have shown superior flocculation properties in comparison with non-aggregated PMBQ. Remarkable difference in water and salt solutions was found in mono- and multilayer formation using polyanion medium molecular weight polystyrenesulfonate (PSS) and PMBQ. Streaming potential measurements, in-situ SPR and AFM were used to follow and quantify layer build up and charge reversal at the solid/liquid interface at the molecular level. SPR measurements, allowing in-situ control of polyelectrolyte adsorption process, have revealed that the layers thickness depended strongly on ionic strength. However, AFM has shown that the layers obtained had very high roughness.

Keywords Layer thickness · Polyelectrolyte layer · SPR · Streaming potential

Introduction

Interfacial interactions between a polyelectrolyte and solid substrate via adsorption from aqueous solution is an important phenomenon in numerous industrial fields which determines high academic and technological interest to this subject. Wide range of polyelectrolytes applications in waste water treatment, paper processing, production of functional polymeric coatings, etc. promotes considerable attention of scientists to the investigation of polymer adsorption at solid-liquid interfaces [1–6]. A deep understanding of this phenomenon is required to control and optimize technological processes involving polyelectrolytes, which are used in many fields to regulate stability and flocculation properties of disperse systems. It is well established that charge and conformation of the adsorbed polyelectrolytes have decisive influence on the stability and flocculation of dispersions [7–10]. Usually, the range of polyelectrolyte concentration providing efficient flocculation is very narrow, and followed by restabilization of pollutants by polymer overdosing due to steric or electrosteric effects. However, a broad flocculation window is highly desirable for industrial applications, since considerable changes in continuous waste stream parameters (solid content, ionic strength, pH, presence of surfactants stabilizing fine suspensions) are possible within one day or even within an hour. If the flocculation range is too narrow, calculation and dosing of the appropriate flocculant amount become very problematic.

Aggregates like polyelectrolyte complexes having positive charges and hydrophobic domains show a broader optimum flocculation concentration range and are considered as new reactive "nanoparticles" [11–14]. Thus, polycations with hydrophobic functionalities represent an interesting class of water-soluble associating polyelectrolytes relevant for controlled stabilization/flocculation of dispersions in numerous industrial applications.

In our investigations we used a new type of high molecular weight flocculant with hydrophobic parts to build up self-assembled thin films. Thin polymer films can be assembled on charged substrates by alternating deposition of polyanions and polycations [15–17]. Each deposition step introduces a reproducible quantity of polyelectrolyte on the oppositely charged surface and reverses the charge on the surface, providing favorable conditions for deposition of the next polymer layer with opposite charge. Repeating the deposition steps allow to build up films from the angstrom up to the micrometer range under precise control of the total film thickness.

Polycations like polyethyleneimine, poly(diallyldimethylammonium chloride) and poly(allylammonium chloride) have mainly been used to build up multilayers. In this paper we followed the layer-by-layer formation of thin films using PMBQ as a cationic component and PSS as an anionic component. The alternating adsorption of PMBQ and PSS was investigated by streaming potential, in-situ SPR, and AFM.

Table 1 Used polyelectrolytes

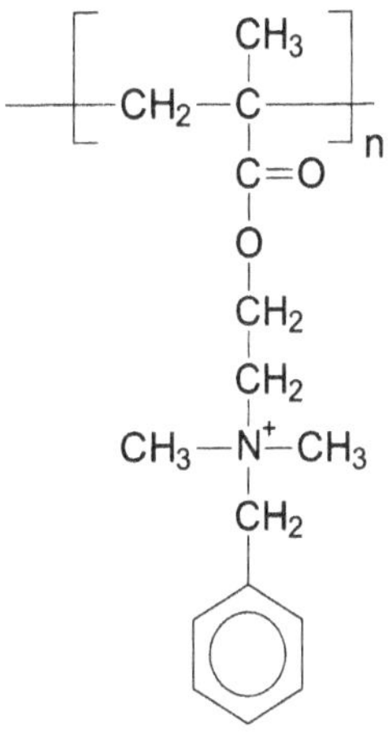

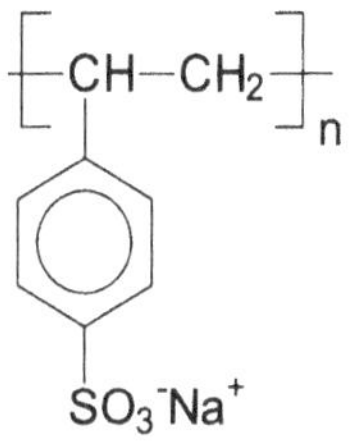

Poly-Methacryl-oyloxyethyl-dimethylbenzyl-amonium-chloride PMBQ	Poly(sodium styrenesulfonate) PSS

Experimental

Materials

The polyelectrolytes used in this study are displayed in Table 1. Hydrophobic flexible polyelectrolyte molecules of poly(methacryloyloxyethyl dimethylbenzylammonium chloride) (PMBQ) with a molecular weight of 4.2 Mio g/mol was synthesized by free radical polymerization in water solution as described elsewhere [18, 19]. Poly(sodium styrenesulfonate) (PSS) with molecular weight of 70 000 g/mol was purchased from Aldrich and was used without further purification. Water purified and deionized (reverse osmosis followed by ion exchange and filtration) by means of Milli-RO 5Plus and Milli-Q Plus systems (Millipore GmbH, Germany) was used as a solvent.

Flocculation Test

Stable model suspensions of kaolin (Sigma) with average diameter of particles 610 nm as determined by means of the flow-type histogram analyzer FPIA-2100 (Sysmex, Japan) based on microscopy principle were used for the flocculation tests. The flocculation studies were carried out using a conventional test equipment for the characterization of the flocculation and sedimentation process under static conditions. After addition of the appropriate amount of polycation solution followed by 10 minutes stirring and 20 minutes sedimentation, the supernatant was separated. The efficiency of the flocculation was estimated as the decrease in turbidity measured as optical density at 500 nm (OD_{500}) by means of spectrophotometer Lambda 900 (Perkin Elmer, UK).

Multilayer Deposition

Glass slides with the dimensions of 76 × 26 × 1 mm were used as supports for the multilayers. The slides were thoroughly cleaned by a mixture of sulfuric acid and potassium dichromate at 80 °C for about 2 hours in an ultrasonic bath prior to the films deposition. The initial concentration c_0 of polyelectrolytes in the solution was 1×10^{-2} mole of the repeating units per liter. The pH value of the solution was about 6. Adsorption was carried out at room temperature in open glass beakers of 100 mL without stirring for 20 min. After every deposition step, the glass slides were rinsed three times for 1 min with Millipore Milli-Q water. The substrate was not dried between the adsorption steps.

For some experiments (surface plasmon resonance), high refractive index glass slides SF10 (Hellma Optic GmbH, Germany) covered by a thin evaporated gold layer with the thickness of approximately 50 nm were used as supports.

Electrokinetic Measurements

Electrokinetic measurements for the polyelectrolyte layers were carried out by means of an Electrokinetic Analyzer device (A. Paar KG, Austria). The glass substrate after polyelectrolyte deposition was several times rinsed with deionized water to ensure that unadsorbed polymers do not contribute to the ζ-potential measured. The values of ζ-potential were calculated according to the formula:

$$\zeta = \frac{\eta}{\varepsilon_0 \varepsilon_r} \frac{\Delta U}{\Delta p} \kappa , \qquad (1)$$

where ΔU is streaming potential measured between two Ag/AgCl electrodes located at the opposite ends of the substrates, η, ε_r, κ are dynamic viscosity, relative dielectric permittivity, and conductivity of the flowing electrolyte solution (0.001 M KCl), respectively. ε_0 is dielectric permittivity of vacuum, and Δp is the pressure applied (150 mbar). Solutions of 0.1 M KOH and HCl were used to change pH of the flowing electrolyte solution in the range from 9 to 3.

Surface Plasmon Resonance

Surface plasmon resonance data for the layers were obtained by means of the equipment consisting of a He-Ne laser with $\lambda = 632.8$ nm (Uniphase, USA), a semi-cylinder made of SF10 glass, a liquid flow cell with the volume of 2.5 ml, and a E10V large area silicon photodiode detector with an integral preamplifier (Linos GmbH, Germany). The laser emitted polarized light (polarization ratio 500 : 1) with power of 3 mW onto a semi-cylinder whose plane face was coupled via an index matching fluid to the substrate examined (SF10 glass slide covered by the gold layer and then by the polyelectrolyte layers). The liquid flow cell was attached to the other side of the substrate and sealed with a rubber O-ring. The light was reflected onto the gold layer to excite surface plasmons. The intensity of the reflected light was measured by the photodiode. Both the semi-cylinder and the detector were mounted in an in-house $\theta/2\theta$ goniometer in such a way that the laser beam was incident on the detector at any angle of incidence. The goniometer and the photodiode were interfaced (Motion-Master 3000, Newport Corp., USA) to a personal computer. An in-house 32-bit software package was used for goniometer control, data acquisition, curve modeling, and curve fitting. For scans over a certain range of incidence angles, a step width of 0.1° was used. The curves obtained were fitted according to Fresnel equations for a four-layer model (glass/metal/dielectric/surrounding medium). Since the refractive index of the adsorbed polymer layer could not be measured at the same time, it was assumed to be 1.5. This is a typical value of a dense polymer.

For the determination of the average thickness of the polyelectrolyte layers the following procedure was applied. At first, the angular dependence of reflectivity (angular scan) was measured for the original slide in the flow cell that was brought into contact with water or 1 M KCl solution. The liquid was continuously circulated by a peristaltic pump through the liquid flow cell at a speed of 10 ml/min. Then, the water or salt solution was replaced by the polycation solution, in water or KCl. After 20 min an angular scan was measured again. The resulting film thicknesses were fitted and plotted as a function of the number of double layers.

Atomic Force Microscopy

The AFM experiments were carried out with a MFP-3D (Asylum Research, Santa Barbara, USA) in the AC mode on air with a scan rate of 0.3 Hz. The tips used for the experiments were silicon probes from Budgetsensors (Bulgaria).

All measurements were carried out at room temperature.

Results and Discussion

The SEM picture (Fig. 1, insert a) confirms that the main part of kaolin particles has size in the range of 300–600 nm, but the distribution is rather broad. This kind of kaolin with rather fine particles has a high specific surface area and a high negative charge density of – 13.75 C/g. Thus, the kaolin suspension remains stable due the strong electrostatic repulsion between particles.

PMBQ was tested for the flocculation as a solution in water as well as in aqueous 1 M KCl solution. The

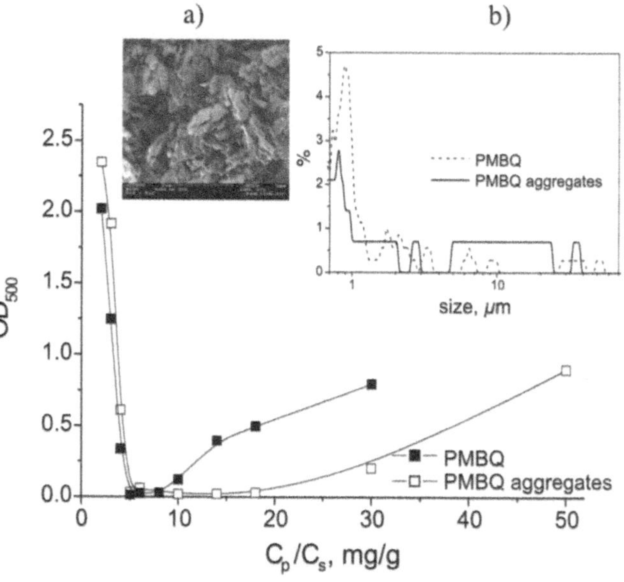

Fig. 1 Optical density of kaolin dispersion at pH 8 as a function of the concentration of PMBQ and PMBQ-aggregates. *Insets*: **a** SEM-picture of kaolin; **b** floc size distribution in kaolin dispersion treated with 4 mg/g PMBQ and PMBQ-aggregates

charge density in water was 3.8 meq/g as determined by polyelectrolyte titration. In 1 M KCl solution the formation of PMBQ aggregates with an average size of 1.2 μm was observed. Here, the high ionic strength resulted probably in the screening of the electrostatic repulsion between the charges along the polymer backbone and, thus, favors short-range hydrophobic attraction leading to the formation of the polyelectrolyte aggregates. The apparent charge density of PMBQ aggregates (freshly prepared) was reduced to 1.8 meq/g. Thus, part of the positively charged groups may be hidden inside the aggregates.

The results of the flocculation investigations with PMBQ and PMBQ aggregates are shown in Fig. 1. PMBQ in aqueous solution is an efficient flocculant in a narrow concentration range: at polymer doses higher than those required for flocculation, redispersion takes place very quickly. That is typical for highly charged polyelectrolytes [20]. However, the PMBQ aggregates formed in 1 M KCl solution exhibited a much broader flocculation window with low residual turbidity and noticeably larger flocs (Fig. 1, insert b) measured by means of flow-type histogram analyzer FPIA-2100 (Sysmex, Japan).

Other potentially important fields of PMBQ aggregate applications are the flocculation of problematic wastes containing nanoparticles with low charge, such as disperse dyes from the textile factories, slurries from microelectronics, or emulsified oil products, which are notoriously difficult to remove using conventional hydrophilic flocculants. For these applications the flocculant PMBQ may be suitable because of the lower charges and the additional hydrophobic parts. Previous investigations showed that as a result of conformation changes and drastic increase of polymer size in salt solution the flocculation properties of PMBQ aggregates significantly differ from that of PMBQ. This was, first of all, illustrated by the fact that no restabilisation of silica dispersions occurred [20].

It was shown [21] that polyelectrolytes having hydrophobic functionalities show higher efficiency for such systems due to the synergetic effect of electrostatic and hydrophobic interactions promoting the polymer adsorption at solid/liquid or liquid/liquid interface.

Since the flocculation efficiency of polyelectrolyte is directly linked to their adsorption behavior it is interesting to investigate PMBQ deposition on planar surfaces. Several substrates varying in surface charge density and hydrophobicity (mica, hydrophilic and hydrophobized silicon wafers) were chosen to elucidate the difference in the adsorbed amounts and in polymer conformations under different experimental conditions. Figures 2a–c show the AFM images of PMBQ adsorbed from water solution. It was previously reported [22, 23] that PMBQ of lower molecular weight adsorbs from water solution on mica as wormlike chains with a height of 0.75 ± 0.05 nm, independent on polymer contour length. Increasing of salt concentration resulted in conformational changes from worm-like to patch-like via partially segregated coil structures. However even at high ionic strength coexistence of several conformations was observed [23]. One can see (Figs. 2a and b) that adsorption of PMBQ of very high molecular weight follows the same trend. Two different conformations – worm-like structures and aggregates could clearly be distinguished for the polyelectrolyte adsorbed from wa-

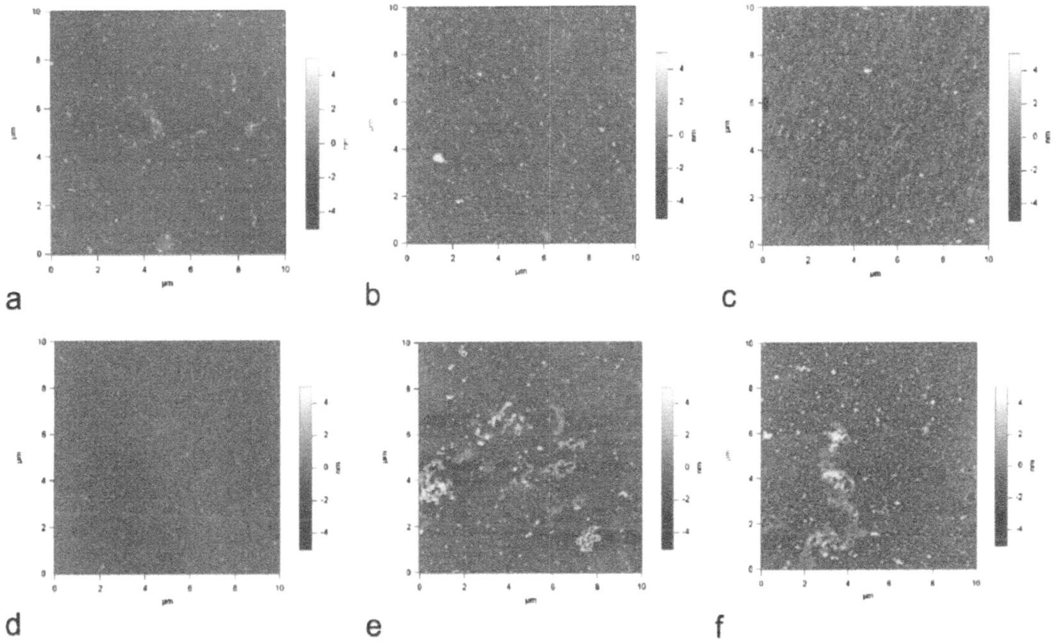

Fig. 2 AFM images after adsorption of PMBQ, Adsorption on different substrates, PMBQ in water: **a** mica, **b** silicon-wafer, hydrophil, **c** silicon-wafer, hydrophob, PMBQ in 1 mol l^{-1} KCl: **d** mica, **e** silicon-wafer, hydrophil, **f** silicon-wafer, hydrophob

ter solution on hydrophilic substrates, mica and silicon-wafer.

In comparison with hydrophilic silicon wafers, remarkably higher PMBQ adsorption from a salt-free solution was found for the hydrophobized silicon-wafer (Fig. 2c). This was related to an additional attraction by hydrophobic interactions between surface sites and hydrophobic PMBQ moieties. The size of the adsorbed aggregates increased with the salt concentration, but in a more profound manner than it was observed for hydrophilic substrate. The AFM micrographs revealed a very high heterogeneity of the adsorbed polymer, with highest values of roughness on hydrophobic substrate.

At high ionic strength significant increase of PMBQ adsorbed amount was observed. Figure 2d shows that mica surface was completely covered by PMBQ aggregates with narrow size distribution, while on silica wafers large and small aggregates coexisted, resulting in very high roughness of the surface (Fig. 2f).

Zeta potential and SPR measurements are sensitive tools to follow changes in surface properties induced by polyelectrolyte adsorption in the wet state. The results obtained by these two methods for PMBQ layer on glass and glass coated with gold were in good agreement, confirming that the higher the ionic strength the higher is the adsorbed amount and the surface coverage. Figure 3 shows that the zeta potentials of the glass surface after deposition of PMBQ were higher than those of the bare glass over all the studied pH range from 3 to 10. Moreover, the zeta potentials of the PMBQ layer adsorbed from 1 M KCl solution was substantially shifted to higher values in comparison with layers deposited from water or 0.1 M KCl. This tendency was confirmed by SPR measurements, where remarkable differences of the reflectivities of the PMBQ layers deposited from solutions with different salt concentrations were detected (Fig. 3). A fully covered surface on glass over the whole investigated range of 2 μm × 2 μm was observed for only one layer PMBQ absorbed but the structure was not flat. The adsorption of polyelectrolyte on glass should be comparable to that on mica because of close charge densities of these substrates at pH 6, while the charge densities of silicon wafer and glass covered with gold are much lower [25].

The adsorption of polyelectrolytes at charged surfaces is used frequently for both charge neutralization and charge reversal. Polyelectrolyte multilayers formed by polycation–polyanion pairs provide additional opportunities to control surface morphology and film thickness [24].

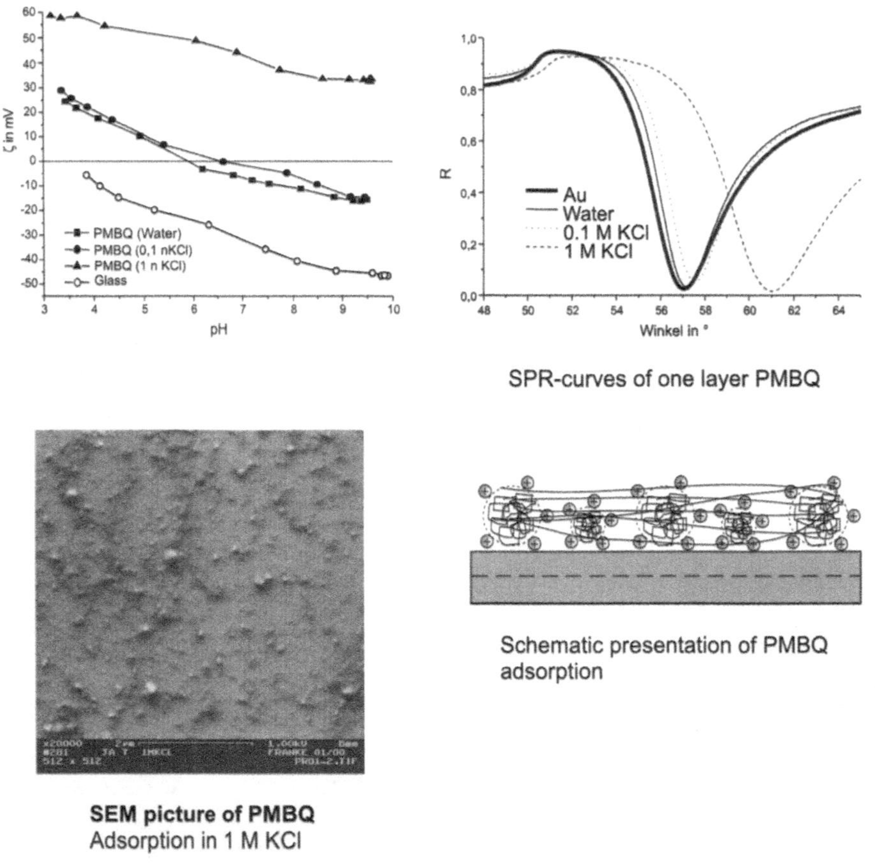

Fig. 3 Zetapotential-pH-profiles and SPR-results for one monolyaer of PMBQ- on glass and glass covered with gold in dependence of salt concentration

To be suitable for the preparation of multilayers polyelectrolytes have to provide reversed surface charge from layer to layer [14], sufficient number of binding sites with the previously adsorbed layer, and certain number of remaining charges as adsorption centers for the next layer deposition. Since the glass surface is negatively charged in the investigated pH range from 3.5 to 9, polycations may be easily adsorbed from solution.

Figure 4 shows that deposition of one layer of PMBQ from solution with pH = 6 resulted in drastic changes of surface properties. This was quantitatively characterized by the shift of surface IEP (isoelectric point) from pH < 3.5 to 6. Since the isoelectric point of the PMBQ solution was higher than 12, it may be assumed that the electrokinetic behavior was still determined by the combination of glass substrate and polycation rather than by a uniform PMBQ layer. This may be a result of the blobwise adsorption of PMBQ under the shear flow during the measurement. The adsorbed PMBQ was used for the deposition of the polyanion PSS. Although polyelectrolytes with high charge densities like PMBQ are expected to adopt mainly a flat conformation in the adsorbed state, it is possible that a lot of loops and tails were extending to the bulk solution due to the very high molecular weight of PMBQ. The low zeta potential values in the range of 6 to 10 indicated a thick and swollen polyelectrolyte layer. It should be mentioned that usually surface properties of multilayer films depend only on the nature of the top layer but not on the absolute number of layers deposited. However, one can see that the zeta potential profile of PMBQ/PSS layers with PSS as the outermost layer were noticeable different from the profile of a plain PSS layer. This illustrates that the surfaces charges can vary strongly in dependence of the underlying surface. One of the possible reasons for this behavior is the extremely high molecular weight of PMBQ (4.2 Mio g/mol) in comparison with that of PSS (70 000 g/mol), which is responsible for extending loops and tails of PMBQ to the top of the film. This resulted in a surface with amphoteric properties.

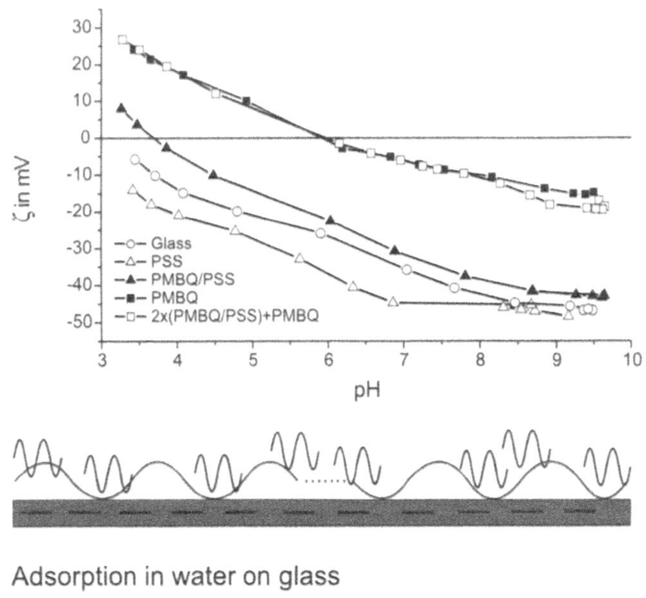

Fig. 4 Zetapotential-pH-profiles of PMBQ and PSS monolayers and multilayers of PMBQ/PSS on glass, Adsorption of polyelectrolytes in water

The hydrophobic character of PMBQ may offer the following advantages for surface modification: electrostatic as well as hydrophobic interactions may contribute to the adsorption. The surfaces were rendered hydrophobic (contact angle 80 in air), and the layers were very stable.

The thickness of adsorbed polyelectrolyte layers on gold-coated glass slides was determined by SPR in situ during the adsorption process in a measuring cell. This allowed measurements in aqueous medium [25]. Figure 5 presents angular scans of the reflectivity for PMBQ/PSS layers in water and in 0.01 M KCl, respectively. Measurements were done after adsorption of each double layer. Each curve showed shift of the SPR minimum due to the adsorption of macromolecules. The thicknesses are averages over the beam spot. The multilayer thicknesses increased linearly with the number of double layers. The

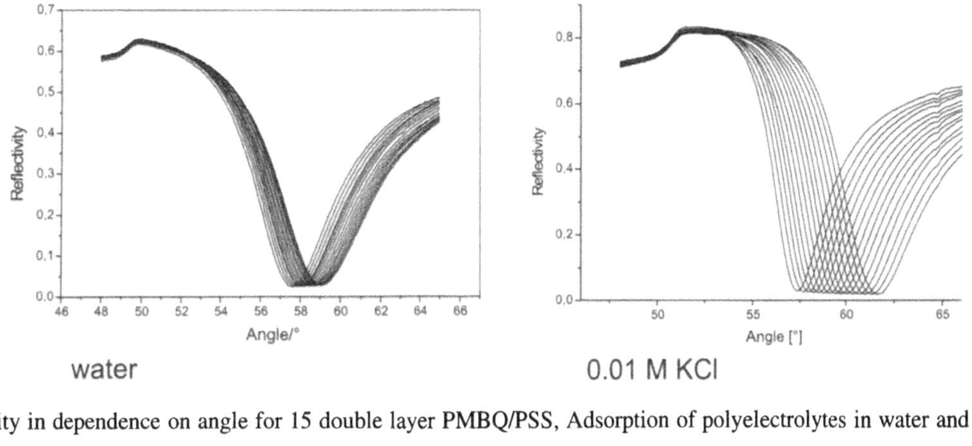

Fig. 5 Reflectivity in dependence on angle for 15 double layer PMBQ/PSS, Adsorption of polyelectrolytes in water and 0.01 M KCl

averaged thickness for one PMBQ/PSS double layer in water was approximately 0.6 nm.

The first steps of the adsorption of a multilayer with 2 double layer PMBQ/PSS were followed by a time drive at which the angle of incidence was fixed to 55.7 and the reflectivity was measured over the time. The time drives show a fast adsorption process for both polyelectrolytes (Fig. 6). The adsorption process was finished after a few seconds, and rinsing with 1 M KCl solution did not lead to a polymer desorption. The comparison of the time drives of the layers made from PMBQ with high molecular weight and PSS with low molecular weight showed that the adsorbed amount for PMBQ is larger. Because at low electrolyte concentration the adsorbed layers are thin and the adsorbed amount hardly depends on molecular weight [26], we can assume that the adsorption of PMBQ is not only driven by electrostatic attraction but also by non-electrostatic forces.

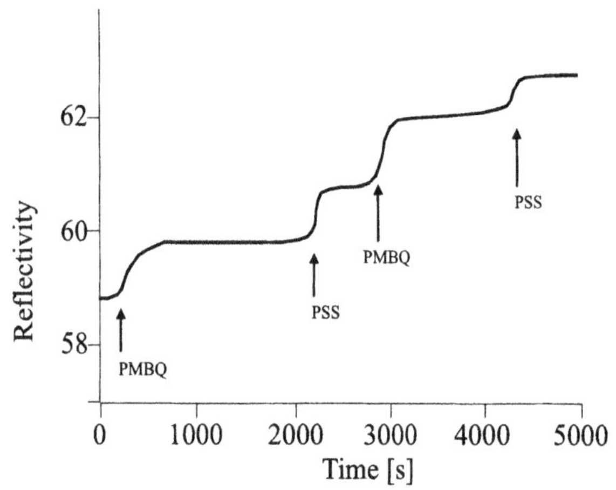

Fig. 6 Time drive for PMBQ/PSS-multilayers, Adsorption of 2× (**PMBQ/PSS(l)**) in 1 M KCl

Summary

PMBQ represents a special class of polyelectrolytes with hydrophobic functional groups, which tends to form aggregates in solution and exhibits a high affinity to hydrophobic surfaces. The presence of hydrophobic parts in PMBQ enhanced its flocculation performance in different dispersions. We have shown that formation of hydrophobic aggregates of PMBQ was responsible for the considerable broadening of the flocculation window that is extremely important and desirable for industrial application.

Study of the mono and multilayer deposition of PMBQ and PSS on planar surfaces were used to characterize the adsorption behavior. The surface properties were investigated by zeta potential measurements, SPR and AFM. It was found that the adsorbed amount of PMBQ depends strongly on the type of substrate (charge density and hydrophobicity) and salt concentration. The conformation of the adsorbed molecules varied from worm-like structures to aggregates of different sizes. The higher surface coverage and larger aggregates were detected for PMBQ adsorption on hydrophobic surfaces. This was attributed to additional hydrophobic interactions between surface sites and polymer hydrophobic functionalities. However, AFM and SEM showed that PMBQ monolayer and PMBQ/PSS multilayers had a very high surface roughness.

Acknowledgement The authors thank Norbert Stiehl and Marina Oelmann for a lot of experimental work. Financial support from AiF under the project 12893 BR is gratefully acknowledged.

References

1. Dautzenberg H, Jaeger W, Kötz J, Phillip B, Seidel Ch, Stscherbina D (1994) Polyelectrolytes. Formation, Characterization and Application. Carl Hanser Verlag, München, p 272
2. Koetz J, Kosmella S (2002) Interactions of Polyelectrolytes with Clay Surfaces. In: Hubbard A (ed) Encyclopedia of Surface and Colloid Science. Marcel Dekker, Inc, New York 1:2620
3. Rojas OJ (2002) Adsorption of Polyelectrolytes on Mica. In: Hubbard A (ed) Encyclopedia of Surface and Colloid Science. Marcel Dekker, Inc, New York 1:517
4. Cohen Stuart MA, Hoogendam CV, de Keizer A (1997) Kinetics of polyelectrolyte adsorption. J Phys, Condens Matter 9:7767
5. Schwarz S, Buchhammer H-M, Lunkwitz K, Jacobasch HJ (1998) Polyelectrolyte adsorption on charged surfaces: study by electrokinetic measurements. Colloids Surf A: Physicochem Eng Aspects 140(1–3):377–384
6. Schwarz S, Eichhorn KJ, Wischerhoff E, Laschewsky A (1999) Polyelectrolyte adsorption onto planar surfaces: a study by streaming potential and ellipsometry measurements. Colloids Surf A: Physicochem Eng Aspects 159:491
7. Schuster Ch, Kötz J, Jaeger W, Kulicke WM (1996) Wechselwirkung zwischen Klärschlammpartikeln und Polyelektrolyten, Chem-Ing-Technik 8:980
8. Bauer D, Buchhammer HM, Fuchs A, Jaeger W, Killmann E, Lunkwitz K, Rehmet R, Schwarz S (1999) Stability of colloidal silica, sikron and polystyrene latex influenced by the adsorption of polycations of different charge density. Coll Surf A 156:291
9. Petzold G, Schwarz S, Lunkwitz K (2003) Higher efficiency in particle flocculation by using combinations of oppositely charged polyelectrolytes. Chemical Engineering and Technology 26:48
10. Solberg D, Wågberg L (2003) Adsorption and flocculation behaviour of cationic polyacrylamide and colloidal silica. Colloids Surf A: Physicochem Eng Aspects 219:161
11. Buchhammer HM, Oelmann M, Petzold G (2001) Flockung von

Dispersionsfarbstoffen in Abwässern mit Polyelektrolytkomplexen, Melliand Textilberichte 5:391
12. Buchhammer HM, Petzold G, Lunkwitz K (2000) Nanoparticles based on polyelectrolyte complexes: Effect of structure and net charge on the sorption capacity for solved organic molecules. Colloid Polym Sci 278:841
13. Petzold G, Nebel A, Buchhammer HM, Lunkwitz K (1998) Preparation and Characterization of Different Polyelectrolyte Complexes and their Application as Flocculants. Colloid Polym Sci 276:125
14. Schwarz S, Dragan S (2004) Nonstoichiometric interpolyelectrolyte complexes as colloidal dispersions based on NaPAMPS and their interaction with colloidal silica particles. Macromolecular Symposia 210(1):185
15. Decher G, Hong JD, Schmitt J (1992) Buildup of ultrathin multilayer films by a self-assembly process: III. Consecutively alternating adsorption of anionic and cationic polyelectrolytes on charged surfaces. Thin Solid Films 210/211:831
16. Hammond PT (2000) Recent Explorations in Electrostatic Multilayer Thin Film Assembly. Current Opinion Colloid Interface Sci 4:430
17. Bertrand P, Jonas A, Laschewsky A, Legras R (2000) Ultrathin Polymer-Coatings by Complexation of Polyelectrolytes at Interfaces – Suitable Materials, Structure and Properties. Macromol Rapid Commun 21:319
18. Zimmermann A, Jaeger W, Reichert KH (1997) Polyelectrolytes – Dispersion polymerization of a water soluble cationic vinyl monomer. Polymer News 22:390
19. Jaeger W, Paulke B-R, Zimmermann A, Lieske A, Wendle U, Bohrisch J (1999) Polymer Prep 40:980
20. Schwarz S, Lunkwitz K, Keßler B, Spiegler U, Killmann E, Jaeger W (2000) Adsorption and Stability of Colloidal Silica. Colloids and Surfaces A 163:17
21. Bolto BA, Dixon DR, Gray SR, Chee H, Harbour PJ, Ngoc L, Ware AJ (1996) The use of soluble organic polymers in waste treatment. Water Sci Technol 34:117
22. Kiriy A, Gorodyska G, Minko S, Jaeger W, Stepanek P, Stamm M (2002) Cascade of Coil-Globule Conformational Transitions of Single Flexible Polyelectrolyte Molecules in Poor Solvent. J Am Chem Soc 124(45):13454
23. Minko S, Gorodyska G, Kiriy A, Jaeger W, Stamm M (2002) Visualization of single polyelectrolyte molecules. Polymeric Materials Science and Engineering 87:185
24. Schönhoff M (2003) Self-Assembled Polyelectrolyte Multilayers. Current Opinion in Colloid and Interface Science 8:86
25. Schwarz S, Nagel J, Jaeger W (2004) Comparison of Polyelectrolyte Multilayers Built Up with Polydiallyldimethylammonium Chloride and Poly (ethyleneimine) from Salt-Free Solutions by in-situ Surface Plasmon Resonance Measurements. Macromolecular Symposia 211:201
26. van de Steeg HGM, Cohen Stuart MA, de Keizer A, Bijsterbosch BH (1992) Polyelectrolyte Adsorption: A Subtle balance of Forces. Langmuir 8:2538–2546

Progr Colloid Polym Sci (2006) 132: 110–116
DOI 10.1007/2882_023

Published online: 10 March 2006

Stefan Spange
Susan Wolf
Frank Simon

Adsorption of Poly(vinyl formamide-co-vinyl amine) (PVFA-co-PVAm) onto Metal Surfaces

Stefan Spange (✉) · Susan Wolf
Department of Polymer Chemistry,
Technical University Chemnitz,
Straße der Nationen 62, 09101 Chemnitz,
Germany
e-mail:
stefan.spange@chemie.tu-chemnitz.de

Frank Simon
Leibniz Institute of Polymer Research,
Hohe Straße 6, 01069 Dresden, Germany

Abstract The formation of stable metal/polymer joints is an enormous challenge in material sciences. Good adhesion requires an interphase which is able to specifically interact with the metal substrate as well as with the second component. Furthermore, the interphase should compensate thermally generated mechanical tensions between the two adhering components. It will be shown that statistic copolymers of poly(vinyl formamide) and poly(vinyl amine) (PVFA-*co*-PVAm) are potential candidates for adhesion promoters. The polyelectrolyte character of the copolymers allows to apply them from its aqueous solutions. The primary amino groups exhibit the copolymers as highly reactive to metal surfaces as well as to the second joint partner. The adsorption mechanism of PVFA-*co*-PVAm onto metal surfaces depends strongly on the metal species and its surface composition. On the other hand, PVFA-*co*-PVAm copolymers offer a wide variety in their number of functional groups may be involved in metal/polymer interactions. Here, the adsorption behavior of PVFA-*co*-PVAm copolymer samples onto different metal species was under investigation. Correlations between the copolymer's composition and the adsorbed amount allowed to get information for tailor-made adhesion promoter systems.

Keywords Metal surface · Poly(vinyl amine) copolymers · Polyelectrolyte adsorption · Solvatochromic measurements · X-ray photoelectron spectroscopy

Introduction

Metals play an important role in technical application and academic research. One challenge in material science is to develop a technology forming permanently stable adhesion joints between various technical materials such as polymers, glass, ceramics or rubber [1–7]. Especially, the joining of metal with technical polymers is of high importance. Clean metal surfaces such as iron, copper, aluminium and zinc are characterized by high surface free energies [8, 9]. Hence, a bare metal surface cannot exist under ambient conditions. Water and oxygen form oxidic layers which are usually contaminated by hydrocarbons lowering the surface free energy. The presence of oxidic layers and other contaminations on each metal surface strongly changes the chemical properties of metal surfaces. Hence, a metal surface produced or stored under environments of technical interests cannot be considered or defined as metal. In this work we will report on the surface functionalization of metal powders and metal sheets using the novel water-soluble polymer Lupamine® provided by the BASF, Ludwigshafen [10–12]. Lupamines® are statistic poly(vinyl formamide-*co*-vinyl amine) (PVFA-*co*-PVAm) copolymers derived from poly(vinyl formamide) by the hydrolysis reaction shown in Scheme 1.

The polymer is applicable as an amine or cationic polyelectrolyte. Depending on the pH value the amino groups

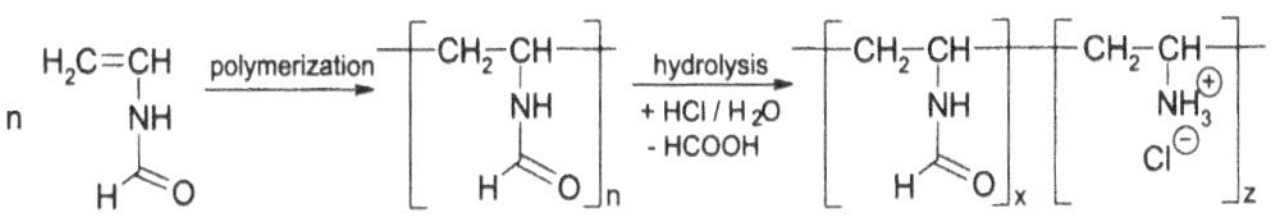

Scheme 1 Synthesis of poly(vinyl formamide-*co*-vinyl amine)

can be present as $-NH_3^+$ or $-NH_2$ group. The number of amino and ammonium groups can be adjusted by the degree of hydrolyzation. Furthermore, the PVFA-*co*-PVAm can be considered as a reactive polymer because the amino groups are suitable to be chemically functionalised in different ways. They can react with epoxides, acid anhydrides, aldehydes and a lot of other reagents [11]. The high chemical potential and the mechanical flexibility of the polymer backbone make the PVFA-*co*-PVAm polymers to promising interphase agents for joining various materials.

The coexistence of amino and formamide groups along the polymer backbone may form amidine groups (Scheme 2).

The amidine formation can take place when polymers with a degree of hydrolysis in the order of 20–80% are used. Furthermore acidic surface groups can also induce the amidinium ion formation.

Adsorption of polymers on inorganic substrates has been subject of numerous investigations because of the simple procedure for application [12–15]. In this work the influence of the amino group content and pH of the PVFA-*co*-PVAm solution on the amount of adsorbed polymer on various metal surfaces has been studied. Metal substrate samples of different size and shape (particles, sheets) have been used which limits the application of special surface sensitive methods such as ellipsometry.

Therefore, X-ray photoelectron spectroscopy (XPS) [12, 16, 17] and diffuse reflection infrared Fourier transform spectroscopy (DRIFT) [17] are used to analyse the chemical composition of the metal surface and the amount of the surface-bonded polymer after the adsorption processes. For the ZnO/PVFA-*co*-PVAm composites solvatochromic dyes were employed to evaluate changes in the surface polarity after the polymer adsorption [18–22].

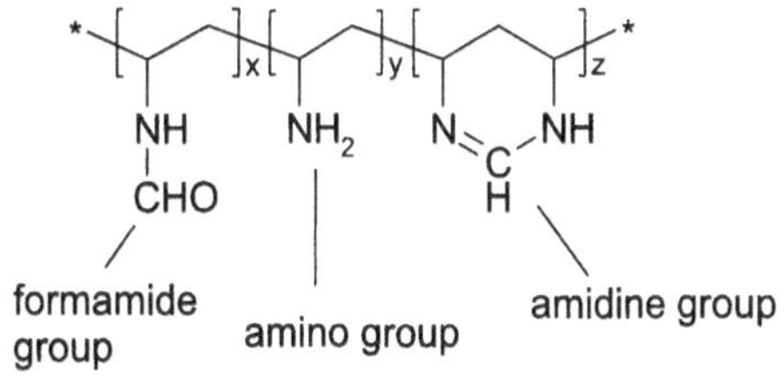

Scheme 2 Functional groups in a Lupamine® polymer with a low number of hydrolysed formamide groups. An amino and formamide group formed an amidine group

Experimental

Materials

Copper (Cu), iron oxide (Fe_2O_3), zinc oxide (ZnO) powders as well as aluminium (Al) and zinc (Zn) sheets were purchased from *Alfa Aesar* (Karlsruhe, Germany). Iron (Fe) particles were bought from *Merck KGaA* (Darmstadt, Germany). The particle sizes and the purity of the metal bulk material are summarized in Table 1.

The aqueous PVAm-*co*-PFA solutions (Lupamines®) were provided by the *BASF-AG* (Ludwigshafen, Germany). The molecular masses and degrees of hydrolysis of the Lupamines® are given in Table 1. From the hydrolyzation reaction technical Lupamines® contain formate ions which can be involved in the adsorption process. For comparison, purified PVFA-*co*-PVAm samples with a hydrolysis degree > 90% were used (named as *desalted*). The purification were carried out by means of dialyses routines.

Table 1 Metals and PVAm-*co*-PFA polymers used

Metals and Metal oxides	Purity [%]	Particle size [mm]
Copper (Cu)	99.00	< 0.04
Iron (Fe)	99.50	0.01
Iron oxide (Fe_2O_3)	99.80	< 0.04
Zinc oxide (ZnO)	99.99	
Aluminium (Al)	99.99	sheets
Zinc (Zn)	99.99	sheets
Polymers	Molar mass [g/mol]	Degree of hydrolysis [%]
Lupamine® 9000	340 000	0
Lupamine® 9030	340 000[a]	30
Lupamine® 9095	340 000[a]	> 90

[a] The molar mass given here is the molar mass of the pre-polymer (PVFA) which was used for hydrolyzation reaction

Adsorption

To modify 1 g of metal or metal oxide powder 50 g of an aqueous polymer or copolymer solution containing 1 w/w % polymer were used. Because the polymer was used in excess, the largest possible amount of polymer should be adsorbed on the surface (adsorption temperature $\vartheta_{ads} = 20\,°C$). Three pH-values were under investigation: pH = 5, 8 and 11. They were adjusted with $0.1\,mol\,l^{-1}$ NaOH or HCl and measured with the pH-equipment by *HANNA-Instruments*.

The powders were dispersed in the aqueous polymer solution for 5 min. Then, the mixture was centrifuged and washed two times with 10 ml of distilled water.

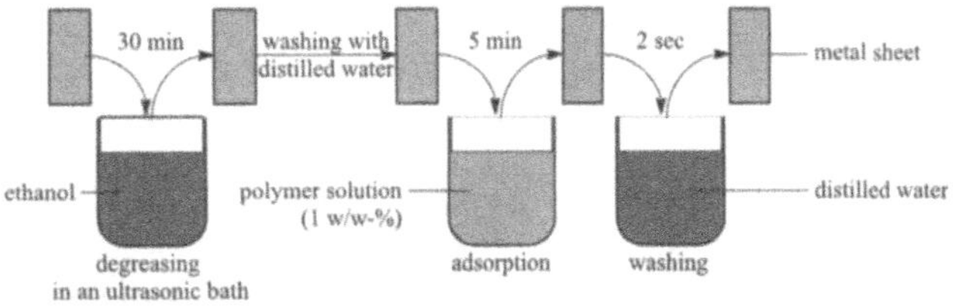

Scheme 3 Preparation of PVAm-*co*-PFA-functionalised metal sheets

Metal sheets were degreased with ethanol in an ultrasonic bath, washed with distilled water, and dipped for 5 min in the 1 w/w % polymer solution. To remove the polymer excess the adsorbates were rinsed for 2 sec with distilled water. The adsorption procedure for sheets is summarized in Scheme 3.

Characterization. All measurements were carried out with dried specimens (particles or sheets). It is established that the thickness of the water-loaded polymer layer is higher than in the dried state, because dried polymer coils collapse [21]. The amount of polymer, however, is not changed after the drying process. Thus an average equilibrium state is always measured.

XPS studies were carried out by means of an AXIS ULTRA photoelectron spectrometer (*KRATOS ANALYTICAL*, Manchester, England). The spectrometer was equipped with a monochromatic Al Kα ($h \cdot \nu = 1486.6$ eV) X-ray source of 300 W at 15 kV. The kinetic energy of the photoelectrons was determined with an hemispheric analyser set to a pass energy of 160 eV for the wide scan spectra and 20 eV for the high-resolution spectra. During all measurements electrostatic charging of the sample was over-compensated by means of a low-energy electron source working in combination with a magnetic immersion lens. Later, all recorded peaks were shifted by the same amount which was necessary to set the C 1s peak to 285.00 eV for saturated hydrocarbons. Quantitative elemental compositions were determined from peak areas using experimentally determined sensitivity factors and the spectrometer transmission function. Spectrum background was subtracted according to Shirley. The high-resolution spectra were deconvoluted by means of a computer routine. Free parameters of component peaks were their binding energy, height, full width at half maximum and the Gaussian-Lorentzian ratio.

The DRIFT spectra were recorded by a FTS 165 spectrometer of BIO-RAD Laboratories (Philadelphia, USA) using the praying-mantis-diffuse-reflection attachment. Unmodified metal or metal oxide particles were used as reference. The spectra were measured in Kubelka-Munk units collecting 32 scans. For data post processing the spectra were transformed in ASCII-files and processed in Origin 5.0. Smoothing was not done.

UV/Vis-measurements were carried out using the MCS 400 spectrometer of *Carl Zeiss Jena GmbH* (Jena, Germany) with a special reflection attachment for particles.

Results and Discussion

As mentioned above, all metal surfaces used in these studies contain an oxidic layer and non-specifically adsorbed surface contaminations. Even the copper surface contains an oxidic layer in the order of 1–2 nm thickness. The differences in the adsorption capacities of a metal and its corresponding metal oxide towards PVFA-*co*-PVAm were compared by means of Zn and ZnO as well as Fe and Fe_2O_3. XPS was employed in order to determine the amount of the polyelectrolyte adsorbed onto the metal or metal oxide surface. XPS is a well established surface sensitive spectroscopic method with an information depth not higher than 10 nm.

The direct determination of the polymer content adsorbed onto the solid surface avoids the systematic measuring error arising from particle-initiated PVFA-*co*-PVAm flocculation which can be important by the determination of the PVFA-*co*-PVAm amount in the supernatant solution after the adsorption [23]. Figure 1 shows a wide-scan XPS spectrum of PVFA-*co*-PVAm (degree of hydrolysis > 90%, pH = 8) adsorbed onto copper.

The adsorbed polyelectrolyte can easily identified by the N 1s peak appearing from the amino and amide groups of the polymer. The substrate is distinguished by the copper photoelectron (e.g. Cu $2p_{3/2}$, Cu $2p_{1/2}$, Cu 3p, Cu

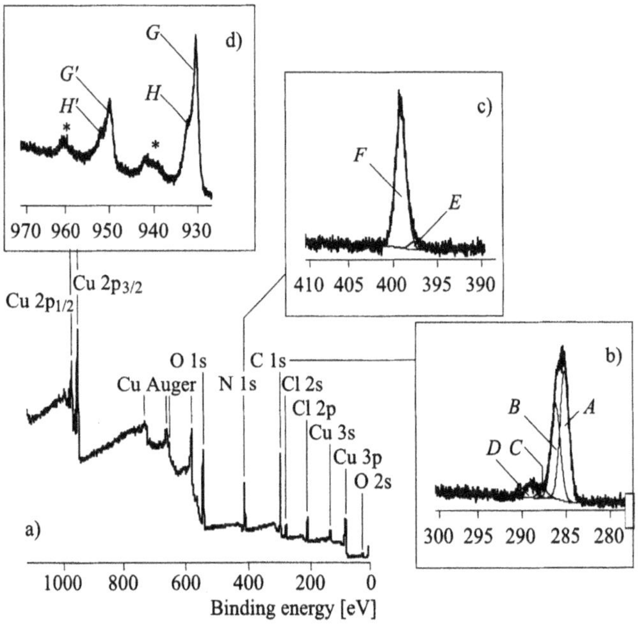

Fig. 1 XPS wide-scan spectrum **a** of copper particles coated with PVFA-*co*-PVAm (molar mass of the PVFA pre-polymer = 340 000 g mol^{-1}, degree of hydrolysis > 90%, adsorption at pH = 8). The *insets* show the C 1s (**b**), N 1s (**c**) and Cu 2p (**d**) element spectra

3s) and Auger peak series. For the determination of the adsorbed amount of PVFA-*co*-PVAm carbon and oxygen cannot be considered as significant elements because surface contaminations can contribute to the carbon C 1s peak and the metal oxidic layer could have a non-stoichiometric composition or traces of water may be incorporated.

The C 1s spectrum in Fig. 1 shows four component peaks indicated by italic letter. The component peak *A* appears from saturated hydrocarbons. Component peak *B* shows C – N bonds of the amino ($C-NH_2$) and formamide (C – NH – CHO) groups. Component peak *C* with an area of ca. 5% of the whole C 1s peak appears from the second carbon of the formamide groups (C – NH – CHO). It is assumed that the component peak *D* is the result of urea formation (N – C(O) – N) by the Brønsted-basic amino groups and atmospheric carbon dioxide [23]. The nitrogen spectrum in Fig. 1c was deconvoluted into two component peaks (*E* and *F*) where the main component *F* represents the amino groups ($C-NH_2$) and the small component peak *E* shows the nitrogen of the residual formamide groups (C – NH – CHO). The two Cu 2p peaks (Cu $2p_{3/2}$ and Cu $2p_{1/2}$, Fig. 1d) show the metallic copper species (indicated by *G* and *G′*) and copper bonded to oxygen (indicated by *H* and *H′*). In addition plasmon resonance peaks of the metallic species were observed (*).

As explained above, only the [N]:[Mt] ratio is of importance to determine the amount of adsorbed polymer (where [N] and [Mt] mean the quantity of nitrogen and the metal, respectively).

For copper, the adsorption behaviour of PVFA-*co*-PVAm seems rather clear. With increasing the amino content of the polymer the amount of adsorbed polymer increases (Table 2). This result is not surprising, because the strong tendency of bonding of copper ions to amines is well known even for copper surfaces [1, 3, 7].

The influence of pH on the amount of adsorbed polymer was studied using a molar mass of 10 000 g mol^{-1}. pH has been varied from 5, 8, to 11. The increase of pH leads to a significant increase of the measured [N]:[Cu] ratio as seen from Table 2. This result is reasonable, because the content of free amino groups increases with pH due to $-NH_3^+$ groups become deprotonated. This result also shows that binding sites of the copper surface to free amino groups play the major role in the adsorption process. It is well known that copper atoms form strong coordinative bonds to amino groups [1, 3, 7].

To investigate the influence of molecular weight on adsorption, a pH of 8 has been chosen. The results are shown in Table 3. The effect of molecular weight on the amount of adsorbed polymer is striking. The [N]:[Cu] ratio increases eight fold from 10 000 to 340 000 g mol^{-1} using the same amount polymer. This result is in line with that observed for the influence of pH and supports the interpretation that coordinative bonds between copper atoms on the surface and free amino groups of the polymer are responsible for the strong adsorption. The more amino groups are present along the polymer backbone the more anchor points are available to fix the polymer at the copper surface [12, 15].

Compared to copper Zn sheets show a contrary adsorption behaviour. With increasing the amount of amino groups, the polymer content on the Zn surface decreases (Table 4).

The adsorption experiments were repeated with ZnO particles as substrate material. For these experiments PVFA-*co*-PVAm samples with a wide range of hydrolyzation were employed. The results are summarized in Table 5.

According to the values in Table 5 there is not a clear relationship between the amount of adsorbed polymer and the number of amino groups (degree of hydrolysis). It is assumed that the adsorption behaviour on ZnO is controlled by a number of structural parameters. In the case of low and moderate degrees of hydrolysis the real number of *free* amino groups does not correspond with

Table 3 The influence of molar masses of PVAm-*co*-PFA on the adsorption of PVAm-*co*-PFA (degree of hydrolysis > 90%, pH 8) onto copper surface

Molar masses [g/mol]	[N]:[Cu] ratio
> 10 000	0.34
45 000	0.82
340 000	2.92

Table 2 The influence of pH on the [N]:[Cu] ratio at adsorption of PVAm-*co*-PFA (molar mass > 10 000 g/mol, degree of hydrolysis > 90%) onto copper surface

pH	[N]:[Cu] ratio
5	0.31
8	0.34
11	0.99

Table 4 Adsorption of PVAm-*co*-PFA onto oxidic metal surfaces. The degree of hydrolysis corresponds with the content of amino groups

Degree of hydrolysis	[N]:[Zn] ratio	[N]:[Cu] ratio
> 90%	1.49	3.25
30%	2.32	0.86
0%	8.16	0.50

Table 5 Adsorption of PVAm-*co*-PFA onto ZnO particle surfaces

Degree of hydrolysis	[N]:[Zn] ratio
< 90%, desalted	0.87
< 90%	0.66
50%	0.60
30%	0.62
10%	1.10
0%	0.04

the degree of hydrolyzation of PVFA because amidinium ions can be formed by reaction of amino with formamide groups (Scheme 2). Formate ions compete with the different functional groups of the PVFA-*co*-PVAm molecule for free adsorption sites on the oxidic metal surface. Hence, the desalted PVFA-*co*-PVAm samples adsorb more effectively. Similar results were found for other oxidic materials [11, 16].

Figure 2 shows the DRIFT spectra of ZnO/PVFA-*co*-PVAm composites. According to Table 5, the degree of hydrolyzation of the polymer was changed.

Advantageously, the DRIFT spectra significantly show the following vibration bands resulting from the different functional groups: The overtone amide-II band was found at $\tilde{\nu} = 3053\,\mathrm{cm}^{-1}$ (*A*), the amide-I band at $\tilde{\nu} = 1675\,\mathrm{cm}^{-1}$ (*B*), the NH-deformation vibration at $\tilde{\nu} = 1590\,\mathrm{cm}^{-1}$ (*C*) and the amide-II band at $\tilde{\nu} = 1540\,\mathrm{cm}^{-1}$ (*D*). Unfortunately, the amidinium ion is difficult to detect, because its vibration band is expected at about $1600\,\mathrm{cm}^{-1}$ and this region is covered by the strong amid-I band and the NH-deformation vibration.

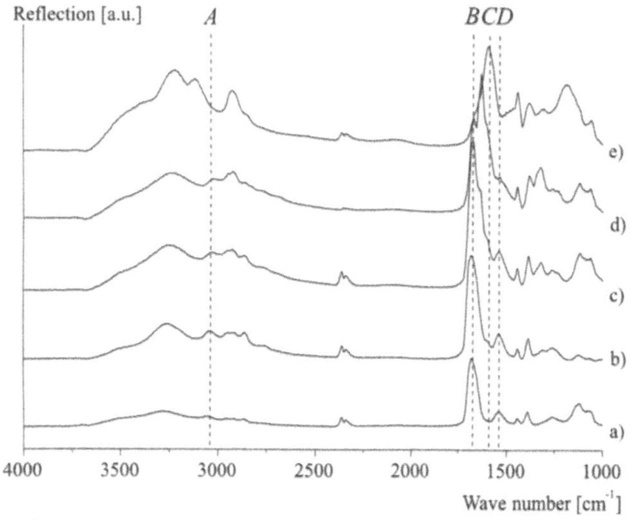

Fig. 2 DRIFT spectra of PVFA-*co*-PVAm modified ZnO particles where the degree of the PVFA-*co*-PVAm hydrolysis was varied: **a** degree of hydrolysis = 0%; **b** degree of hydrolysis = 10%; **c** degree of hydrolysis = 30%; **d** degree of hydrolysis = 50%; and **e** degree of hydrolysis > 90%, desalted. The molar mass of the non-hydrolysed PVFA polymer was $340\,000\,\mathrm{g\,mol}^{-1}$, during the adsorption the pH was pH = 8. The spectra show the overtone amide-II band at $\tilde{\nu} = 3053\,\mathrm{cm}^{-1}$ (*A*), the amide-I band at $\tilde{\nu} = 1675\,\mathrm{cm}^{-1}$ (*B*), the NH deformation vibration at $\tilde{\nu} = 1590\,\mathrm{cm}^{-1}$ (*C*), the amide-II band at $\tilde{\nu} = 1540\,\mathrm{cm}^{-1}$ (*D*). ($\tilde{\nu}$ = wave number, a.u. = arbitrary units according to Kubelka and Munk)

The lowest intensities of the amide-I (*B*) and amide-II (*D*) bands of non-hydrolysed PVFA confirm the results of the XPS investigations. It supplies that the smallest amount of adsorbed polymer on the ZnO surface is observed for non-hydrolysed PVFA.

The strong adsorption of PVFA-*co*-PVAm onto ZnO can be also evidenced by measuring the surface polarity of ZnO/PVFA-*co*-PVAm composites using co-adsorbed solvatochromic dyes [18]. According to earlier studies of silica/PVFA-*co*-PVAm composites [21, 22], we applied a *tert*-butyl substituted Reichardt's dye [19–21] as surface polarity indicator. Employing UV/Vis spectroscopy the absorption maxima λ_{max} of the adsorbed Reichardt's dye was measured (Fig. 4). Its position is strongly correlated with the surface polarity that can be expressed by the $E_T(30)$ polarity parameters (Eq. 1) [21].

$$E_T(30)[\mathrm{kcal\cdot mol^{-1}}] = \frac{30338.5\,[\mathrm{nm\cdot kcal\cdot mol^{-1}}]}{\lambda_{max}} - 1.9185\,[\mathrm{kcal\cdot mol^{-1}}] \quad (1)$$

Figure 3 shows a series of UV/Vis spectra of *tert*-butyl substituted Reichardt's dye loaded ZnO particles coated with PVFA-*co*-PVAm of different degrees of hydrolyses.

Table 6 shows with increasing the amino content the $E_T(30)$ polarity parameter of the ZnO/PVFA-*co*-PVAm

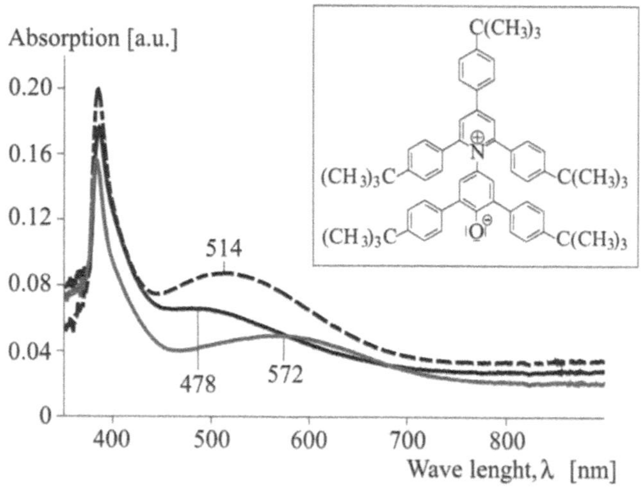

Fig. 3 UV/Vis spectra of the *tert*.-butyl-substituted Reichardt's dye (shown inset) adsorbed onto PVFA-*co*-PVAm modified ZnO particles: ZnO/Lupamine® 9000 (*grey solid line*), ZnO/Lupamine® 9030 (*black dotted line*) and ZnO/Lupamine® 9095 (*black solid line*)

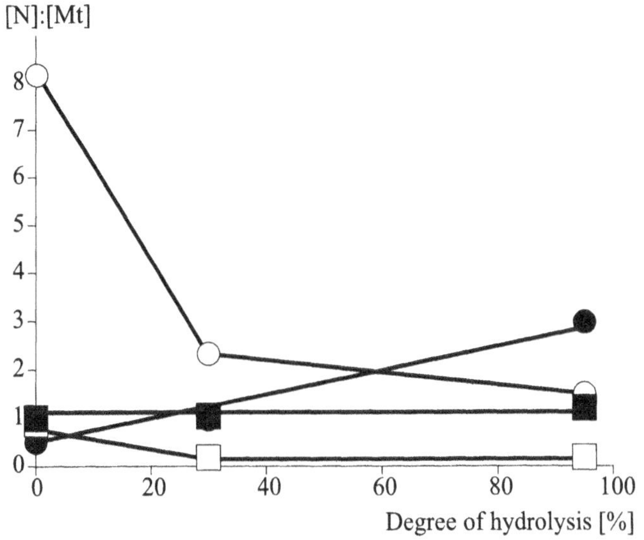

Fig. 4 [N]:[Mt]-ratio of modified Cu (•), Fe (■), Fe_2O_3 (□) particles and Zn sheets (○) as function of degree of hydrolysis of the adsorbed PVFA-*co*-PVAm

Table 6 Measured UV/Vis absorption maxima λ_{max} of Reichardt's dye adsorbed onto bare and PVFA-*co*-PVAm modified ZnO particles and the corresponding $E_T(30)$ values [22] calculated according to Eq. 1

Sample	λ_{max} [nm]	$E_T(30)$ [kcal · mol^{-1}]
bare ZnO	488.9	60.1
ZnO/Lupamine®9000	478.2	61.5
ZnO/Lupamine®9030	513.6	57.2
ZnO/Lupamine®9095	571.6	51.0

composites decreases. These results indicate that the surface acidity of ZnO is compensated by the adsorbed polymer layer, because the $E_T(30)$ parameter reflects the surface acidity as well as dipolarity/polarizability of the environment [19].

Compared to the oxidic surfaces of copper and zinc, the adsorption process of PVFA-*co*-PVAm onto Fe and Fe_2O_3 powders was weakly dependent on the composition of the polymer. The results are given in Table 7.

Summarizing all results, Fig. 4 shows the content of adsorbed polymer expressed by the [N]/[Mt] ratio as function of degree of hydrolyses of PVFA-*co*-PVAm.

Table 7 Adsorption of PVAm-*co*-PVFA onto Fe and Fe_2O_3 particle surfaces

Degree of hydrolysis	[N]:[Fe] ratio for Fe powders	[N]:[Fe] ratio for Fe_2O_3 powders
> 90%	1.25	0.21
30%	0.95	0.19
0%	0.95	0.88

Conclusion and Outlook

The results of these studies showed that PVFA-*co*-PVAm adsorb well onto various surface oxidized metals and their corresponding oxides. The adsorbed polymer amount is significantly higher for metal powders while oxides show a lower adsorption capacity. The oxide layer may act as the mediating environment, but it decreases the high surface energy of the bare metal surface by lowering the Van der Waals interaction between the solid surface and the adsorbing polymer. This phenomenon corresponds with the well known difference of the Hamaker constants, which are usually higher for metals compared with the corresponding metal oxides [24–26]. Furthermore, the formation of an oxide layer prevents coordinative interactions between metal atoms and the freely accessible amino groups. The origin of the driving force of the adsorption of PVFA-*co*-PVAm from its aqueous solution onto various metals and metal oxides and the role of water requires still further studies. For the adsorption of highly hydrolysed PVFA-*co*-PVAm onto silica (amorphous SiO_2) we discussed an adsorption process driven by the adsorption enthalpy [23]. In the case of transition metals or its metal oxides additional coordinative bonds can likely contribute to the enthalpic term. Functional surface groups of metal oxides (e.g. Mt – OH) can be involved in the adsorption mechanism by the formation of specific acid-base interactions to the Brønsted-basic PVFA-*co*-PVAm polymer, but can also be an anchor for hydrogen bonded water layers which must be removed during the adsorption. The release of surface adsorbed water would contribute to the entropy term of the adsorption free energy.

Acknowledgement Financial support by the *Deutsche Forschungsgemeinschaft* and the *Fond der Chemischen Industrie* is gratefully acknowledged. We thank Dr. Schornick and Dr. Hähnle from the *BASF AG*, Ludwigshafen for providing the poly(vinyl formamide-*co*-vinyl amine)s and their helpful scientific correspondence.

References

1. Jensen WB (1991) The Lewis acid-base concepts: Recent results and prospects for the future. In: Mittal KL, Anderson HR (eds) Acid-Base-Interactions. VSP Utrecht, p 3–23
2. Fowkes FM, Sun CY, Joslin ST (1980) Enhancing polymer adhesion to iron surfaces by acid/base interaction. Corrosion Control by Organic Coatings Symposium Lehigh Univ, Edited Proc NACE 1981, 1

3. Barthes-Labrousse MG, Joud JC (2000) Acid-Base Characterization of metallic materials and the use of model molecules in the study of adhesion mechanism. Acid-Base Interactions: Relevance to Adhesion Science and Technology, Mittal KL (ed) VSP Utrecht, 453
4. Cain SR (1991) Quantum-mechanical approach to understanding acid-base interactions at metal-polymer interfaces Jensen. In: Mittal KL, Anderson HR (eds) Acid-Base-Interactions. VSP Utrecht, p 47
5. Berger EJ (1991) A method of determining the surface acidity of polymeric and metallic materials and its application to lap shear adhesion. In: Mittal KL, Anderson HR (eds) Acid-Base-Interactions. VSP Utrecht, p 207
6. Huang YC, Fowkes FM, Loyd TB (1991) Acidic and basic nature of ferric oxide surfaces. Adsorption, adhesion, zeta potential and dispersibility in magnetic inks for hard disks. In: Mittal KL, Anderson HR (eds) Acid-Base-Interactions. VSP Utrecht, p 363
7. Preuss M, Schmidt WG, Bechstedt F (2005) Physical Review Letters 94:236102
8. Spencer MJS, Hung A, Snook IK, Yarovsky I (2002) Surface Science 515:464
9. Spencer MJS, Hung A, Snook IK, Yarovsky I (2003) Surface Reviews and Letters 10:169
10. Chen CW, Tano D, Akashi M (1999) Coll Polym Sci 277:488
11. Pinschmidt RK, Renz WL, Carroll WE, Yacoub K, Drescher J, Nordquist AF, Chen N (1997) J Macromolar Science 10:1885
12. Spange S, Meyer T, Voigt I, Eschner M, Estel K, Pleul D, Simon F (2003) Adv Polym Sci 165:43
13. Fleer GJ, Lyklema J (1983) In: Parfitt GD, Rochester CH (eds) Adsorption from Solution at the Solid/Liquid Interface. Academic Press, London
14. Amjad Z (1998) Water soluble Polymers: Solution Properties and Application, Plenum Press 1998, ISBN 0-306-45931-0
15. Raymond SF, Paul L (1999) Colloid Polymer Interactions: From Fundamentals to Practice, Wiley ISBN 0-471-24316-7
16. Fleer GJ, Stuart MAC, Scheutjens JMHM, Cosgrove T, Vincent B (1998) Polymers at Interfaces. Chapman & Hall, London
17. Wolf S, Simon F, Spange S (2005) to be publish
18. Horner MR, Boerio FJ, Clearfield HM (1992) An XPS investigation of the adsorption of aminosilanes onto metal substrates. In: Mittal KL (ed) Silanes and Other Coupling Agents. VSP Utrecht, p 241
19. Macquarrie DJ, Tavener SJ, Gray GW, Heath PA, Rafelt JS, Saulzet SI, Hardy JJE, Clark JH, Sutra P, Brunel D, di Renzo F, Fajula F (1999) New J Chem 23:725
20. Reichardt C, Harbusch-Görnert E (1983) Justus Liebigs Ann Chem 721
21. Reichardt C (1994) Chem Rev 94:2319
22. Voigt I, Simon F, Estel K, Spange S (2001) Langmuir 17:308
23. Spange S, Reuter A, Lubda D (1999) Langmuir 15:2103
24. Bucatariu F, Simon F, Spange S, Dragan ES (2005) Reactive nano-structured thin films on silica particles. Construction and characterization, Proceedings of the 21st Annual Meeting of the Polymer Processing Society, Leipzig, in print
25. Mahany J, Ninham BW (1976) Dispersion Forces. Academic Press, New York
26. Israelachvilli JN (1974) Quart Rev Biophys 6:34
27. Bergström L (1997) Adv Coll Interf Sci 70:125

Progr Colloid Polym Sci (2006) 132: 117–123
DOI 10.1007/2882_022

Published online: 10 March 2006

N. Elsner
F. Dubreuil
R. Weinkamer
M. Wasicek
F.D. Fischer
A. Fery

Mechanical Properties of Freestanding Polyelectrolyte Capsules: a Quantitative Approach Based on Shell Theory

N. Elsner · F. Dubreuil · A. Fery (✉)
Abteilung Grenzflächen, Max Planck Institut für Kolloid- und Grenzflächenforschung, Am Mühlenberg, 14424 Potsdam, Germany
e-mail: andreas.fery@mpikg.mpg.de

R. Weinkamer
Abteilung Biomaterialien, Max Planck Institut für Kolloid- und Grenzflächenforschung, Am Mühlenberg, 14424 Potsdam, Germany

M. Wasicek · F.D. Fischer
Institute of Mechanics, Montanuniversität Leoben, and Erich Schmid Institute of Material Science, Austrian Academy of Sciences, and Materials Center Leoben, GmbH, Austria

Abstract In this paper we report on AFM force spectroscopy measurements on hollow polymeric spheres of colloidal dimensions made from polyelectrolyte multilayers of polyallylamine and polystyrenesulfonate in water. We find that the shells show a linear force-deformation characteristic for deformations of the order of the shell wall thickness. This experimental outcome is discussed in terms of analytical results of continuum mechanics, in particular the scaling behaviour of the shell spring constant with wall thickness, shell radius and speed of the deformation is analysed. The experimental results agree well with the predictions of Reissner for thin shells and allow us to rescale our stiffness data such that a master curve of shell stiffness is obtained. The result of Reissner is strictly valid only for point like loading situations, while in our experiments a more extended plate like load is applied. Experimentally we find indeed little influence of the probe geometry on the shell spring constant. This result agrees well with finite element (FE) calculations that show that the Reissner result is a good approximation also for non point like loading situations, as long as small deformations are considered.

Keywords Polyelectrolytes · Multilayers · Mechanical properties · Deformation · Shells · Capsules

Introduction

Polymeric hollow spheres of colloidal dimensions are of considerable interest for various applications like encapsulation of agents in medicine for targeting or for sustained release (drug delivery), cosmetics or food design to mention a few examples only. At the same time, examples of hollow capsules of colloidal dimensions are numerous in biology, like the shells of single cell organisms or virus capsids [1]. Thus, artificial polymeric shells can also serve as (biomimetic) model systems for these more complex biological counterparts. The mechanical properties of shells play an important role in many processes and an understanding, therefore, is indispensable for their application. The mechanical properties govern the deformation behaviour under external loads or pressures and, thus, not only the stability of shells, but also the extent of deformations that are induced by these external stimuli and whether the deformations are reversible (elastic) or irreversible (plastic). The adhesion behaviour is also closely linked to the deformation properties, since the size of adhesion areas that are developed as a result of adhesive interactions is given by the balance between energy gain due to the work of adhesion and energy cost due to the deformations involved in the process.

The development of novel methods that allow probing the mechanics of colloidal scale shells is a prerequisite for gaining understanding of the physico-chemical mechanisms that govern shell mechanics. This however requires also an in depth treatment of the shell deformation problem under the specific circumstances of the experiment. Both aims are subject of this paper for the particular case of polyelectrolyte multilayer shells investigated with the colloidal probe AFM technique. While the experimental results are specific for this system, the mechanical results concerning the deformation process and the interpretation of the AFM force distance data are generally applicable for other shell systems like the examples mentioned above.

Within artificial polymeric capsules, polyelectrolyte multilayer capsules (PMCs) like we use them in this study offer several advantages as model systems, since they can be produced with well defined wall thickness, radius and low polydispersity. The shell production process is based on the layer-by-layer electrostatic self assembly (lbl-esa), which was first introduced by Decher and co-workers as a surface coating technique [2, 3]. Polyelectrolyte multilayers can be built up on charged surfaces by alternating adsorption of positively and negatively charged polyelectrolytes from aqueous solution and for many polyelectrolyte combinations (in particular the one used here, polyallylamine (PAH)/polystyrenesulfonate (PSS)), the total thickness of the surface coating is linear in the number of deposition cycles, with a thickness increase per polyelectrolyte layer pair on the order of nanometres. Furthermore, since the deposition process is based on non-covalent interactions, restrictions on the type of substrate and polyelectrolyte are few. Hollow capsules can be produced by a two step process [4]: First colloidal particles are coated using lbl-esa, than the particles are dissolved under conditions that do not destroy the multilayer. During the core dissolution process, the core material has to permeate through the membrane, which limits the choice of the core particle. So far, the process has been demonstrated for a number of colloidal particles, like melamin-formaldehyde [4], non-crosslinked polystyrene latices, silica or polylactic acid particles [5]. In these cases, the composition and thickness of the shell walls can be controlled with a precision that is comparable to the case of solid supported lbl-esa films. Additionally, the capsules have the same shape and monodispersity as the template particles, which makes them one of the best defined capsule systems with respect to these parameters that is currently available. A more complete overview on polyelectrolyte multilayers can be found in [6].

The mechanical properties of polyelectrolyte multilayer capsules have been subject of several studies using different methods. Bäumler and co-workers [7] have used the micropipette technique and found that PMCs are not conserving their volume if pressure differences are applied between inside and outside of the shell. This is expected, since the shells can only be formed in first place because the membrane is permeable to low molecular weight species, the core dissolution products. They found no deformation up to a critical pressure followed by an irreversible collapse, showing that shells deform not elastically but plastically for large deformations. First quantitative estimates of the Young's modulus of the shell material were obtained by Gao and coworkers, using osmotic pressure differences between inside and outside of the shell [8, 9]. These authors monitored the onset of the buckling instability as a function of shell radius and thickness and found values around 1.5 GPa for the Young's modulus of PAH/PSS multilayers. They concluded that the multilayer material is in a glassy state. In contrast, Vinogradova and coworkers found much lower elastic constants for the same system using AFM force spectroscopy (1–150 MPa) and concluded that the wall material is in an elastomeric state [10] (see [11] for a complete overview). But some of the assumptions made in this work (volume conservation during deformation, elasticity at large deformations) are not in line with previous findings of others, which can explain the underestimation as compared to Gao [12]. We have independently applied colloidal probe AFM force spectroscopy, but limited our analysis to deformations on the order of the shell wall thickness only, to avoid problems occurring at larger deformations and, additionally, no assumptions concerning the permeability are necessary [12–14]. Our results of the Young's modulus are between the values found by Vinogradova and those of Gao and coworkers and are on the same order as those found in recent experiments based on capsule swelling [15] of 200–400 MPa.

In the present paper we extend our analysis of the experimental results obtained from this small deformation regime and we show that the result found by Reissner for the deformation of shallow spherical caps represents an excellent analytical approximation for the interpretation of the measurements. This result is varified by finite element modelling (FEM) and by experimental variation of the force probe geometry and radius as well as wall thickness of the studied capsules. This result is also applicable for other capsule deformation measurements, since it is independent of the specific Young's modulus. Furthermore, we report on speed dependent measurements that indicate the glassy nature of PAH/PSS multilayers.

Experimental

Materials

Polyallylamine ($M_w = 70\,000$ g/mol) and polystyrene sulfonate ($M_w = 70\,000$ g/mol) were obtained from Aldrich. PAH was used as received, while PSS was purified from low molecular weight impurities by dialysis (Polyethersulfone membranes, MW cut off 10 000 g/mol, Millipore) against ultra pure water and freeze dried. All water used for preparing solutions or for dialysis was purified by a Purelab Plus UV/UF, Elga LabWater system and had a resistivity smaller than 0.055 μS/cm and a total organic content between 2 and 12 ppb (parts per billion).

Part of the PAH was fluorescently labelled using Rhodamin-B-isothiocyanate (RBITC, Fluka) and the procedure described in Richter et al. [16]. The labelled PAH (R-PAH) was dialyzed subsequently against ultra pure water (regenerated cellulose membrane, MW cut off =

15 000 g/mol, Spectrum Laboratories). The labelling degree was not larger than 1% to avoid dye induced changes of the interactions. Polyethyleneimine (branched, $M_w = 25\,000$ g/mol, Aldrich), which was used in this work to coat the glass substrates to promote adhesion of the microcapsules, was obtained from Aldrich.

Capsule Preparation and Characterization

All capsules were produced by lbl-esa coating of non-crosslinked polystyrene latex particles (Microparticles GmbH, particle radii 7.8 and 9.6 microns). Adsorption was carried out from aqueous solutions containing 0.5 moles NaCl (Merck) per litre added and a polyelectrolyte concentration of 10 mM based on the monomer unit. Adsorption was allowed to take place for 20 minutes and was followed by three washing steps to remove non adsorbed polyelectrolyte in which the particle solutions were centrifuged and the supernatant was replaced with ultrapure water. Dissolution was carried out in tetrahydrofurane (THF, stabilized with 2,6-Di-tert.-butyl-4-methylphenole, p.a., Aldrich). Again several washing steps with THF were carried out before the capsules were re-transferred to water. In order to achieve labelling of the capsules, one of the PAH layers was formed from a 1 : 1 mixture of PAH and R-PAH. Capsules of different wall thickness were formed as described in the results part, but in each case, the terminating layer was PSS.

Shape and uniformity of the capsules as well as their diameter were controlled after dissolution and during the following force spectroscopy measurements with fluorescence microscopy. Capsules that showed indentations or deformations were not used in the experiments. The capsule wall thickness was determined for each batch by AFM imaging of dried capsules on using a Nanoscope IIIA AFM (Veeco Instruments) in Tapping mode® using Silicon cantilevers (Nanosensors NCH-W, $k_c = 31$–50 N/m, $f_c = 302$–350 kHz, NanoWorld Services). This method was also used to ensure that the core dissolution was complete and no residuals were found inside the capsules. The dry thickness of the capsules was found to be between 4.5–5 nm per layer pair for the different batches, which is in line with findings of others using the same preparation conditions [17].

Force spectroscopy measurements were carried out on a MFP 1D and a MFP 3D setup (Atomic Force, Germany) mounted onto an inverted optical microscope (Olympus IC71). We used tipless cantilevers (contact silicon cantilevers CSC12/tipless/50, $f_c = 7$ bis 95 kHz, $k_c = 0.01$ to 0.45 N/m, Mikromash) onto which glass-beads (diameter 30 to 50 μm, Polyscience Inc, USA) were glued using epoxy glue (UHU plus 300) with the help of a micromanipulator. All spring constants were determined by both the method of Sader [18] before gluing and the thermal noise method directly during the experiment [19] and cantilevers showing differences greater than 10 percent were discarded.

For the capsule measurements, glass substrates coated with polyethyleneimine were used on which the PSS-terminated capsules showed adhesion. The capsule force deformation characteristics of individual capsules were obtained by placing a capsule under the colloidal probe and pressing onto it. Special care was taken to press onto the pole of the capsule as described in more detail in [13]. Reference curves on hard substrates were obtained before and after the experiments to determine the inverse optical lever sensitivity (InvoLS) and to derive the absolute deformation of the capsules. In rare cases when the InvoLS was found to change during the measurement, the measured data was discarded.

Results and Discussion

Figure 1 shows a representative force deformation characteristic as obtained from the measurements of a capsule made from PAH/PSS in water. The dried thickness of the capsule was 25 nm and the radius 7.9 microns. For deformations on the order of 1–3 times the shell wall thickness, a linear force deformation characteristic is found. For higher deformations discontinuities in the force deformation characteristic are observed, which are separating quasi-linear sections. The position of these discontinuities as well as their shapes scattered a lot between different shells and the shells showed plasticity in this deformation regime. We avoided this regime in the measurements and obtained the results exclusively from a detailed analysis of the linear regime. Based on classical thin shell theory [20], one would expect a linear force deformation characteristic for deformations up to a few times the wall thickness (fit indicated as dotted line). The onset of buckling should lead to a deviation from the linear dependency, like dis-

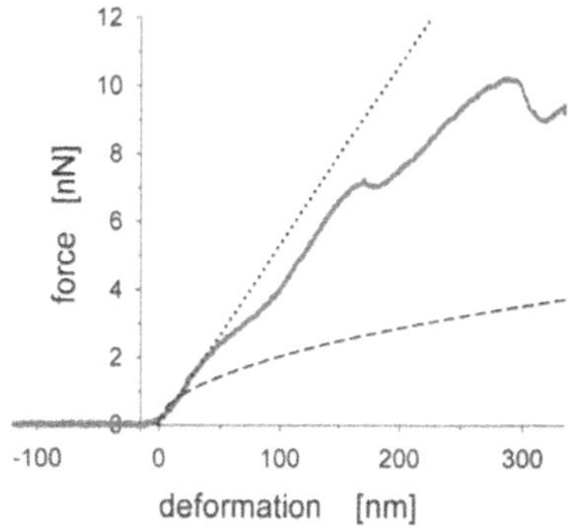

Fig. 1 *Solid line:* experimental force deformation characteristic of a PAH/PSS capsule of 25 nm dry thickness and 7.9 μm radius in water. *Dotted line:* Reissner's result for thin, shallow shells (see Eq. 1). *Dashed line:* square root dependency between deformation and force as expected from Pogorelov's work (see Eq. 2)

cussed within the approach of Pogorelov [21] (indicated by the dashed line in Fig. 1).

In previous work [13], we have shown that the discontinuities observed in the force deformation characteristic are indeed connected to buckling instabilities and we have focused therefore on the linear part at small deformations in the characterization of the shells mechanical properties. We sticked to this approach also for the work presented here. Introducing the capsule spring constant k_{shell}, the relation between the force F and deformation d can be written as $F = k_{\text{shell}}\, d$. Figure 2a shows a histogram of slopes ($= k_{\text{shell}}$) as measured from a large number (35) of different capsules of the same batch and Fig. 2b the corresponding histogram of dried wall thicknesses of capsules of the same batch.

The distribution of capsule spring constants, although broad, clearly shows a maximum at a distinct value of about 0.26 N/m. Basically two effects can explain the scatter. First, the thickness of the capsule walls is not uniform within one batch as shown in Fig. 2b (for a quantitative discussion of the thickness dependency see below). Second, the loading situation is not completely identical for the different experiments, although we attempted to deform all shells axisymmetrically by hitting them exactly at the pole. Lateral positioning, however, was only possible with an accuracy of $\pm 0.5\ \mu$m. This explains why even for repeated deformation of the same shell a scatter of about 10 percent of the shell spring constant was found. It is important, however, to notice that for a deformation range of a few times the wall thickness, repeating the experiment on the same shell up to 100 times did not produce a systematic stiffening or softening of the shell spring constant or any plastic deformation.

To quantify the mechanical properties of the shells from these measurements, we use the analytical solution obtained by Reissner as a starting point [22, 23]. This result describes the deformation of shallow spherical caps under point like loads. It is a special case of the more general solution found by Koiter for the deformation of a thin elastic shell under point like loads at the poles [24]. Already Koiter pointed out that both results are equivalent within the accuracy of thin shell theory. The applicability of thin shell theory is limited to shells with a ratio between shell thickness and radius smaller than 1/20 [25] which is definitely fulfilled in the AFM experiments. Further, for both results it is assumed that the shell is made of a material of isotropic elasticity that can be described by a Young's modulus E and a Poisson ratio ν. This assumption is justified, at least for the particular case of PAH/PSS, since the different polyelectrolyte species are strongly interdigitated [6]. Under these assumptions, Reissner predicts the following relation between the deformation d and the force F [22, 23]:

$$d = \frac{\sqrt{3\left(1-\nu^2\right)}}{4} \frac{R}{h^2} \frac{F}{E}\,, \tag{1}$$

where h denotes the wall thickness and R the radius of the shell.

Pogorelov described the post-buckling case [21], where a dimple is created by a geometrical inversion of the spherical cap around the force. Under the main assumption that all the deformation energy is concentrated in the rim around the dimple, proportionality between the square root of d and F is obtained, or more explicit

$$\sqrt{d} = \frac{1-\nu^2}{1.68} \frac{R}{h^{5/2}} \frac{F}{E}\,. \tag{2}$$

The intersection point between the two results (Fig. 1) should provide a rough indication for the onset of buckling. The intersection occurs at a value well below the wall thickness h, where numerical and experimental results show buckling for deformations of a few shell thicknesses [26].

In the solution of Reissner (Eq. 1) the shell spring constant k_{shell} should scale $\propto h^2$ and $\propto R^{-1}$. To check this scaling relation, we have prepared shells with different wall thickness and different radius. Figure 3 shows a plot of k_{shell} multiplied with the radius versus the square of the shell wall thickness, so that the slope of the plot reflects only the fixed material properties.

For both data sets, the shell spring constants scale indeed with the square of the wall thickness, h^2. In both cases, the fit to the data produces only a small offset for the limit of zero thickness. Thus, contributions to the shell spring constant from long range interactions or a surface tension, which both would not be expected to strongly depend on the wall thickness, can be ruled out, since they would result in an offset. While the rescaling of the spring constant with R shows the predicted tendency, the match is not as convincing as for the h dependency. The smaller shells show a slightly higher stiffness compared to the larger ones, although the deviation is not dramatic (within 10 percent error). Therefore, the scaling analysis is in line with the interpretation of the data within the Reissner model.

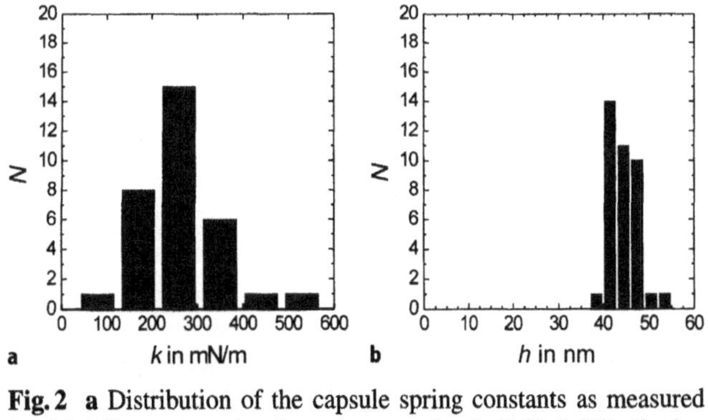

Fig. 2 **a** Distribution of the capsule spring constants as measured for 35 different capsules from the same batch (*left*). **b** Distribution of wall thicknesses in dry state of the same batch (*right*)

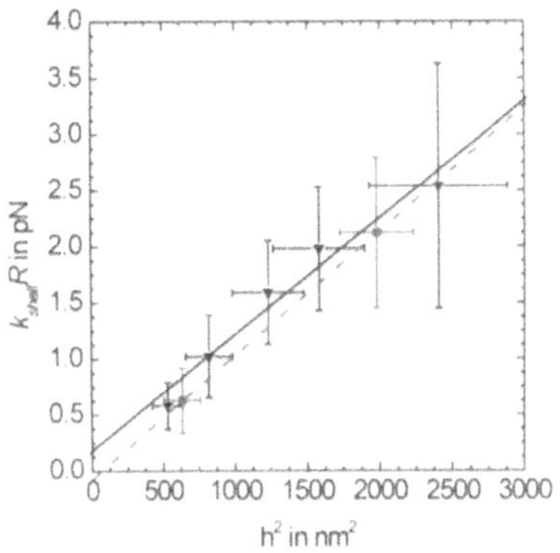

Fig. 3 Shell spring constant multiplied with the shell radius, which makes the data from shells of different radius collapse onto a master curve, and plotted against the square of the shell thickness. *Black triangles:* shells with 9.6 micrometer radius, *grey squares:* shells with 7.85 micrometer radius. *Full lines* correspond to linear regressions performed on each data set separately. In both cases the predicted scaling of the shell spring constant proportional to h^2 is found in the experiment

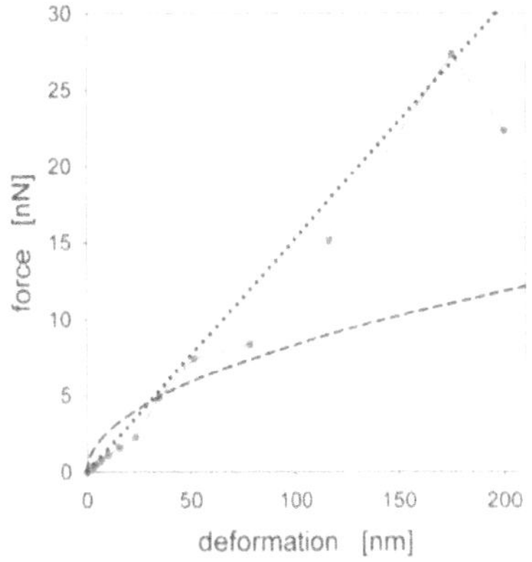

Fig. 4 Comparison of FEM results (*filled circles connected by solid lines*) for the deformation of a spherical shell with a ratio thickness to radius as in the experiment by a rigid spherical colloid. *Dotted and broken lines* are the analytical results from Reissner (Eq. 1) and Pogorelov (Eq. 2). In the FEM, $R = 10\,\mu\text{m}$, $h = 50\,\text{nm}$, $E = 250\,\text{MPa}$ und $\upsilon = 1/3$ were used which is close to the experimental situation

However, a question concerning the quantitative applicability of Reissner's solution is, how far the analytical result is suited for the deformation situation that we have in the present problem. The capsule is not compressed by a point like load but rather by a second sphere with a radius about four times larger than the shell radius. The same assumption is made in the framework of Pogorelov's approach, which is frequently used at higher deformations [15]. In order to answer this question, we have performed finite element (FE) calculations for the particular situation under investigation (loading by a rigid sphere of radius four times the shell radius instead of a point like force). A large deformation analysis has been performed using Abaqus (http://www.abaqus.com/) under the assumption of a frictionless contact between the two spheres. Figure 4 shows the comparison of the FEM results with the analytical predictions from Reissner (dotted line) and Pogorelov (broken line). Note that these are predictions using the same material parameters and dimensions of the shell in both analytical results and not simply fits performed to the FEM data.

Figure 4 allows several conclusions: First of all, the prediction of the Reissner result matches the FEM result in the small deformation region well, which shows that the finite size of the contact area does not lead to strong deviations from the analytical theory based on point like contact. This justifies an application of this result to derive elastic constants of the shell material from the measurements at small deformations.

Second, this is not the case concerning the result of Pogorelov, which should predict the deformation after the onset of buckling. Rather, similar to the situation observed in the experiment (Fig. 1), the force-deformation characteristic follows closer to the Reissner deformation scheme even at deformations of about more than three shell thicknesses were buckling already occurred (which is found both in the experiment and in the FEM calculations for those deformations). This surprising effect should be attributed to the finite contact area between probe and shell. After buckling has occurred, the contact between the two bodies is not point like, but a contact circle between the sphere and the rim like top of the buckled sphere. This increased contact area results in an increase of the shell spring constant with the effect that the Reissner solution stays a reasonable solution far beyond its actual range of validity. It is noteworthy that this result depends only on geometrical properties of the shell, i.e., the shell thickness and the shell radius, and the assumption that the material is deforming linear elastically. Thus, this result is quite generally applicable for deformation problems of spherical shells in colloidal probe or other (even macroscopic) deformation problems.

This allows us to derive the elastic properties of the capsules from the measured shell spring constants. The most precise way to do so is to use the combination of the data in Fig. 3, since the slope of the rescaled data contains the Young's modulus of the material and is an average over all measurements. However, since the measurements were carried out in water and not in dried state, like the thickness measurements, the thickness has to be modified to account for the swelling of PAH/PSS. Kügler et al. showed that the thickness of the multilayers depends on the degree of humidity [27]. According to their study, PAH/PSS multilayers swell by 15% between multilayers dried under ambient conditions and immersed into water. Similar results were found by Lösche et al. [28], who

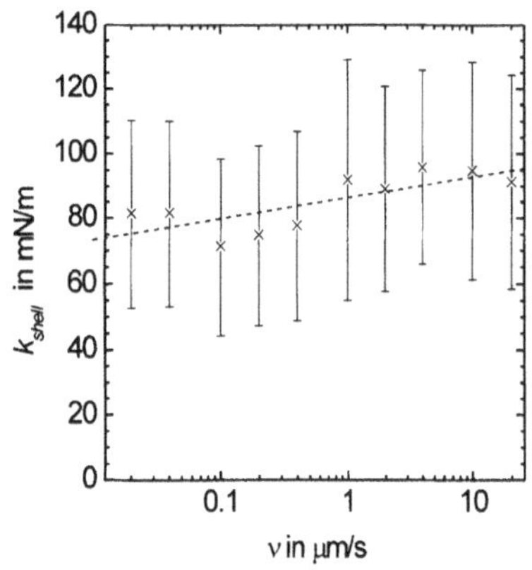

Fig. 5 Dependency of the measured shell spring constant on speed: the speed of the AFM probe deforming the surface is varied over 3 orders of magnitude which results, within the accuracy of the measurement, in no change of the elastic constants

measured an increase of 22% between fully dehydrated and rehydrated PAH/PSS multilayers. We verified this result by AFM imaging of the scratch height of PAH/PSS multilayers that were formed under the same conditions as the capsules on silicon wafer substrates. Transfer from air into water was found to lead to a multilayer thickness increase of the same order (30%). Therefore, we have used 20 percent higher thicknesses for calculating the Young's modulus. Assuming a Poisson ratio $\nu = \frac{1}{3}$, this results in a Young's modulus E of 294 ± 32 MPa, which agrees with the results of Heuvingh from capsule deformation [14] and Vinogradova [15] from capsule swelling with the respective errors. The value and the error of this value is derived from the linear regression in Fig. 3. Our own estimate in a previous paper based on a simpler version of shell theory [13] as well as the estimate of Gao are higher and the estimate of Vinogradova based on large deformation AFM studies is lower. In the case of Gao, we believe that the discrepancy stems from the fact that larger deformations are probed and that the material is probably showing strain hardening. In the case of Vinogradova, this is most likely due to problems with handling the finite permeability of the shells in the proposed model, that would explain an underestimation of the stiffness. Apart from the magnitude of the Young's modulus, the question whether the material is in a glassy or in a viscoelastic state is of importance. To answer this question, we have carried out speed dependent measurements, whose results that are shown in Fig. 5.

Figure 5 clearly shows that there is no speed dependency of the shells elasticity when the speed of deformation is varied over 3 orders of magnitude. This clearly rules out a viscoelastic material, which is also consistent with the (high) order of magnitude of the derived Youngs modulus.

Conclusions

In this paper we have reported on both experimental and theoretical results on the deformation behaviour of polymeric shells with microns diameter and membrane thickness on the order of 20–50 nanometers. AFM force spectroscopy allowed to directly determining the force versus deformation relation of individual capsules made from multilayers of polyallylamine and polystyrenesulfonate. We found good agreement of the experimental results with the analytical Reissner solution of the shell deformation problem which predicts a linear force deformation relation for deformations comparable to the capsule wall thickness. We compared these analytical findings with finite element modelling, in which a deformation of the shell by a plate can be modelled, which is closer to the experimental situation, while the Reissner solution assumes a pointlike deformation. The FEM data shows that the effect of using a non-pointlike probe to deform the capsule is to prolong the linear part of the force deformation, which is also found in the experiment. The FEM data shows therefore a better agreement with the data, but is identical to the analytical theory in the regime of interest of small deformations. This result is not depending on the particular system as long as the capsule wall material is reacting linear elastically. Therefore, it can serve as a guide for the measurement of other capsule like systems with similar techniques as the one used here.

Concerning the particular system, the Youngs modulus was determined with a higher accuracy then previously possible to be 294 ± 32 MPa, assuming a Poisson ratio of $\nu = \frac{1}{3}$. Together with the non detectable impact of the speed of deformation which was varied over 3 orders of magnitude, this makes us believe that the system is in a glassy state. Temperature dependent measurements which will be published elsewhere show a pronounced softening of the material upon temperature increase for systems comparable to the one investigated here. Therefore we conclude that the strong ionic interactions are leading to the formation of a glassy material. An important result is the finding that the capsule spring constant can be adjusted by the radius and the wall thickness over a large range and that the analytical Reissner solution predicts the scaling of the thickness correctly. This shows that lbl-esa capsules can be precisely controlled in their compliance. This property is unique amongst shell systems and opens perspectives towards controlling deformability and adhesion in these systems via their mechanical properties.

Acknowledgement We would like to acknowledge financial support from the German Science foundation (DFG) within the project "Adhäsion und Mechanik von Polyelektrolyt-Hohlkörpern" and the Max Planck Society. We are grateful for stimulating discussions with Prof. Helmuth Möhwald.

References

1. Ivanovska IL, de Pablo PJ, Ibarra B, Sgalari G, MacKintosh FC, Carrascosa JL, Schmidt CF, Wuite GJL (2004) Proc Natl Acad Sci 101:7600
2. Decher G, Hong JD, Schmitt J (1992) Thin Solid Films 210/211:831
3. Decher G (1997) Science 277:1232
4. Donath E, Sukhorukov GB, Caruso F, Davis SA, Möhwald H (1998) Angewandte Chemie 110:2324
5. Shenoy DB, Antipov AA, Sukhorukov GB, Möhwald H (2003) Biomacromol 4:265
6. Decher G, Schlenoff JB (2003) Multilayer Thin Films. Wiley-VCH
7. Bäumler H, Artmann G, Voigt A, Mitlöhner R, Neu B, Kiesewetter H (2000) J Microencapsul 17:651
8. Gao CY, Leporatti S, Moya S, Donath E, Möhwald H (2001) Langmuir 17:3491
9. Gao C, Donath E, Moya S, Dudnik V, Möhwald H (2001) Eur Phys J E 5:21
10. Lulevich VV, Andrienko D, Vinogradova OI (2004) J Chem Phys 120:3822
11. Vinogradova OI (2004) J Phys: Condensed Matter 16:R1105
12. Fery A, Dubreuil F, Möhwald H (2004) New J Phys 6:13
13. Dubreuil F, Elsner N, Fery A (2003) Eur Phys J E 12:215
14. Heuvingh J, Zappa M, Fery A (2005) Langmuir 21:3165
15. Vinogradova OI, Andrienko D, Lulevich VV, Nordschild S, Sukhorukov GB (2004) Macromol 37:1113
16. Richter B, Kirstein S (1999) J Chem Phys 111:5191
17. Schwarz B, Schönhoff M (2002) Langmuir 18:2964
18. Sader JE, Chon JWM, Mulvaney P (1999) Rev Sci Instrum 70:3967
19. Hutter JL, Bechhoefer J (1993) Rev Sci Instrum 64:1868
20. Niordson FI (1985) Shell Theory, North-Holland Series in Applied Mathematics and Mechanics, Vol. 29, North-Holland
21. Pogorelov AV (1988) Bendings of surfaces and stability of shells. Am Chem Soc, Providence
22. Reissner E (1946) J Math Phys 25:80
23. Reissner E (1946) J Math Phys 25:279
24. Koiter WT (1963) in Progress in Applied Mechanics Macmillan: New York 1963 Vol. Prager Anniversary, p 155
25. Gregory RD, Milac TI, Wan FYM (1999) SIAM J Appl Math 59:1080
26. Pauchard L, Rica S (1998) Phil Mag B 78:225
27. Kügler R, Schmitt J, Knoll W (2002) Macromol Chem Phys 203:413
28. Lösche M, Schmitt J, Decher G, Bouwman WG, Kjaer K (1998) Macromolecules 31:8893

Progr Colloid Polym Sci (2006) 132: 124–130
DOI 10.1007/2882_027

Published online: 10 March 2006

Andra Dedinaite
Joseph Iruthayaraj
Natalija Gorochovceva
Ricardas Makuska
Per M. Claesson

Interfacial Properties of Chitosan-PEO Graft Oligomers: Surface Competition with Unmodified Chitosan Oligomers

Andra Dedinaite · Joseph Iruthayaraj ·
Per M. Claesson (✉)
Department of Chemistry, Surface Chemistry, Royal Institute of Technology, 100 44 Stockholm, Sweden
e-mail: per.claesson@surfchem.kth.se

Natalija Gorochovceva · Ricardas Makuska
Department of Polymer Chemistry, Vilnius University, Naugarduko 24, 03225 Vilnius, Lithuania

Abstract Oligomers of chitosan carrying 45 units long poly(ethylene oxide), PEO, chains grafted to the C-6 position of the sugar units were prepared using a novel synthesis route. The graft density was high, close to one poly(ethylene oxide) chain grafted to each sugar unit of the chitosan oligomer but a small fraction of unreacted chitosan remained in the sample. The molecular weight distribution of the sample was determined using GPC. The interfacial properties of the chitosan-PEO graft oligomers were evaluated using X-ray photoelectron spectroscopy and surface force measurements. It was found that the small fraction of unreacted chitosan was significantly enriched at the solid-solution interface on negatively charged muscovite mica surfaces. The interactions between chitosan-PEO oligomer coated surfaces were found to be dominated by the extended PEO chains, and at high coverage the measured forces were consistent with those expected for polymer brushes. Addition of salt up to 10 mM did not result in any significant desorption of preadsorbed oligomer layers.

Keywords Chitosan · Graft copolymers · Poly(ethylene oxide) · Polymer brushes · Surface forces

Introduction

Block copolymers in selective solvents can be designed to be efficient steric stabilizers provided one block, the anchor block, adsorbs strongly to the adsorbent whereas the other block, the buoy block, experiences good solvency conditions and thus extends into solution forming a steric barrier against flocculation [1]. Comb copolymers with adsorbing backbone can to some extent be seen as block copolymers covalently bound together via their anchor group. The general formula can be written as $A(B)_n$ where a number of n buoy blocks are grafted to one anchoring chain. The density of grafted side-chains is determined by the distance between grafting points along the main chain. Surprisingly little is know about the interfacial behaviour of comb copolymers, and nearly all of the experimental and theoretical work reported in the literature is concerned with comb copolymers of rather low graft density (see e.g. [2–4]). Theoretical work indicates that comb homopolymers for entropic reasons adsorb preferentially with the backbone towards the surface [3]. It has also been shown that the thickness of an adsorbed comb homopolymer layer is thinner than that of a linear homopolymer of the same molecular weight and at the same surface coverage. This is due to the fact that even though the number of tails in the comb homopolymer is large, their length is short compared to those formed by adsorbing linear homopolymers [3]. The structure of an adsorbed layer of a comb copolymer with an adsorbing backbone is similar to that of a comb homopolymer. On the other hand, for the case when the grafted chains are adsorbing whereas the backbone is not, the backbone will form many loops extending from the surface, the length of which will depend on the graft density.

When comb copolymers with a high graft density adsorb with the main chain closely packed on the adsorbent, the grafted chains will form a tight brush layer that is expected to give rise to strong steric stabilization, good lubricating properties and, by proper choice of grafted chains, protein and particle repellency. In fact, in a rather recent article excellent protein repellent properties were reported for a comb copolymer consisting of a polylysine backbone and poly(ethylene oxide), PEO, grafts [5]. Similar results have previously been obtained by using PEO grafted to poly(ethylene imine), PEI [6]. The use of comb copolymers as stabilizers and flocculation aids for particle dispersions has also been explored. For instance, it has been shown that titanium dioxide particles can be stabilized by comb copolymers of methacrylic acid (backbone) and PEO (side-chains) [7]. However, the polymer samples used in that study was not well defined and thus it was not possible to draw any conclusions about the relation between the polymer architecture and the stabilizing power of the polymer. Another area of interest is to use comb copolymers as lubricants. The need for thin, firmly anchored molecular lubricants for construction of micro- and nanodevices has recently been stressed [8].

In this report we address interfacial properties of cationic oligomeric chitosan, having close to one grafted 45 units long EO-chain on each sugar unit, by studying its adsorption to negatively charged surfaces and the forces generated by the adsorbed layers. We have previously investigated the interfacial properties of chitosan on negatively charged surfaces [9, 10], and found that it adsorbs in a very flat conformation when the ionic strength is low. We have also investigated interfacial properties of poly(ethylene imine) with grafted PEO chains [11]. However, this study is different from the previous ones since now a very short molecular backbone is used, and the adsorption competition between chitosan with grafted PEO chains and unmodified chitosan oligomers is in focus. One motivation for this choice of oiligomer is that the chitosan backbone is obtained from a naturally occurring polysaccharide, chitin, and show good biodegradability and low toxicity. The grafted side-chains of PEO are motivated by the low affinity of proteins and soils to this compound, which makes PEO interesting in applications ranging from biomedical to household care. The use of oligomers rather than polymers is sometimes advantageous due to their more rapid diffusion to interfaces, and due to that long-range attractive bridging forces, which may cause unwanted flocculation or adhesion between soil particles and surfaces, are avoided. When preparing high graft density copolymers, or oligomers, some unreacted backbone segments are often present in the final product. Since, excessive purification procedures are both time consuming and expensive it is of interest to know if a low concentration of unreacted and adsorbing backbone material has a significant effect on the composition of the adsorbed layer.

Materials and Methods

Synthesis of Chitosan-EO_{45}

The chitosan sample with 45 units long grafted poly (ethylene oxide) side-chains, henceforth referred to as chitosan-EO_{45} oligomers, were prepared according to the following scheme in Fig. 1.

Chitosan (M_r 400 000 g/mol, degree of deacetylation (DD) 72%) and poly(ethylene glycol) monomethyl ether (MPEG) (M_r 2000 g/mol) were obtained from FLUKA. Synthesis of methoxypoly(ethylene glycol) iodide (MPEGI) was done as described in our recent paper [12]. 2-*N*-phthaloyl chitosan (**2** in Scheme 1), 2-*N*-phthaloyl-6-*O*-triphenylmethyl chitosan (**3** in Scheme 1), 2-*N*-phthaloyl-3-*O*-acetyl-6-*O*-triphenylmethyl chitosan (**4** in Scheme 1) and 2-*N*-phthaloyl-3-*O*-acetyl chitosan (**5** in Scheme 1) were synthesized from initial chitosan by a method described earlier [13]. Ag_2O was synthesized from $AgNO_3$ and KOH just before use.

The last step in the synthesis is novel and here a mixture of 2-*N*-phthaloyl-3-*O*-acetyl chitosan (1.0 g, 3.01 mmol), MPEGI (6.4 g, 3.01 mmol) and Ag_2O (0.73 g, 3.01 mmol) in 50 ml DMF was heated at 60 °C with magnetic stirring for 16 hours. Then hydrazine monohydrate (40 ml) and water (80 ml) were added to the reaction vessel and the mixture was heated under magnetic stirring for 15 hours at 90 °C. To remove hydrazine monohydrate the mixture was evaporated using rotary evaporator until a viscous solution was obtained. The solution was diluted with water and evaporated again until a solid residue was formed. This procedure was repeated three times. Then the residue was washed several times by acetone and the obtained product was dried in a vacuum oven to give 5.8 g of chitosan-6-*O*-MPEG graft oligomers. Intermediate and final derivatives of chitosan were characterized by FTIR and ^{1}H-NMR spectroscopy.

The product obtained was analyzed by gel permeation chromatography, GPC, using a Viscotek TDA 302 Triple Detection System with Light Scattering, Viscosity and RI Detection. The columns used were 2x ViscoGel GMPWxl (Mixed Bed), and the eluent was aqueous 0.8 M $NaNO_3$ at pH $\sim$ 7. The eluent flow rate was 0.6 ml/min. and the oligomer concentration approximately 1 mg/ml. The temperature was 35 °C.

Other Materials

The potassium bromide and sodium chloride, with > 99.5% purity, were obtained from Merck and used as received. The water used in the experiments was first pre-treated with a Milli-RO 10 Plus system and further purified with a Milli-Q PLUS 185 system. When used in the surface force apparatus it was deaerated for at least one hour using a water jet pump. All glassware was cleaned by Hellmanex

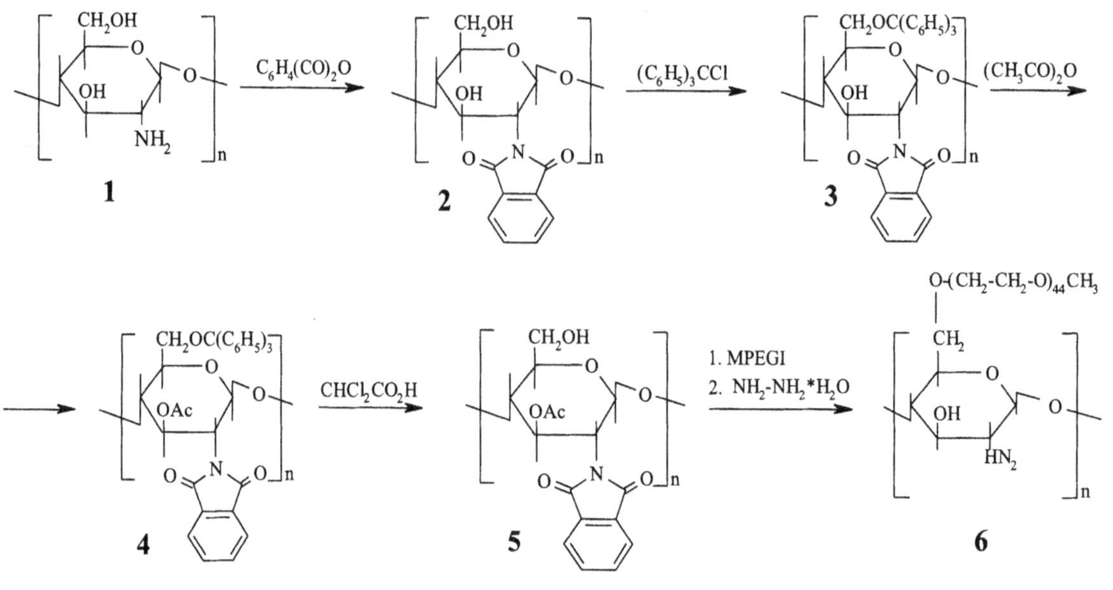

where: $C_6H_4(CO)_2O$ - phthalic anhydride; $(C_6H_5)_3CCl$ - triphenylmethylchloride; MPEGI - methoxypoly(ethylene glycol) iodide; $(CH_3CO)_2O$ - acetic anhydride; $CHCl_2CO_2H$ - dichloroacetic acid

Fig. 1 A scheme of the synthesis of chitosan-EO_{45} oligomers of regular structure

solutions for 24 hours. It was then rinsed with purified water and finally blow-dried with nitrogen gas.

X-ray Photoelectron Spectroscopy

The adsorbed amount of the oligomer on mica surfaces was determined by means of a Kratos AXIS-HS X-ray photoelectron spectrometer (XPS also called ESCA) equipped with a hemispherical analyzer. Photoelectron emission was induced by non-monochromatic X-rays (Al Kα, 1486.6 eV) that were emitted from a dual anode. Electrostatic lenses were used to collect the photoelectrons that were emitted normal to the surface plane. The quantification was carried out using the sensitivity factors for K2p, C1s and N1s provided by Kratos. The samples used for XPS analysis were prepared in the following way. Freshly cleaved mica pieces were immersed in 10 mL beakers containing 0.1 mM NaCl and the appropriate amount of the chitosan-EO_{45} oligomer for 40 minutes. Next, the beakers with the surfaces were immersed in 2.5 liters of 0.1 mM NaCl and withdrawn from this solution. The surfaces were then dried by a nitrogen jet. All surface preparation steps were carried out inside a dust-free laminar flow cabinet. Desorption of the oligomer from surfaces precoated with chitosan-EO_{45} was also investigated in NaCl solutions of different ionic strengths. The surfaces were left in these solutions for up to 12 hours, and then withdrawn and dried as described above.

Surface Force Aparatus

The forces acting between muscovite mica surfaces coated with chitosan-EO_{45} were studied with a surface force apparatus, SFA, using the Mark IV [14] model. The technique is described in detail in [15]. The results of the surface force measurements are plotted as force normalized by the geometric mean radius, F/R, as a function of surface separation, D. The distance resolution is about 2 Å, and the F/R detection limit is about 1×10^{-5} N/m. We note that the force normalized by radius for the crossed cylinder geometry used in the experiments is related to the free energy of interaction per unit area (G_f) between flat surfaces ($F/R = 2\pi G_f$). This relation is valid when the radius of the surfaces (≈ 2 cm in our experimental set-up) is much larger than the range of the measured forces, and when the surfaces remain undeformed. The force measuring system is unstable once the slope of the force, dF/dD, exceeds the spring constant. When this occurs a "jump" to the next mechanically stable region of the force curve takes place.

Results and Discussion

The average structure of the chitosan-EO_{45} oligomers is depicted in Fig. 2.

The degree of O-substitution of MPEG to the monosaccharide residues of chitosan (DS, %) was calculated by two

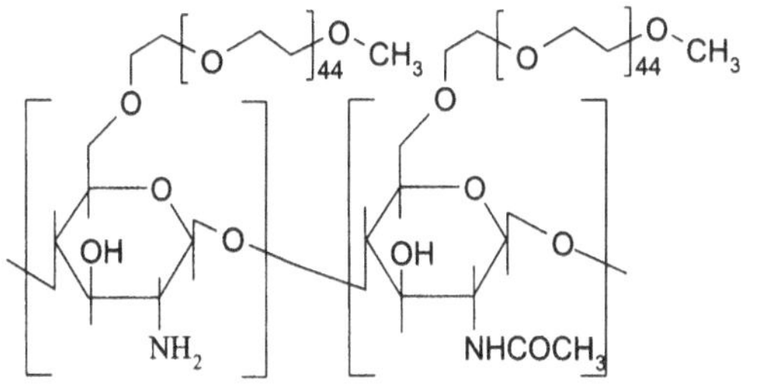

Fig. 2 The structure of the chitosan-EO_{45} oligomers

independent methods, according to the content of primary amino groups and according to the content of PEO (PEG) units in the oligomers determined experimentally [12]. The data characterizing the composition of the chitosan-EO_{45} oligomers are: content of primary amino groups 0.55%, content of PEO units 92%, DS 96%. Thus, 72% of the monosaccharide residues are deacetylated and the obtained product carries one grafted PEO chain on close to each C-6 position of the sugar units present in chitosan. The oligomers have well defined regular structure because of regioselective modification of chitosan through C-6 position only. This was achieved by the use of effective protection of both amino and C-6 hydroxyl groups in the course of the multistep synthesis (see Fig. 1, Scheme 1). The amino groups of the initial chitosan remained unchanged during the modification process thus providing certain cationicity to the chitosan-EO_{45} oligomers.

The GPC chromatogram of the chitosan-EO_{45} oligomers is illustrated in Fig. 3. Analysis of the graft oligomers by contemporary GPC technique revealed that the average molecular weight of the sample is very low. The oligomers consist mainly of conjugates of glucoseamine and MPEO as well as of dimers and trimers of these conjugates, but some larger species with up to 15 chitosan monosaccharide residues with grafted MPEG are present in the sample. A small peak corresponding to unmodified chitosan oligomers is also noticeable.

The main parameters from GPC with triple detection characterizing the chitosan-EO_{45} are as follows: the average molecular weights $M_n = 2770$ g/mol and $M_w = 3790$ g/mol, the hydrodynamic radius of the graft oligomers is 1.8 nm, and the intrinsic viscosity is 0.105 dl/g. The breakdown of the chitosan backbone is consistent with earlier findings [16], and thought to occur due to the amine group protection – deprotection procedure. We suspect that residual Ag_2O has an additional effect on the degradation of the chitosan backbone explaining why oligomers only are received in this case.

X-ray Photoelectron Spectroscopy, XPS

The cationic chitosan-EO_{45} oligomers are expected to adsorb by electrostatic attraction to the negatively charged mica surface. Thus, the chitosan unit will be oriented towards the mica surface with the PEO chains facing outwards. Mica surfaces coated with adsorbed layers of chitosan-EO_{45} were analyzed by XPS as a function of the chitosan-EO_{45} concentration used during the adsorption step. The nitrogen N1*s* peak emanating from the aminosugar groups was used for quantifying the amount of such groups adsorbed to the mica surface, and the carbon C1*s* peak was used for quantifying the amount of poly(ethylene oxide) chains associated with the surface. The method used for quantifying the adsorbed amount is based on using the lattice potassium ions within the mica crystal as an internal standard as described in detail by Rojas et al. [17]. We note that in the analysis the carbon content of the sugar units was considered.

The results obtained from the XPS analysis are illustrated in Fig. 4. The results clearly demonstrate that

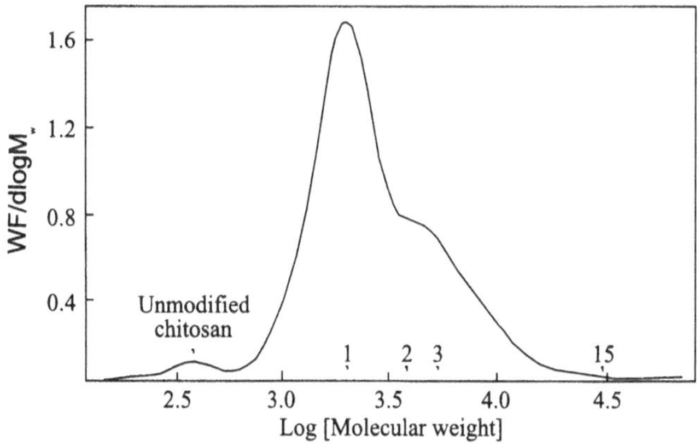

Fig. 3 The molecular weight distribution of chitosan-EO_{45} as determined by GPC. The peak corresponding to unmodified chitosan as well as the peaks corresponding to monomers, dimers, and trimers and 15-mers of conjugates of glukosamine and MPEO of chitosan-EO_{45} are indicated

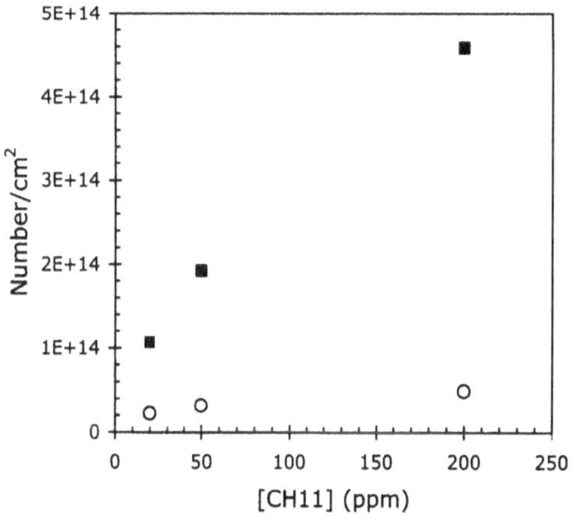

Fig. 4 The number of nitrogen atoms, which equals the number of sugar units, per cm^2 (*filled squares*) and the number of poly(ethylene oxide) chains per cm^2 (*unfilled circles*)

the number of sugar units per unit surface area significantly exceeds the number of PEO chains. This result demonstrates a strong preference for adsorption of the unmodified chitosan fraction over that of the chitosan-EO_{45} oligomers. Further, this preference increases as the surface coverage increases even though also the number of chitosan-EO_{45} oligomers increases with the concentration of oligomers in the adsorbing solution. This strongly indicates that repulsion between the adsorbed PEO chains disfavors the adsorption of the grafted cooligomer over that of the unmodified chitosan.

We note that XPS measurements may result in some degradation of the adsorbed layer due to the X-ray bombardment. However, repeated measurements at the same position provided the same results, indicating that this effect is unimportant and cannot explain that significantly more sugar groups than PEO chains are detected on the surface.

The radius of gyration of the PEO chain in bulk solution can be calculated from [18]:

$$R_g = 0.0215 M_w^{0.583} (\mathrm{nm}) \,. \tag{1}$$

This gives a radius of gyration of the PEO chain of about 1.8 nm, meaning that an undistrurbed coild would cover a surface area of $\pi R_g^2 \approx 10\,\mathrm{nm}^2$. The ratio between the area of the undisturbed PEO coil and the area per PEO chain adsorbed on the surface is a measure of the extension of the chain away from the surface, where a value larger than one indicates that the interaction between neighboring adsorbed PEO chains forces the chain to adopt a stretched conformation away from the surface. The results of this calculation are illustrated in Fig. 5.

Clearly, the repulsion between the adsorbed PEO chains is significant which forces them to adopt strongly stretched conformations, a brush structure. The degree of

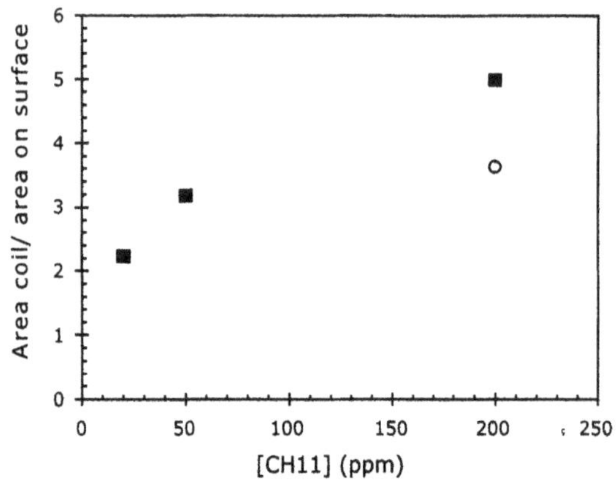

Fig. 5 The area for an undisturbed PEO coil divided by the area per adsorbed PEO chain as a function of the concentration of the polymer sample in the solution. The *unfilled circle* illustrates the result after allowing a preadsorbed layer to desorb in 100 mM NaCl for 12 hours

Fig. 6 A schematic illustration of the adsorbed layer. The *circles* represent sugar units and the *thick lines* the stretched PEO chains

stretching increases with increasing polymer concentration in solution. These results support the hypothesis that it is the steric intralayer repulsion between the PEO chains that causes the strong preferential adsorption of the unmodified chitosan. However, we note that the adsorption of chitosan-EO_{45} oligomers is sufficiently large to give rise to a brush structure, as schematically shown in Fig. 6.

Desorption Experiments

Desorption of the adsorbed layer in salt solutions were also quantified by XPS. It was found that no measurable amount of oligomer was desorbed by immersing a preadsorbed layer in NaCl solutions up to a concentration of 10 mM, whereas immersion in 100 mM NaCl for 12 hours resulted in removal of about 23% of the polymer. Hence, the chitosan-EO_{45} oligomer is surprisingly strongly attached to the mica surface, which indicates a significant non-electrostatic surface affinity between the sugar group and the mica surface, as also observed for other sugars groups [19]. One would expect that such an adsorbed layer would act as an efficient steric stabilizer. This aspect can be further investigated by surface force measurements.

Surface Force Measurements

The forces acting between two mica surfaces across a 0.1 mM KBr solution before and after addition of 20 ppm of chitosan-EO_{45} oligomers are illustrated in Fig. 7. Before addition of the EO_{45} oligomers a long-range repulsive force is dominating the interactions. The decay length of this force is consistent with that of a double-layer force and it is thus due to overlap of the diffuse ionic clouds outside the negatively charged mica surfaces. At shorter separations, from about 5 nm, the attractive van der Waals force becomes dominant and the this force causes the surfaces to jump into contact, which is due to a mechanical instability of the force measuring system. These results are in agreement with those reported for similar systems before, see e.g. [20, 21].

The forces experienced after adding 20 ppm of chitosan-EO_{45} oligomers are quite different. Firstly, the repulsive double-layer force is significantly weaker, which is due to adsorption of cationic groups on the negatively charged surface. Secondly, at distances below 10 nm, the

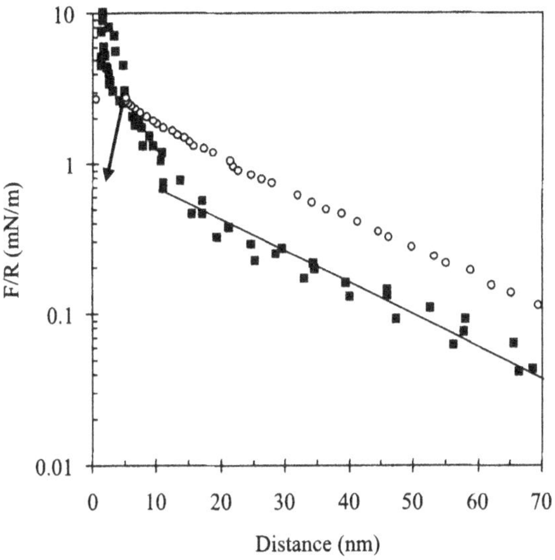

Fig. 7 Force normalized by radius as a function of surface separation. The forces were measured between muscovite mica surfaces across a 0.1 mM KBr solution (*unfilled circles*) and in the same solution after addition of 20 ppm chitosan-EO_{45} oligomers. The *arrow* represents the jump into contact to $D = 0$ due to the action of the attractive van der Waals force

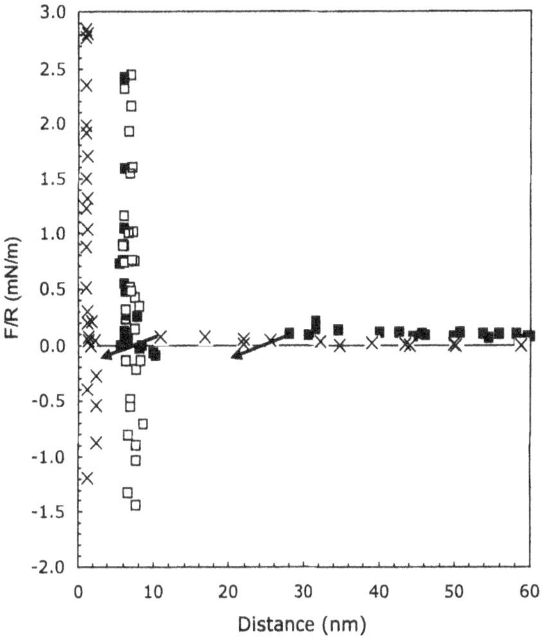

Fig. 8 Force normalized by radius as a function of surface separation. The forces were measured between muscovite mica surfaces across a 0.1 mM KBr solution containing 50 ppm chitosan-EO_{45} oligomers. *Filled* and *unfilled symbols* represent forces measured on approach and on separation, respectively. The forces measured between mica surfaces coated with polymeric chitosan are represented by *crosses*. The *arrows* represent inward jumps due to the action of an attractive force

forces are dominated by a steric repulsion due to compression of the extended PEO chains. The final separation obtained under a high compressive force is about 4 nm, corresponding to a 2 nm thick layer on each surface.

An increase in the chitosan-EO_{45} oligomer concentration to 50 ppm results in a decrease in the repulsive double-layer force, and the surfaces now appears to be close to uncharged, see Fig. 8.

The suppression of the repulsive double layer force makes a weak attractive force component measurable. This force causes the surfaces to jump from a separation of about 30 nm to about 8 nm. The origin of this attraction is at present unclear, but it may be suggestive of a weak positive charge on one surface and a weak negative charge on the other. The force encountered on further compression is very steep. On separation a weak adhesive minimum is observed at a separation of 10 nm, i.e. the uncompressed layer thickness on each surface is about 5 nm. For comparison, the forces measured between mica surfaces coated with a chitosan polymer (M_w 150.000 g/mol) are shown. The strong electrostatic affinity between the polymer and the surface results in a very thin adsorbed layer, significantly thinner than that obtained in presence of the chitosan-EO_{45} oligomer, demonstrating that the steric barrier in the latter case is due to the adsorbed PEO-chains.

A further increase in the chitosan-EO_{45} oligomer concentration to 200 ppm, results in a further increase in the short-range repulsion, as illustrated in Fig. 9.

The repulsion commences at a separation of just below 25 nm, and the compressed layer thickness is about 7–8 nm. No attraction is observed upon separation. The results obtained with XPS indicates that the PEO chains anchored to the surface should be quite extended at this bulk polymer concentration and it is therefore of interest to compare the measured forces with those expected for polymer brushes interacting in a good solvent.

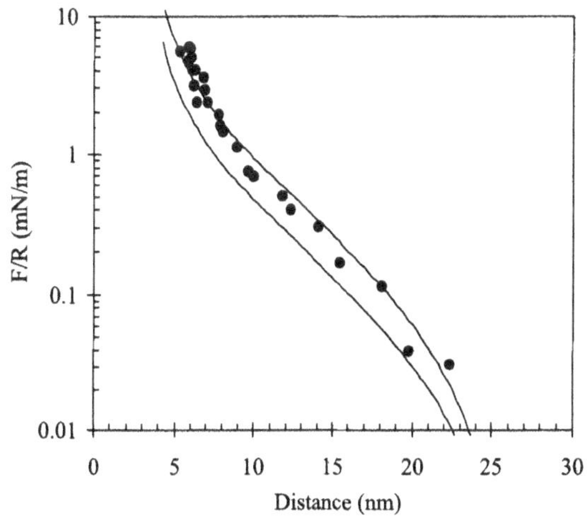

Fig. 9 Force normalized by radius as a function of surface separation. The forces were measured between muscovite mica surfaces across a 0.1 mM KBr solution containing 200 ppm chitosan-EO_{45} oligomer. The *lines* are theoretically calculated force curves, for details see text

One way is to use scaling theory that predicts that the pressure $P(D)$ between two flat surfaces coated with polymer brushes in a good solvent as given by [22]:

$$P(D) \propto \frac{kT}{s^3}\left[\left(\frac{D^*}{D}\right)^{9/4} - \left(\frac{D}{D^*}\right)^{3/4}\right], \qquad (2)$$

where Eq. 2 is valid provided the separation, D, is less than D^* (where D^* is twice the length of the polymer tail), and s is the linear distance between the anchored chains on the surface. The scaling theory does not provide the value of the prefactor, but it is expected to be of the order of unity. For the interactions between two crossed cylinders, which is the experimental geometry, this relation is, according to the Derjaguin approximation [23, 24], modified to:

$$\frac{F(D)}{R} = -2\pi \int_{D^*}^{D} P(D')\mathrm{d}D' . \qquad (3)$$

When calculating the theoretical force curves shown in Fig. 9 we used the XPS results to obtain the value of s and twice the length of the polymer brush was set to 23 nm. The anchoring points for the grafted PEO-chains were set at a distance 1.5 nm away from the mica surface. This value was chosen to be equal to the undisturbed layer thickness of the adsorbed chitosan polymer without grafted side-chains. When the prefactor is set equal to 1, the lower curve in Fig. 9 is obtained, and when a prefactor equal to two is used the upper curve is obtained. Clearly, the agreement between theory and experiment is satisfactory.

The interactions between surfaces coated with a high density of chitosan-EO_{45} oligomers, as illustrated in Fig. 9, are similar to those obtained in previous studies using lysine end-capped PEO chains with similar chain length adsorbed to mica [25], and for the (butylenes oxide)$_8$–(ethylene oxide)$_{41}$ copolymer adsorbed to non-polar surfaces [26].

Conclusions

Positively charged chitosan-EO_{45} oligomers have been prepared using a novel synthesis route. These oligomers adsorb readily to negatively charged surfaces. XPS analysis have shown that the small fraction of unreacted chitosan is strongly enriched in the surface layer, which is due to the strong steric intralayer repulsion between the PEO chains of the chitosan-EO_{45} oligomers. Nevertheless, the surface density of PEO chains are sufficiently high to create a brush structure that generates strong steric repulsive forces, similar to those observed for tightly packed copolymers on non-polar surfaces. Thus, the chitosan-EO_{45} oligomers can be used for modifying the surface properties of negatively charged surfaces and as a steric stabilizer.

Acknowledgement This work was carried out as a part the Marie Curie Network "Self-assembly" under confinement, "SOCON" and travel support for Natalija Gorochovceva is acknowledged from the Swedish Royal Academy of Sciences. PC and AD acknowledge financial support from the Swedish Research Council, VR.

References

1. Fleer GJ, Cohen Stuart MA, Scheutjens JMHM, Cosgrove T, Vincent B (1993) Polymers at Interfaces. Chapman & Hall, London
2. Bijsterbosch HD, Cohen Stuart MA, Fleer GJ (1998) Macromolecules 31:8981
3. van der Linden CC, Leermakers FAM, Fleer GJ (1996) Macromolecules 29:1000
4. De Sousa Delgado A, Léonard M, Dellacherie E (2001) Langmuir 17:4386
5. Huang N-P, Michel R, Voros J, Textor M, Hofer R, Rossi A, Elbert D, Hubbel JA, Spencer ND (2001) Langmuir 17:489
6. Brink C, Österberg E, Holmberg K, Tiberg F (1992) Colloids Surf 66:149
7. Liang W, Bognolo G, Tadros TF (2000) Langmuir 16:1306
8. Tsukruk VV (2001) Adv Mater 13:95
9. Claesson PM, Ninham B (1992) Langmuir 8:1406
10. Dedinaite A, Ernstsson M (2003) J Phys Chem B 107:8181
11. Claesson PM, Blomberg E, Paulson O, Malmsten M (1996) Colloids Surf A 112:131
12. Gorochovceva N, Makuska R (2004) European Polymer J 40:685
13. Nishimura S, Kohgo O, Kurita K (1991) Macromolecules 24:4745
14. Parker JL, Christenson HK, Ninham BW (1989) Rev Sci Instrum 60:3135
15. Israelachvili JN, Adams GE (1978) J Chem Soc Faraday Trans 1 74:975
16. Holappa J, Nevalainen T, Savolainen J, Soininen P, Elomaa M, Safin R, Suvanto S, Pakkanen T, Masson M, Loftsson T, Järvinen T (2004) Macromolecules 37:2784
17. Rojas O, Ernstsson M, Neuman RD, Claesson PM (2000) J Phys Chem B 104:10032
18. Devanand K, Selser JC (1991) Macromolecules 24:5943
19. Claesson PM, Christenson HK, Berg JM, Neuman RD (1995) J Colloid Interface Sci 172:415
20. Pashley RM (1981) J Colloid Interface Sci 83:531
21. Pashley RM (1981) J Colloid Interface Sci 80:153
22. de Gennes PG (1987) Adv Colloid Interface Sci 27:189
23. Derjaguin B (1934) 69:155
24. Israelachvili JN (1991) Intermolecular and Surface Forces. Academic Press, London
25. Claesson PM, Gölander CG (1987) J Colloid Interface Sci 117:366
26. Schillén K, Claesson PM, Malmsten M, Linse P, Booth C (1997) J Phys Chem 101:4238

Progr Colloid Polym Sci (2006) 132: 131–137
DOI 10.1007/2882_036

Published online: 8 March 2006

Anetta Walawska
Edward Rybicki
Barbara Filipowska

Physicochemical Changes on Wool Surface after an Enzymatic Treatment

Anetta Walawska (✉) · Barbara Filipowska
Institute of Textile Architecture,
276 Piotrkowska St., 90-950 Lodz, Poland
e-mail: iat@iat.com.pl,
anetta_walawska@poczta.onet.pl

Edward Rybicki
Department of Textile Finishing, Technical University of Lodz, 116 Zeromski St., 90-924 Lodz, Poland

Abstract A new ecological method has been proposed for the shrink-proof finishing of wool, based on an enzymatic pre-treatment and chitosan deposition, ensuring dimensional stability after multiple washing processes. It was demonstrated that an enzymatic treatment using highly selected proteolitic enzymes causes a distinct decrease in the shrinkage of wool fabric. The present work is aimed at analyzing the physicochemical changes in the wool surface following an enzymatic treatment. In order to characterise these changes measurements of the electrokinetic potential of wool samples and the enzymatic bath were performed. An increase in the electronegative PCD potential of wool samples and the enzymatic bath during the process was observed. Also, the enzymatic treatment causes a change in the isoelectric point (IEP) value of wool, as obtained by the electrokinetic measurements. The amount of the individual amino acids released during the enzymatic treatment of wool was monitored by the HPLC method. Moreover, an XPS analysis of enzymatic-modified wool fabric samples and contact angle measurements were performed. The data obtained by the XPS method allowed comparison of the changes in the elemental concentration on the wool surface after enzymatic treatment. The results of the contact angle measurements demonstrated an increase in the wettability of the modified wool surfaces.

The results obtained proved that very good correlations exist between the value of electrokinetic potential, HPLC results, XPS measurements, contact angle values and changes in the isoelectric point (IEP).

Keywords Chitosan · Electrokinetic measurements · Enzymatic treatment · Surface properties · Wool

Introduction

Wool is a natural protein fiber characterized by the scaly structure of its external surface-cuticle (Fig. 1). This structure, i.e. the stiffness of cuticle and smoothness of the epicuticle as well as the ability of wool to contract, causes the shrinking of wool fabric during mechanical washing processes.

The oxidative plasma treatment of wool, at low temperature and pressure, modifies to a large extent the cuticle surface of the fibres, improving its surface wettability, dyeability, fibre cohesion and shrink-resistance [3–8]. The use of plasma treatment in wool finishing is an environmentally acceptable alternative to the conventional chlorination finishing process. However, the shrink-resistant properties obtained by plasma treatment do not impart a machine-washable finish – one of the end-users' demands.

The use of enzymes to achieve wool shrink resistance, better whiteness, and improved handle is of considerable

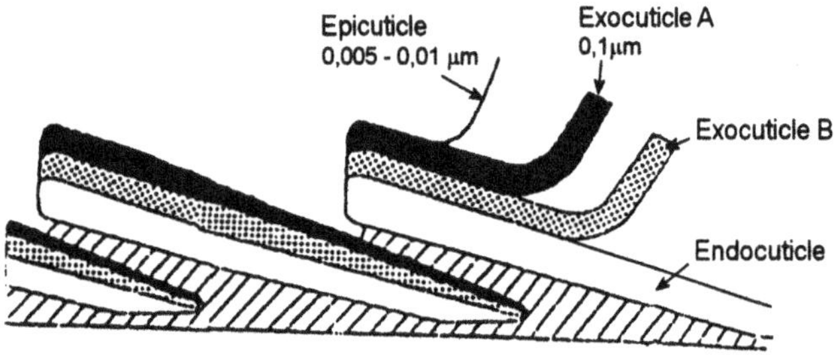

Fig. 1 The schemata of wool fibre cuticle; thickness 0.3–1 μm [1, 2]

interest [9–12]. However, if enzymes are applied at levels that provide the required shrink resistance, wool fibres are often unacceptably damaged [13–15].

Hence, a new ecological method for the shrink-proof finishing of wool based on an enzymatic pre-treatment and chitosan deposition has been proposed. It has been demonstrated that an enzymatic treatment of highly selected proteolytic enzymes causes a distinct decrease in the shrinkage of wool fabric. Chitosan, deposited on the wool fabric surface, stabilises the obtained shrink-resistant effect and restores the "full handle" typical of standard finished wool fabrics [16–21]. Moreover, the deposition of chitosan, a cationic polymer, causes a change in the wool surface charge from electronegative to electropositive. This phenomenon affects the changes in the mechanism of wool dyeing using anionic dyestuffs, soiling, and dirt removal [20, 21].

The existing knowledge concerning the character of physicochemical changes on wool surface as a result of an enzymatic action is still insufficient.

In this paper, studies of enzyme-modified wool fabric surfaces using the PCD potential, XPS analysis, HPLC analysis of amino acids, and contact angle measurements are presented.

Materials and Methods

Raw wool fabric samples of a plain-weave (155 g/m^2), supplied by the Wool Company "9 May" S.A., Lodz, Poland, were used in this research. These wool samples were thoroughly extracted with a mixture of ethanol/benzene (32.4/67.6 wt %) for 8 hours using Soxhlet apparatus. The extracted wool fabrics were washed twice with 96% ethanol and rinsed twice with deionised water. Finally, the wool fabrics were dried in an oven at 50 °C for 1 hour and were then air dried.

For the enzymatic treatment, a selected mixture of proteolytic enzymes was used – Perizym LAN (Dr Petry, GmbH, Germany). Its optimum activity was at pH 8–8.5 and a temperature of 50–55 °C. The optimum treatment time was in the range of 30–40 min, as advised by the manufacturer. The mean proteolytic activity determined by the Anson method [22] was 2.676 units/cm^3. A liquor ratio of 10 : 1 was applied. To deactivate the enzymes, a bath with wool samples was heated to 85 °C and kept at this temperature for 20 min. The concentration of enzymes in the bath is given in weight per cent related to the weight of fabric – e.g. 4% of weight of fabric (o.w.f.) of Perizym LAN.

Chitosan (*CHITON* Co. Ltd., Gdynia, Poland), having a deacetylation degree of 83% and viscosity (1% water solution of acetic acid, pH 3.5–4) of 20 mPas, was deposited on wool fabric pre-treated with enzymes using the pad-dry-fix method [19].

The shrink-resistance testing was carried out in a domestic washing-drying machine, using a non-ionic washing agent solution (LAVAN JET, Textilcolor AG) at 60 °C, pH 5.3. The changes in dimensions of wool samples after multiple washing were assessed in warp – the longitudinal direction of the fabric, and weft – the transverse direction of fabric, in accordance with the standard procedure described elsewhere [23, 24].

Scanning electron micrographs (SEM) were obtained using a JSM 5500 LV (Jeol, Japan) electron microscope. The observations were performed in a secondary electron (SE) and in a backscattering electron (BSE) mode at a low vacuum pressure of 12 kPa.

The most commonly used approach to determine the contact angle of a liquid drop to a solid surface is the sessile drop method, in which a goniometer is used to measure the angle given by the line appearing from the solid surface and the tangent arising from the three-phase point. The accuracy of the method, however, is limited [25]. Practically, all the contact angle measurement methods proposed in the presence of fibrous materials should be used with caution and the results regarded as an approximation rather than as absolute values. Nevertheless, the sessile drop method was often applied to determine the contact angles of keratin fibres [26–29]. The water contact angle (Θ) of shrink-proof finished wool fabric surface was measured on a sessile drop. 1 μl of distilled water was dropped on the surface of modified wool fabric samples. The contact angle value of one sample was calculated as the arithmetic mean of 10 measuring results.

The streaming potential (proportional to the zeta potential) of wool fibres was determined using the Particle Charge Detector (PCD), a product of Mütek GmbH, Germany. The induced streaming potential was measured using two gold electrodes connected with an electronic analyser. The potential was recorded by a measuring millivoltmeter. Powdered fabric samples (wool yarn placed in the microtome and then cut into portions of about 40–60 μm) were suspended in distilled water (100 mg/100 ml). Following careful stirring, the samples were placed in a cylindrical vessel of the apparatus in order to determine the value of the streaming potential. The determination of an isoelectric point (IEP) of modified wool surface was also carried out using the Particle Charge Detector (Mütek GmbH, Germany). The changes in the PCD potential values were determined in the pH range from 2.30 to 8.00. In order to adjust the pH values, water solutions of 0.1 M HCl and 0.1 M NaOH were used.

The studies of the composition and state of chemical elements on enzyme-modified wool surfaces (carbon, nitrogen, oxygen, sulphur) as compared to untreated ones were performed by means of XPS analysis (Vacuum System Workshop Ltd., England) using non-monochromatized AlKα radiation with energy 1486.6 eV, 10 kV and 200 W. The base pressure in the analysis chamber was 3×10^{-6} Pa [30]. XPS spectra were acquired in the constant analyser transmission mode with energy of electron transmission 22 eV.

The HPLC (High Performance Liquid Chromatography) analysis of enzymatic bath was performed using a chromatographic column C18 DABS (Beckman, USA) modified to amino acid analysis.

Results and Discussion

As can be seen from Fig. 2, the finishing of wool fabrics based on an enzymatic pre-treatment and chitosan deposition causes a decreased tendency in the wool samples to shrink during the mechanical washing process. The enzymatic pre-treatment greatly improves the shrink-proof properties of wool fabric. The treated fabric becomes more pleasant to the touch, but at the same time it is too flabby. The amount of chitosan deposited on wool fabric pre-treated with enzymes is practically of no essential consequence to its shrink-resistance properties, as it is also the case for fabric without enzymatic pre-treatment, but a slight decrease in shrinkage was observed. Chitosan deposited on wool fabric preliminarily pre-treated with enzymes restores its "full handle" and improves the appearance of the wool fabric surface after washing [16, 18].

Figure 3 presents SEM pictures of untreated, enzyme-treated, and wool fibres after enzymatic treatment and chitosan deposition. The characteristic scaly structure of the

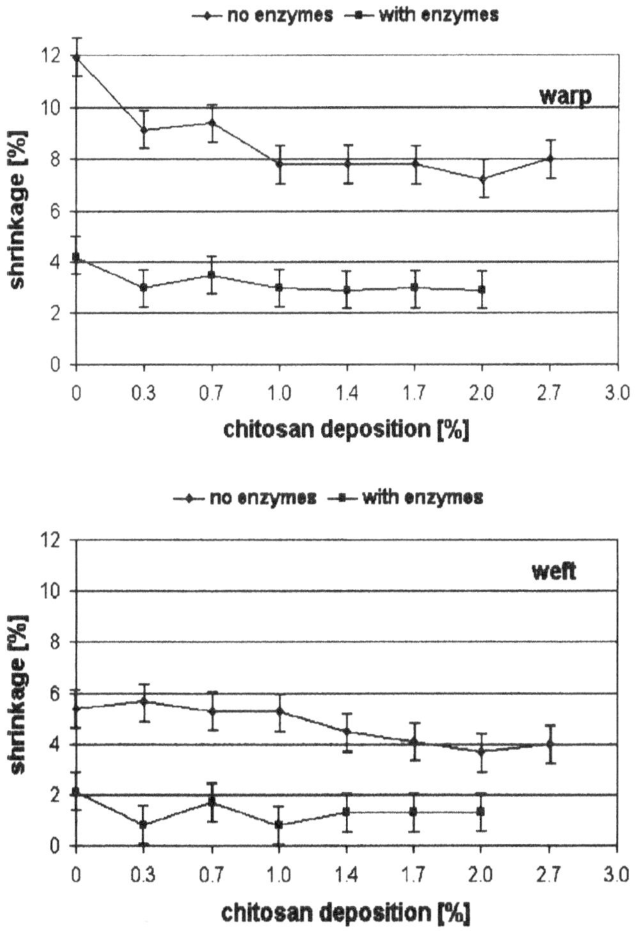

Fig. 2 Changes in the linear dimensions of wool fabrics after three washings at 60 °C versus chitosan deposition

wool fibre surface is evident in the SEM micrographs of blank wool fibre samples (Fig. 3a). Enzymatic treatment of wool leads to a rougher surface, with a reduced scales' height. Chitosan application seems to lower the roughness; however SEM observations did not show visible chitosan on the fibre surface even when high chitosan amounts had been applied (up to 1.7%). This implies that chitosan forms a very thin film on wool fibres which is sufficient

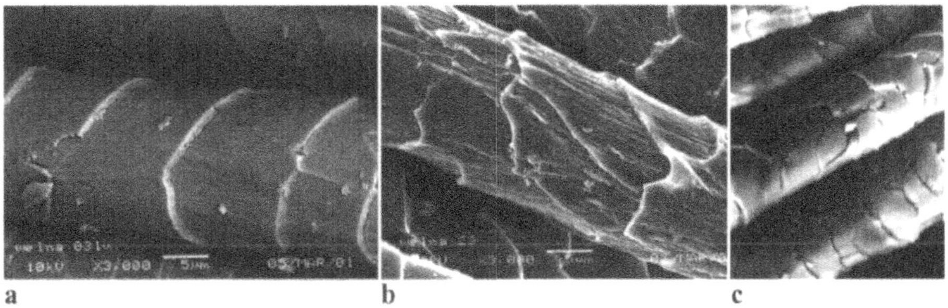

a b c

Fig. 3 SEM micrographs: **a** blank wool fibre – magnification 3000×; **b** wool fibre after enzymatic treatment – 4% o.w.f. of Perizym LAN, 30 min – magnification 3000×; **c** wool fibre after enzymatic treatment – 4% o.w.f. of Perizym LAN, 30 min and 1.7% o.w.f. of chitosan deposition – magnification 2000×

to provide an additional shrink resistant effect, as has been also previously reported [7, 31].

As a consequence of the interaction of enzymes with wool, changes in the physicochemical properties of modified wool surface are observed.

Figure 4 shows an increase in the electronegative PCD potential, both for the enzymatic bath (from −70 mV to about −310 mV) and the wool fabric sample (from −180 mV to −670 mV), as a function of the time of enzymatic treatment. This increase in PCD potential can be ascribed to an increased wool surface acidity, which occurs as a result of an increased acid groups' exposition at the wool surface.

The results of the PCD measurements confirm the acidic character of wool surface and an enzymatic bath and demonstrate that the optimum time of enzymatic treatment amounts to about 35–45 min. After that period a plateau in the PCD potential for wool fibres can be seen.

Also, the enzymatic treatment causes a change in the isoelectric point (IEP) value of wool, as demonstrated by the electrokinetic measurements (Fig. 5). With an increase in enzyme concentration the IEP value shifts in the direction of more acidic pH (from 3.8 for untreated wool to 3.7 for wool after 1% enzymatic treatment, and 3.45 for wool

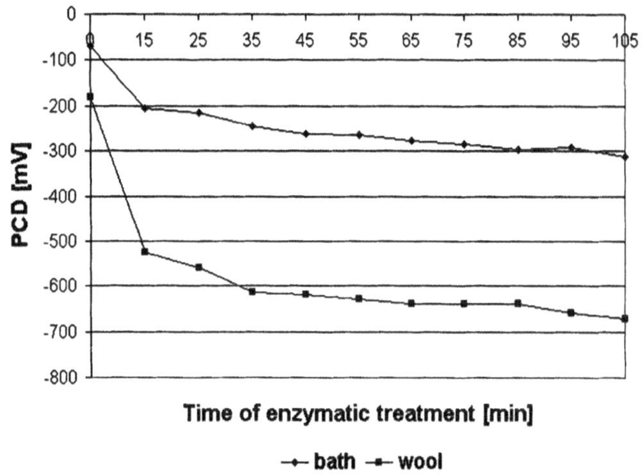

Fig. 4 Changes in the PCD potential during the enzymatic treatment of wool (4% o.w.f. of Perizym LAN)

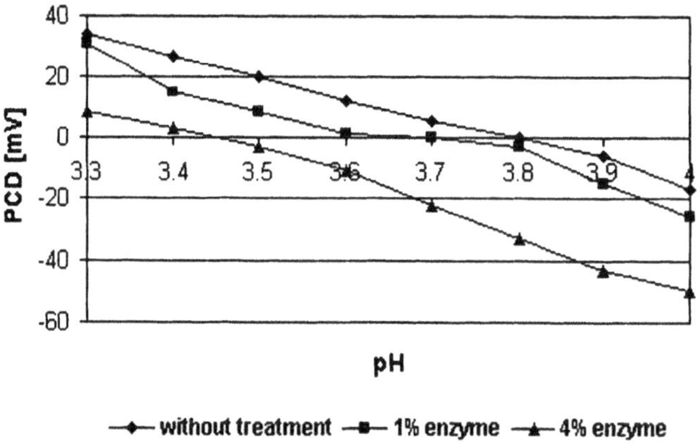

Fig. 5 Changes in the IEP value of wool after enzymatic treatment (1%, 4% o.w.f. of Perizym LAN, 60 min)

Table 1 Changes in the elemental concentration for untreated and enzymatic-treated samples (4% o.w.f. of Perizym LAN, 60 min) assessed by XPS method

		Untreated wool				Wool after enzymatic treatment			
		Binding energy [eV]	FWHM * [eV]	Elemental concentration [%]	Share of line [%]	Binding energy [eV]	FWHM * [eV]	Elemental concentration [%]	Share of line [%]
C 1*s*	C−C	284.60	2.22	78.6	76.9	284.60	2.14	79.4	77.4
	C−O	286.10	2.22		10.5	286.10	2.14		12.7
	C=O	287.60	2.22		11.6	287.60	2.14		8.4
	O−C=O	289.10	2.22		1.0	289.10	2.14		1.5
N 1*s*	**	399.92	2.35	6.5	100.0	399.56	2.48	5.3	100.0
S 2*p*	Sulphide	163.00	1.70	2.1	37.3	163.00	1.70	1.6	52.3
	Sulphide	164.00	1.70		50.9	164.00	1.70		28.1
	SO_2, $-SO_3$, $-SO_4$	168.00	1.70		11.8	168.00	1.70		19.6
O 1*s*	OH	531.68	2.29	12.7	70.0	531.54	2.30	13.8	60.8
	OH_2	533.28	2.29		21.7	533.00	2.30		27.0
	O^{2-}	530.18	2.29		8.3	530.04	2.30		12.2

* full width at half maximum

** width of the spectral line show the presence of many groups containing *N*, but the shape of the spectrum does not permit to isolate them; typical width of the spectral line amounts to 1.7 eV

after 4% enzymatic treatment). This can be attributed to the better exposition of acid groups at the wool surface after enzymatic treatment.

A similar IEP value for untreated wool (3.6) was recently obtained by Jocic et al. [32], using EKA streaming current measurements.

Most likely the changes in IEP value, as well as in PCD potential, during the enzymatic treatment of wool are the result of enzyme-initiated oxidation reactions. As can be seen from the XPS results specified below (Table 1), a slight increase in SO_2, SO_3, SO_4 groups' concentration, from 0.248% (0.248% = 11.8% of 2.1% of total elemental concentration) for untreated sample to 0.314% (0.314% = 19.6% of 1.6% of total elemental concentration) can be observed.

The changes in PCD potential for enzymatic bath, shown in Fig. 4, correspond well with the results obtained by the HPLC method (Fig. 6). In the enzymatic bath 18 amino acids were analysed. Aspartic acid, as well as methionine and cysteine, two sulphur containing amino acids showed increased concentrations (Fig. 6).

The increase in the electronegative PCD potential can be explained by the increase in the aspartic acid concentration in the enzymatic bath released during the interaction of wool with enzymes. However, the changes in the methionine and cysteine concentration may suggest changes in the structure of the cuticle cells.

The data obtained by the XPS method (Table 1, Fig. 7) show the changes in the elemental concentration of C, N, S, and O on the wool surface after enzymatic treatment, as compared to untreated wool. The increase in the atomic concentration of carbon, from 78.6% for untreated wool to 79.4% for wool after enzymatic treatment, and the decrease in the atomic concentration of nitrogen from 6.5% for untreated wool to 5.3% for wool after enzymatic

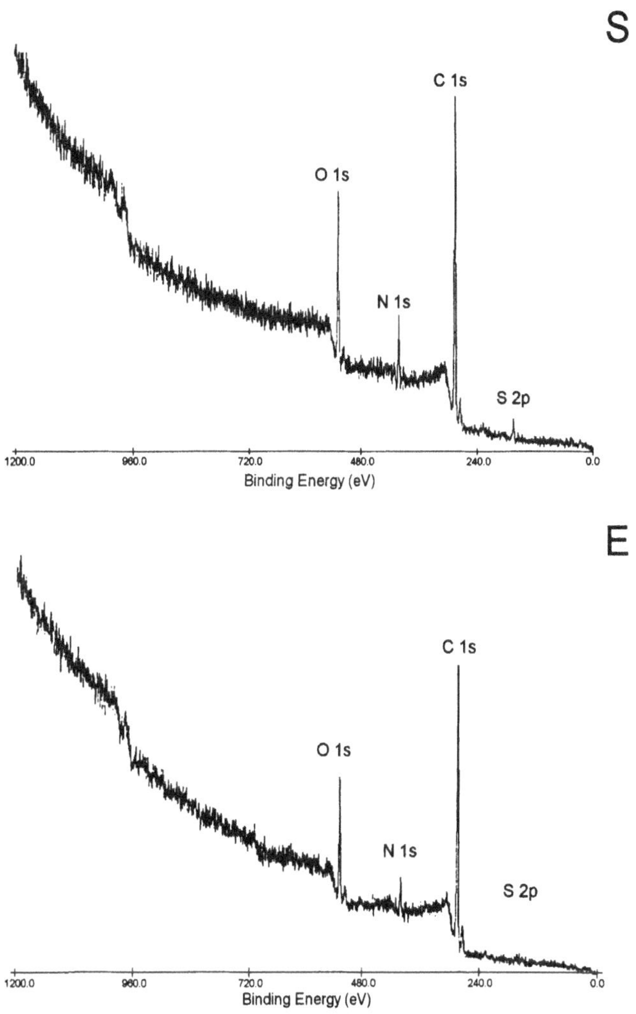

Fig. 7 The review spectrum of untreated wool fabric samples (**S**) and samples after enzymatic treatment (**E**) (4% o.w.f. of Perizym LAN, 60 min) obtained by XPS-method

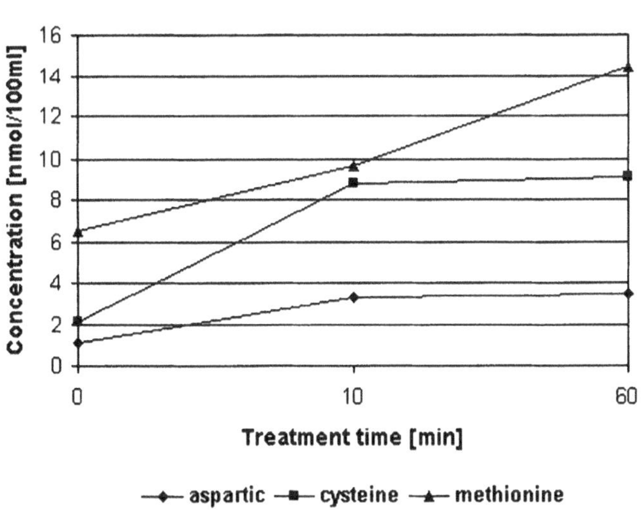

Fig. 6 Changes in amino acid concentration during the enzymatic treatment of wool (4% o.w.f. of Perizym LAN)

treatment, suggest an enrichment of acid groups on the wool surface, which may be confirmed by the C/N ratio, which increased from 12.1 for untreated to 15.0 for enzyme treated wool. The reduction in the sulphur content (from 2.1% to 1.6%) on the wool surface following enzymatic treatment corresponds well with the increase in the concentration of amino acids (containing sulphur) in the enzymatic bath (Fig. 6).

The increase in the atomic concentration of oxygen (Table 1) suggests an increased amount of hydrophilic groups on the wool surface. This fact can be confirmed by the results of the contact angle measurements (Table 2). Following enzymatic treatment the contact angle is distinctly decreased, from 122.5° for untreated wool to 108° and even 102°, but the wool fabric surface is still hydrophobic ($\Theta > 90°$). The contact angle value obtained for wool after an intensive treatment (4% o.w.f. of enzyme) equals the one for wool samples with chitosan deposited.

Table 2 Changes in the contact angle (with water) of wool modified by enzymatic treatment (1%; 4% o.w.f. of Perizym LAN, 40 min) and chitosan (deposition 1.7% o.w.f.)

Wool samples	Contact angle Θ [deg]
Untreated	122.5 ± 4
After 1% enzymatic treatment	108 ± 2
After 4% enzymatic treatment	102 ± 2
After 1% enzymatic treatment and 1.7% chitosan deposition	101.5 ± 2

The hydrophobic nature of wool is known to be due to the specific properties of its exocuticles and epicuticles. The exocuticle is hydrophobic because it contains a high degree (~ 35%) of disulphite crosslinkage in the A-layer. The epicuticle surrounds each cuticle cell of the wool fibre, and it consists of fatty acid (~ 25% by mass), the main component of which is 18-methyleicosanoic acid. Fatty acid chains of 18-methyleicosanoic acid are covalently bound to the wool fibre as a thioester to cysteine and are oriented away from the fibre to produce a "polyethylene-like" layer at the fibre surface, thus making the epicuticle hydrophobic [7, 33]. Even after the removal of grease by aqueous scouring or solvent extraction, wool maintains its hydrophobic character. The change in the contact angle for the enzyme-modified wool and chitosan deposition (101.5° ± 2) corresponds well with the data obtained by Jocic et al. (100° ± 2) [7] using the Wilhelmy balance method.

Summary and Conclusions

Based on the experimental results obtained during this work, the following conclusions can be drawn.

1. Changes in the topography of wool fibres, confirmed by SEM micrographs, can be observed as a result of the enzymatic treatment of wool fabric.
2. An enzymatic treatment causes the release of a significant amount of amino acids containing sulphur (methionine and cysteine) as well as an aspartic acid from the wool surface.
3. As a consequence of enzymatic treatment, an increase can be observed in the acidity of the wool surface, which is demonstrated by a distinct increase in the electronegative PCD potential of wool fabric, the reduction of IEP value, and by XPS results.
4. Moreover, an enzymatic treatment of wool causes an increase in the amount of hydrophilic groups at the surface. This is proved by the increase in the elemental concentration of oxygen, as obtained by XPS, and the increase in the wettability of enzyme-modified wool surfaces.
5. The results obtained prove that the PCD potential measurement is a very sensitive method for determination of the charge on a wool surface. In connection with other methods (XPS, HPLC), it allows the explanation of the phenomena which occur on modified wool fibre surfaces.

Acknowledgement This work was financially supported by the Polish State Committee for Scientific Research (research project No 7 T09B 035 20).

References

1. Virnich A, Thomas H, Blankenburg G, Hoecker H (1990) Melliand Textilberichte 11:887
2. Przygocki W, Wlochowicz A (1983) Textile Review (Przeglad Wlokienniczy) 7–8–9:279
3. Gregorski KS, Pavlath AE (1980) Textile Res Journal 50:42–46
4. Rakowski W (1989) Melliand Textilberichte 70:780–785
5. Ryu J, Wakida T, Takagishi T (1991) Textile Res Journal 61:595–601
6. Byrne KM, Godau J (1995) Proceedings of the 9th Int Wool Text Res Conf, vol IV, p 415–421
7. Jocic D, Jovancic P, Petrovic ZL, Bertran E, Navarro A, Julia MR, Erra P (2002) Proceedings of 2nd AUTEX Conference, p 299–312
8. Chi-wai K, Kwong Ch, Chun-wan Y (2003) AUTEX Research J 3(4):194–205
9. Haefely HR (1989) Textilveredlung 24:271
10. Fornelli S (1994) Melliand Textilberichte 74:120
11. Heine E, Hoecker H (1995) Rev Prog Colour 112:57
12. Cegarra J (1996) J Soc Dyers Colour 112:326
13. Riva A, Cegarra J, Prieto R (1993) J Soc Dyers Colour 109:210
14. Jovancic P, Jocic D, Dumic J (1998) J Text Inst, Part 1, Text Sci Techn 89(2):390
15. Jovancic P, Jocic D, Erra P, Molina R, Julia MR (1998) European Microscopy and Analysis 51:15
16. Rybicki E, Filipowska B, Walawska A, Wlochowicz A (1999) Polish Pat Appl No. P-33 23 27
17. Rybicki E, Filipowska B, Walawska A (1999) International Seminar Textile Science for 21st Century. June 8–9, Guimaraes, Portugal
18. Filipowska B, Walawska A, Rybicki E (2000) L'Industrie Textile 1317:21–22
19. Rybicki E, Filipowska B, Walawska A (2000) Fibres & Textiles in Eastern Europe 2–3:62–65
20. Filipowska B, Rybicki E, Walawska A (2002) In: Chitosan in Pharmacy and Chemistry, Muzzarelli RAA, Muzzarelli C (eds) ATEC Italy, p 263–270

21. Rybicki E, Filipowska B, Walawska A (2001) Proceedings of the 40th International Detergency Conference, April 30th–May 3rd, Strasbourg, France
22. Anson ML (1938) J Gen Physiol 22:79–82
23. Polish-European Standard PN-EN 25077:1998
24. Polish – European Standard PN-EN ISO 3759:1008
25. Allen JS (2003) J Colloid Interface Sci 261(2):281
26. Hoecker H (2002) Pure Appl Chem 74(3):423–427
27. McCord M (2001) National Textile Center Annual Report: November C99–S09:1–12
28. Smith JR, Eaton PJ, Waddington K, Alexander C, Tsibouklis J, Swift JA (2000) The Proceedings of The 10th Int Wool Tex Res Conf HH-2, p 1–9
29. Dupres V, Camesano T, Langevin D, Checco A, Guenoun P (2004) J Colloid Interface Sci 269:329–335
30. Scofield JH, Hartree-Slater (1976) J Electron Spectroscopy and Related Phenomena 8:129–137
31. Davidson RS, Xue Y (1994) JSDC 110:24–29
32. Radetic M, Jocic D, Jovancic P, Petrovic ZL, Thomas H (2003) The Proceedings of The 3rd AUTEX Conference Book I, p 309–314
33. Kalkbrenner K, Koerner A, Hoecker H, Rivett DE (1990) The Proceedings of the 8th Int Wool Tex Res Conf Christchurch

Progr Colloid Polym Sci (2006) 132: 138–144
DOI 10.1007/2882_026

Published online: 9 March 2006

Astrid Roosjen
Willem Norde
Henny C. van der Mei
Henk J. Busscher

The Use of Positively Charged or Low Surface Free Energy Coatings versus Polymer Brushes in Controlling Biofilm Formation

Astrid Roosjen · Henny C. van der Mei · Henk J. Busscher (✉)
Department of Biomedical Engineering, University Medical Center Groningen and University of Groningen,
Antonius Deusinglaan 1,
9713 AV Groningen, The Netherlands
e-mail: H.J.Busscher@med.umcg.nl

Willem Norde
Laboratory of Physical Chemistry and Colloid Science,
Wageningen University, Dreijenplein 6,
6703 HB Wageningen, The Netherlands

Abstract Biofilm formation on biomaterials implant surfaces and subsequent infectious complications are a frequent reason for failure of many biomedical devices, such as total hip arthroplasties, vascular catheters and urinary catheters. The development of a biofilm is initiated by the formation of a conditioning film of adsorbed macromolecules, such as proteins, followed by adhesion of microorganisms, where after they grow and anchor through secretion of extracellular polymeric substances. Adhesion of microorganisms is influenced by the physico-chemical properties of the biomaterial surface. Positively charged materials stimulate bacterial adhesion, but prevent growth of adhering bacteria. The use of low surface free energy materials did not always reduce in vitro adhesion of bacteria, but has been found beneficial in in vivo applications where fluctuating shear forces prevail, like on intra-oral devices and urine catheters. Polymer brushes have shown a very high reduction in in vitro adhesion of a great variety of microorganisms. However, for clinical application, the long term stability of polymer brushes is still a limiting factor. Further effort is therefore required to enhance the stability of polymer brushes on biomaterial implant surfaces to facilitate clinical use of these promising coatings.

Keywords Biofilm · Biomaterial centered infection · Microbial adhesion · Polymer brushes · Surface free energy

Introduction

Biomaterials are materials foreign to the human body that are used in medicine to replace, support or restore body function. Applications range from central venous and urinary catheters to more complex devices such as prosthetic joints and heart valves. The risk of biomaterial centered infection (BCI) is a key factor limiting their use [1]. The incidence of this type of infections varies for each application for instance 4% for hip prostheses [2] and 10–20% for urinary catheters (see Table 1). In BCI microorganisms are present in close association with the biomaterial surface forming a so-called biofilm. Different species of microorganisms are found in BCI that are often commensals of the skin or the intestines, for instance *Staphylococcus epidermidis*, often found in hip prosthesis infections and *Escherichia coli* in urinary catheter infections [2] (see Table 1). BCI can cause severe problems, from disfunctioning of the implanted device to lethal sepsis of the patient. Furthermore, treatment of BCI is complicated, as microorganisms in a biofilm are more resistant to antibiotics [3] than their planktonic counterparts [4]. As a consequence, the only remedy for a BCI is removal of the infected implant at the expense of considerable costs and patients suffering. A more convenient way to deal with this problem is to prevent the development of an infectious biofilm on the biomaterial surface. To achieve this, a thorough understanding of the development of biofilms is necessary.

Table 1 Incidences of infection of different biomedical implants and devices adapted from Dankert et al. [2]

Body site	Implant or device	Incidence (%)	Commonly causative bacterial species
Urinary tract	UT catheters	10–20	*Escherichia coli*
Percutaneous	CV catheters	4–12	*Staphylococcus epidermidis* *Staphylococcus aureus*
	Temporary pacemaker	4	*Staphylococcus epidermidis* *Staphylococcus aureus*
	Peritoneal dialysis catheters	3–5	*Staphylococcus epidermidis*
	Orthopedic pins*	50	*Staphylococcus aureus*
Subcutaneous	Cardiac pacemaker	1	*Staphylococcus epidermidis*
Soft tissue	Mammary prosthesis	1–7	*Staphylococcus aureus*
	Intraocular lenses	0.13	*Pseudomonas* *Staphylococcus epidermidis*
Circulatory system	Prosthetic heart valve	1.88	*Staphylococcus aureus* viridans streptococci
	Vascular graft	1.5	*Staphylococcus epidermidis* *Staphylococcus aureus* Gram negative bacteria
Bones	Prosthetic Hip	2.6–4.0	*Staphylococcus epidermidis* *Staphylococcus aureus*
	Total knee	3.5–4	*Staphylococcus epidermidis* *Staphylococcus aureus*

* Data obtained from [5] and [6]

Biofilm Formation

Although the function and appearance of biofilms in various environments may be different, all biofilms are formed according to the following basic sequence of events [7] (see Fig. 1).

1. Formation of a conditioning film of adsorbed macromolecular organic components (i.e. proteins and other organic molecules) on the substratum surface prior to microbial deposition.
2. Transport of microorganisms towards the substratum surface through diffusion, convection, sedimentation, or by intrinsic bacterial motility.
3. Initial microbial adhesion.
4. Strong attachment or anchoring of microorganisms to the substratum surface through the production of extracellular polymeric substances (EPS), mostly composed of polysaccarides [8] and proteins [9].
5. Surface growth of adhering microorganisms and continued secretion of EPS.
6. Localized detachment of isolated clumps of microorganisms caused by occasionally high fluid stress or other detachment forces operative in the environment of the biofilm.

Microbial adhesion is mediated by generic physico-chemical interactions forces as well as by specific interaction forces between cell surface structures and molecular groups on a substratum surface [10]. Generic, non-specific interaction forces include Lifshitz–Van der Waals forces and electrostatic forces, which both operate over a long range, and hydrophobic and acid-base interactions that act over a shorter range [11]. Specific interactions result in fact from non-specific forces acting on highly localized regions of the interacting surfaces over distances smaller than 1.5 nm [12].

Upon approach, organisms will be attracted or repelled by the biomaterial surface, depending on the resultant of the various interaction forces. Thus, the physico-chemical surface properties of the biomaterial, with or without conditioning film, and those of the microorganisms play a decisive role in this process. Because the size of microorganisms is in the μm range, adhesion can be described in terms of colloid science. Indeed, for several strains and species physico-chemical models like the Derjaguin-

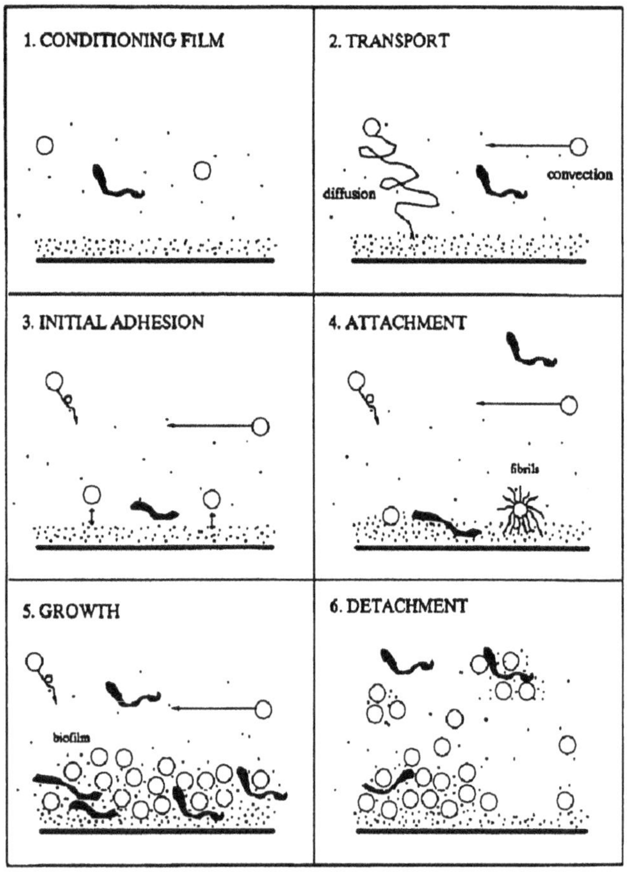

Fig. 1 Schematic, sequential presentation of the steps involved in biofilm formation. Reprinted from Meth. Enzymol. 310, Models for studying initial adhesion and surface growth in biofilm formation on surface, 523–533, Copyright 1999, with permission from Elsevier

Landau-Verwey-Overbeek (DLVO) theory of colloidal stability have been successful in qualitatively explaining microbial adhesion to solid substrata [8, 13].

Prevention of Biofilm Formation

Surgeons take considerable effort in preventing the contamination of implants with microorganisms during implantation. Although application of prophylactic antibiotics and better operation hygiene have reduced the incidence of BCI over the last four decades, still a significant number of patients suffer from such infections [2, 14].

Different strategies may prevent biofilm formation and thus BCI. In general, it is aimed to reduce the attractive force between microorganisms and a biomaterial surface by optimizing the physico-chemical surface properties of the biomaterial. For instance more negatively charged biomaterials [15], biomaterials coated with albumin [16], heparin [17, 18] or polysaccharide [19, 20] have shown to attract less bacteria. However, the most promising and most extensively studied methods to prevent biofilm formation are positively charged coatings, low surface free energy coatings and surfaces covered with a polymer brush. These three types of surfaces will be evaluated and compared in this review.

Positively Charged Coatings

Despite the fact that bacteria adhere more readily to positively charged surfaces than to negatively charged ones [21, 22], there are some aspects of biofilm formation on positively charged surfaces that deserve their further consideration. *E. coli* and *P. aeruginosa* hardly showed any growth after their adhesion to positively charged poly(methacrylate) surfaces [21], although *S. aureus* and *S. epidermidis* were able to grow on these surfaces. However, on negatively charged poly(methacrylate) surfaces, growth was found for all four species.

These observations may be explained by the strong binding of the negatively charged bacteria [23] to positively charged surfaces through attractive electrostatic interactions, impeding elongation and division needed for bacterial growth. It has indeed been demonstrated that as the binding strength of *P. aeruginosa* AK1 to substrata increases, their surface growth reduces [24].

Also in vivo, positively charged surfaces have appeared promising in certain applications. Differently charged poly(methacrylate) coated discs have been seeded with *E. coli* or *P. aeruginosa* and implanted in rats [25]. After 48 h, only 50% of the positively charged disks contained viable *E. coli*, while on all negatively charged disks viable bacteria were found. *P. aeruginosa*, however, was isolated from both positively and negatively charged surfaces, probably because this bacterium can circumvent the effect of the positive charge through production of extracellular polymeric substances.

Low Surface Free Energy Coatings

The surface free energy (s.f.e.) of a material is a measure for the work required to enlarge its surface (mJ m^{-2}). At constant pressure, constant temperature and if the surface composition remains constant, the s.f.e. equals the surface tension (γ_{SV}) of the material against its vapor [26]. To determine the value of γ_{SV} and hence of the s.f.e. of a solid, contact angles with several liquids with known s.f.e. (γ_{LV}) are required. Thomas Young's equation relates the surface free energies and the contact angle (θ) based on the force balance at the three-phase boundary [27]:

$$\gamma_{SV} = \gamma_{SL} + \gamma_{LV} \cos\theta \qquad (1)$$

Equation 1 can only be solved using different methods for which additional assumptions are required. These methods are subject to serious controversy in surface science [26] and lead to different γ_{SV} values for one and the same surface.

Therefore, literature values of γ_{SV} are mainly used as a relative measure for the s.f.e.

The type of forces determining the s.f.e. are Lifshitz–Van der Waals interactions and non-dispersive interactions like hydrogen bonding, stacking between π-electrons and ion pairing [26]. The value of γ_{SV} is determined mainly by the chemical nature and structure of the surface. Values for some constituent groups decrease in the order CH_2 ($36\,mJ\,m^{-2}$) > CH_3 ($30\,mJ\,m^{-2}$) > CF_2 ($23\,mJ\,m^{-2}$) > CF_3 ($15\,mJ\,m^{-2}$) [29]. The low s.f.e. of materials containing the CF_2 of CF_3 group may be explained by the highly electron-negative F atom that is able to withdraw electrons from the carbon backbone leading to a very inert, noble gas like configuration [30]. In water minimal interaction possibilities between this inert surface and a bacterium are expected, which should lead to minimal bacterial adhesion and easy removal of attached bacteria. From a thermodynamic viewpoint, it has also been predicted that bacteria with a s.f.e. higher than water, which constitute the majority of the bacterial population, in for instance dental plaque [31], have the lowest interaction energy with low s.f.e. surfaces [32].

Indeed fouling resistant gorgonian corals were found to have low s.f.e. ($23–27\,mJ\,m^{-2}$) [33]. Also, the first bacterial adhesion tests on materials with different s.f.e. showed lowest bacterial adhesion on materials with the lowest surface free energies [34]. In in vitro experiments, Everaert et al. [35] showed that fluoro-alkylsiloxane layers chemisorbed to silicone rubber surfaces reduced both yeast (*Candida albicans* and *Candida tropicalis*) and bacterial (*Streptococcus salivarius*, *S. epidermidis*) adhesion by 50 to 77% as compared to original silicone rubber. Furthermore, microorganisms were more easily detached when passing an air-liquid interface. Tsibouklis et al. [36] developed films of poly(perfluoroacrylate) and poly(methylpropenoxyfluoroalkylsiloxane) with surface free energies as low as 12.2 and $5.6\,mJ\,m^{-2}$ respectively. Retention of *Pseudomonas* and *S. aureus* after a washing step showed more than 95% reduction after 6 weeks of incubation as compared to glass controls. Adhesion in a stationary state of one of the initial colonizers of human teeth, *Streptococcus sanguis*, was determined on low s.f.e. fluoroethylenepropylene (FEP) and high s.f.e. glass [37]. On FEP adhesion was 0 to 94% lower than on glass depending on the buffer concentration and shear rate. Furthermore bacteria were more easily detached from the FEP surface through shear forces.

On the other hand, for low s.f.e. bacteria, a preference for low s.f.e. surfaces is predicted by the thermodynamic approach of Absolom et al. [32]. This effect may be explained mechanistically by hydrophobic interaction between the low s.f.e. surface and the low s.f.e. bacterium. Adhesion is driven by release of water from both the surface and the bacterium thus leading to an increase in entropy [38].

Indeed a bacterial preference for low s.f.e. surfaces has been shown by several authors [31, 32, 39]. However, in an in vitro study, low density poly(ethylene) films were treated with tetrafluoromethane plasma, which resulted in a dramatic reduction in s.f.e., but which did not show a significant reduction in bacterial adhesion [40].

In vivo studies mostly showed a beneficial effect of a low s.f.e. material. For instance in a study by Quirynen et al. [41], four surfaces with surface free energies ranging from 20 to $88\,mJ\,m^{-2}$ were attached to the teeth of healthy humans. The low s.f.e. surfaces attracted significantly less microorganisms after 9 days and those attached were less tightly associated with the surface as compared to bacteria adhering on high s.f.e. surfaces. Treatment of enamel surfaces with silicone oil, which lowered its s.f.e., resulted in a significant reduction in plaque formation [42]. Another study indicated, that after three months the exposed areas of low s.f.e. FEP coated abutments (transmucosal dental implants) displayed lower bacterial colonization and a lower plaque maturation as compared to bare titanium [43]. However, in the areas covered by the gingiva, where less mechanical shear forces are present, this effect was not apparent. Finally, modification of a silicone rubber voice prosthesis with perfluoroalkylsiloxane of 8 fluorocarbon units resulted in reduced biofilm formation in vivo [44].

In the in vivo examples mentioned above, low adhesion numbers were obtained when low s.f.e. surfaces were applied in combination with high shear forces, induced by high flow or passage of an air-liquid interface. Thus, in medical applications where high shear forces are operative, like oral devices, urine catheters and voice prostheses, low s.f.e. surfaces are promising. Fluoropolymers are especially suitable for use as coatings on biomaterials as they are known for their chemical and thermal stability [30]. However, applications that are completely internal like artificial hips, artificial veins and intraocular lenses are still prone to biofilm buildup and thus need a more general anti-adhesive coating, like e.g. a polymer brush.

Polymer Brushes

Polymer brushes are polymer chains that are attached to a surface and stretch out into the surrounding medium. Brushes that are designed to prevent biofilm buildup are usually made from poly(ethylene oxide) (PEO), which are highly water soluble and non-toxic [45]. As the PEO chains are highly mobile [46] and attain extremely large exclusion volumes [47], they make the surface difficult to approach by incoming proteins or bacteria. Penetration or compression would lead to an increase in the local concentration of PEO, which, in turn, would lead to a repulsive osmotic interaction. Therewith, a PEO-brush forms a steric barrier preventing close approach, thus keeping the protein or bacterium at a distance where the attractive Lifshitz–Van der Waals interaction is relatively low. The weak residual attraction generally leads to low adsorption and adhesion, if occurring at all, and to easy removal of biological matter from a brushed surface.

Protein adsorption on PEO brushes has been extensively studied and has recently been reviewed [48]. In general, PEO brushes greatly reduce and sometimes even completely prevent protein adsorption. Easy removal of proteins from PEO coatings has also been described [49, 50].

One of the first to study bacterial adhesion to PEO grafted surfaces were Bridgett et al. [51], who used copolymers of PEO and polypropylene oxide (PPO) of different lengths, where the PPO block anchored the PEO chains to the surface. All copolymers induced significant adhesion reductions (up to 97%) of three clinical isolates of *S. epidermidis*. Also another strain of *S. epidermidis* and the skin borne bacteria *Serratia marcescens* were reduced by about 90% applying a coating of this copolymer with 99 ethylene oxide (EO) units [52]. Coating with a copolymer of poly(*L*-lysine) and PEO with 47 EO units reduced adhesion of *Staphylococcus aureus* by 89 to 93% on titanium surfaces [53]. Self-assembled monolayers of only 6 EO units even showed more than 99.5% reduction of both a medical *S. epidermidis* strain and a marine *Delaya marina* strain [54]. Covalent attachment of PEO with 66 EO units to polyurethane resulted in 90 and 95% reduction of adhesion of *S. epidermidis* and *E. coli*, respectively [55]. Adhesion of *Pseudomonas* sp. was reduced by more than 99% by covalently attaching PEO chains with 110 PEO units to PET surfaces [56]. Reductions obtained in above mentioned studies are much higher than obtained with low surface energy coatings. In nearly all papers cited, the adhesion methodology employed some kind of washing step, which may cause detachment and contribute to the low adhesion numbers [57]. However, in a study by Razatos et al. [58], adhesion of *E. coli* was determined in situ without washing steps and still a reduction of more than 99% was found.

In our studies on well characterized covalently attached PEO brushes, adhesion of *S. epidermidis*, *S. aureus*, *S. salivaris* and *E. coli* [59] was reduced by more than 94%. Reductions of *Pseudomonas aeruginosa* and the yeast strains *Candida albicans* and *Candida tropicalis* amounted about 80%, thus indicating that different microbial species can adhere in different numbers to the same PEO brush. More hydrophobic microorganisms showed more adhesion and stronger adhesion of the yeast strains could be attributed to stronger Lifshitz–Van der Waals forces, due to their larger size. We have also shown that a PEO brush is effective at both 20 °C and 37 °C and that those microorganisms that do adhere can very easily be removed [60]. As for protein adsorption [61–64], the longest PEO chains were found to be most effective in preventing adhesion of bacteria and yeast [60]. This confirms the theory of attenuation of particle-substratum interaction forces by a brush [65–68].

Despite these excellent in vitro results, little progress has been made in vivo. For instance, under clinical conditions, blood proteins have been demonstrated to adsorb extensively to a PEO coated polymer, in contrast to in vitro results [69]. Also, in vivo research on PEO coatings in the oral cavity showed poor results [70], despite excellent in vitro reductions in salivary protein adsorption and oral bacterial adhesion to PEO coatings on glass and hydroxyapatite [70, 71]. Possibly, the durability of the thin layer of grafted PEO chains in the oral cavity was not sufficient over a clinically relevant time scale.

Recently we have developed a method to easily assess the stability of a PEO coating, by using a marker bacterium to directly determine the effectivity of a PEO coating [72]. We showed that our coating was only stable for at most 48 hours depending on the biological fluid it was exposed to. In another study, a PEO coating remained stable for less than a month [73]. In both cases, degradation most probably does not occur at the alkyl or ether bonds of the PEO polymer itself, as these can only be degraded by aggressive chemicals [74], high temperatures [75] or specific bacteria [76]. The grafting of the PEO chain to the surface is, in general, the weak point of a PEO coating. For instance, the often used organosilane linkage is susceptible to hydrolytic cleavage [77]. In our study, the PEO chains were coupled by a Si – O – C linkage to the glass substratum, which may readily hydrolyze as well [78]. In case the polymer chains are grafted at the surface by physical adsorption, they can be expected to desorb over time, therewith losing their effectiveness.

Conclusions

This review has shown that positively charged materials, low s.f.e. materials and polymer brushes show potential in preventing BCI. Positively charged materials have shown to be effective in preventing growth of certain strains and species, while other stains readily adhere and grow on them. Possibly, positively charged surfaces can only be applied in situations where there is little supply of new organisms, such as for totally implanted surfaces, because a large supply of new organisms will cause adhesion of viable organisms on the organisms killed or hampered in their growth by the positive charge. Also low s.f.e. materials are not generally applicable as some bacterial strains are able to adhere to them or even show preference for these materials. However, for applications where fluctuating shear forces are operative, excellent in vivo results have been obtained. Theoretically, polymer brushes should be able to reduce adhesion of any bacterium and, indeed, have shown a very high effectiveness in preventing bacterial adhesion in vitro. Whether or not those bacteria that do successfully adhere to a polymer brush will also grow and form a biofilm, has not yet been established. Problems concerning durable attachment of the polymer chains to the surface preventing successful in vivo applications with these coatings could be overcome by developing a stable coupling between a surface and the brush.

References

1. Gristina AG (1987) Science 237:1588
2. Dankert J, Hogt AH, Feijen J (1986) Crit Rev Biocompat 2:219
3. Costerton JW, Stewart PS, Greenberg EP (1999) Science 284:1318
4. Gilbert P, Das J, Foley I (1997) Adv Dent Res 11:160
5. Lee-Smith J, Santy J, Davis P, Jester R, Kneale J (2001) J Orthop Nurs 5:37
6. Mahan J, Seligson D, Henry SL, Dobbins J (1991) Orthopedics 14:305
7. Escher A, Characklis WG (1990) Modeling the initial events in biofilm accumulation. In: Characklis WG (ed) Biofilms. Wiley, New York, p 445–486
8. Azeredo J, Visser J, Oliveira R (1999) Colloids Surf B 14:141
9. Dufrene YF, Vermeiren H, Van der Leyden J, Rouxhet PG (1996) Microbiology 142:855
10. Christensen GD, Baddour LM, Hasty DL, Lowrance JH, Simpson WA (1989) Microbial and foreign body factors in the pathogenesis of medical device infections. In: Bisno AL, Waldvogel FA (eds) Infections Associated with Indwelling Medical Devices. American Society of Microbiology, Washington, DC, p 27–59
11. van Oss CJ (1991) Biofouling 4:25
12. Busscher HJ, Weerkamp AH (1987) FEMS Microbiol Rev 46:165
13. Lyklema J, Norde W, van Loosdrecht MM, Zehnder AB (1989) Colloids Surf 39:175
14. Donlan RM (2001) Emerg Infect Dis 7:277
15. Hogt AH, Dankert J, Feijen J (1986) J Biomed Mater Res 20:533
16. Keogh JR, Eaton JW (1994) J Lab Clin Med 121:537
17. Paulsson M, Gouda I, Larm O, Ljungh A (1994) J Biomed Mater Res 28:311
18. Nomura S, Lundberg F, Stollenwerk M, Nakamura K, Ljungh A (1997) J Biomed Mater Res 38:35
19. Morra M, Cassineli C (1999) J Biomater Sci Polym Ed 10:1107
20. Holland NB, Qiu YX, Ruegsegger M, Marchant RE (1998) Nature 392:799
21. Gottenbos B, Grijpma DW, Van der Mei HC, Feijen J, Busscher HJ (2001) J Antimicro Chemother 48:7
22. Harkes G, Feijen J, Dankert J (1991) Biomaterials 12:853
23. Jucker BA, Harms H, Zehnder AB (1996) J Bacteriol 178,18:5472
24. Gottenbos B, Van der Mei HC, Busscher HJ (2000) J Biomed Mater Res 0:208
25. Gottenbos B, Van der Mei HC, Klatter F, Grijpma DW, Feijen J, Nieuwenhuis P et al. (2003) Biomaterials 24:2707
26. Norde W (2003) Interfacial tension. In: Norde W (ed) Colloid and interfaces in life sciences. Marcel Dekker Inc., New York, p 47–61
27. Young T (1805) Phil Trans Roy Soc 95:65
28. Sharma PK, Rao KH (2002) Adv Colloid Interface Sci 98:341
29. Tsibouklis J, Nevell TG (2003) Adv Mater 15:647
30. Thomas RR (1999) Material properties of fluoropolymers and perfluoroalkyl-based polymers. In: Hougham G, Cassidy PE, Johns K, Davidson T (eds) Fluoropolymers 2: Properties. Plenum Press, New York, p 47–65
31. Weerkamp AH, Quirynen M, Marechal M, Van der Mei HC, Van Steenberghe D, Busscher HJ (1989) Microb Ecol Health Dis 2:11
32. Absolom DR, Lamberti FV, Policova Z, Zingg W, Van Oss CJ, Neumann AW (1983) Appl Environ Microbiol 46:90
33. Vrolijk NH, Targett NM, Baier RE, Baier AE (1990) Biofouling 2:39
34. Dexter SC, Sullivan JD, Williams IIJ, Watson SW (1975) Appl Microbiol 30:298
35. Everaert EPJM, Van der Mei HC, Busscher HJ (1998) Coll Surf B:Biointerfaces 10:179
36. Tsibouklis J, Stone M, Thorpe AA, Graham P, Peters V, Heerlien R et al. (1999) Biomaterials 20:1229
37. Busscher HJ, Sjollema J, Van der Mei HC (1990) Relative importance of surface free energy as a measure of hydrophobicity in bacterial adhesion to solid surfaces. In: Doyle RJ, Rosenberg M (eds) Microbial Cell Surface Hydrophobicity. American Society for Microbiology, Washington DC, p 335–359
38. Norde W (2003) Water. In: Norde W (ed) Colloid and interfaces in life sciences. Marcel Dekker Inc., New York, p 47–61
39. Bakker DP, Huijs FM, De Vries J, Klijnstra JW, Busscher HJ, van der Mei HC (2003) Coll Surf B:Biointerfaces 32:179
40. Olde Riekerink MB, Engbers GHM, Van der Mei HC, Busscher HJ, Feijen J (2001) Microbial adhesion onto superhydrophobic fluorinated low density poly(ethylene) films. In: Olde Riekerink MB (ed) Thesis: Structural and Chemical Modification of Polymer Surfaces by Gas Plasma Etching. Printpartners Ipskamp, Enschede, p 65–82
41. Quirynen M, Marechal M, Busscher HJ, Weerkamp A, Arends J, Darius PL (1989) J Dent Res 68:796
42. Rolla G, Ellingsen JE, Herlofson B (1991) Biofouling 3:175
43. Quirynen M, Van der Mei HC, Bollen CML, Van den Bossche LH, Doornbusch GI, Van Steenberghe D et al. (1994) J Periodontology 65:162
44. Everaert EPJM, Mahieu HF, Van de Belt-Gritter B, Peeters AJGE, Verkerke GJ, Van der Mei HC et al. (1999) Arch Otolaryngol Head Neck Surg 125:1329
45. Harris JM (1992) Poly(ethyleneglycol) Chemistry: Biotechnical and Biomedical Aplications. Plenum Press, New York
46. Harris JM (1992) Introduction to Biotechnical and biomedical applications of poly(ethylene glycol). In: Harris JM (ed) Poly(ethyleneglycol) Chemistry:

Biotechnical and Biomedical Aplications. Plenum Press, New York, p 1–13
47. Ryle AP (1965) Nature 206:1256
48. Currie EPK, Norde W, Cohen Stuart MA (2003) Adv Coll Interf Sci 100:205
49. Zhu B, Eurell T, Gunawan R, Leckband D (2001) J Biomed Mater Res 56:406
50. Huang NP, Michel R, Voros J, Textor M, Hofer R, Rossi A et al. (2001) Langmuir 17:489
51. Bridgett MJ, Davies MC, Denyer SP (1992) Biomaterials 13:411
52. Marsh LH, Coke M, Dettmar PW, Ewen RJ, Havler M, Nevell TG et al. (2002) J Biomed Mater Res 61:641
53. Harris LG, Tosatti S, Wieland M, Textor M, Richards RG (2004) Biomaterials 25:4135
54. Ista LK, Fan H, Baca O, Lopez GP (1996) FEMS Microbiol Lett 142:59
55. Ki DP, Young SK, Dong KH, Young HK, Eun HBL, Hwal S et al. (1998) Biomaterials 19:851
56. Kingshott P, Wei J, Bagge-Ravn D, Gadegaard N, Gram L (2003) Langmuir 19:6912
57. Gomez-Suárez C, Busscher HJ, Van der Mei HC (2001) Appl Environ Microbiol 67:2531
58. Razatos A, Ong YL, Boulay F, Elbert DL, Hubbell JA, Sharma MM et al. (2000) Langmuir 16:9155
59. Roosjen A, Kaper HJ, Van der Mei HC, Norde W, Busscher HJ (2003) Microbiology 149:3239
60. Roosjen A, Van der Mei HC, Busscher HJ, Norde W (2004) Langmuir 20:10949
61. Golander C-G, Herron JN, Lim K, Cleasson P, Stenius P, Andrade JD (1992) Properties of immobilized PEG films and the interaction with proteins. In: Harris JM (ed) Poly(ethylene glycol) chemistry, Biotechnological and Biomedical Applications. Plenum Press, New York, p 221–245
62. Prime KL, Whitesides GM (1993) J Am Chem Soc 115:10714
63. Gombotz WR, Guanghui W, Horbett TA, Hoffman AS (1991) J Biomed Mater Res 25:1547
64. Lee J, Martic PA, Tan JS (1989) J Coll Interf Sci 131:252
65. Jeon SI, Lee JH, Andrade JD, Gennes PG (1991) J Coll Sci 142:149
66. Szleifer I (1997) Biophys J 72:595
67. Halperin A (1999) Langmuir 15:2525
68. Leckband D, Sheth S, Halperin A (1999) J Biomater Sci Polym Ed 10:1125
69. Park KD, Kim SW (1992) PEO-modified surfaces – In vitro, ex vivo andin vivo blood compatibility. In: Harris JM (ed) Poly(ethyleneglycol) chemistry: biotechnical and biomedical aplications. Plenum Press, New York, p 283–301
70. Olsson J, Van der Heijde Y, Holmberg K (1992) Caries Res 26:428
71. Olsson J, Carlen A, Holmberg K (1990) J Dent Res 69:1586
72. Roosjen A, De Vries J, Van der Mei HC, Norde W, Busscher HJ (2005) J Biomed Mater Res Part B: Appl Biomater 73:347
73. Branch DW, Wheeler BC, Brewer GJ, Leckband DE (2001) Biomaterials 22:1035
74. Bhatt MV, Kulkarni SU (1983) Synthesis 4:249
75. Han S, Kim C, Kwon D (1997) Polymer 38:317
76. Kawai F (2002) Appl Microbiol Biotechnol 58:30
77. Hong HG, Jiang M, Sligar SG, Bohn PW (1994) Langmuir 10:153
78. Maas JH, Cohen Stuart MA, Sieval AB, Zuilhof H, Sudholter EJR (2003) Thin Solid Films 426:135

Progr Colloid Polym Sci (2006) 132: 145–151
DOI 10.1007/2882_025

Published online: 10 March 2006

Pradeep Pareek
Hans-Jürgen P. Adler
Dirk Kuckling

Tuning the Swelling Behavior of Chemisorbed Thin PNIPAAm Hydrogel Layers by N,N-Dimethyl Acrylamide Content

Pradeep Pareek · Hans-Jürgen P. Adler · Dirk Kuckling (✉)
Institut für Makromolekulare Chemie und Textilchemie, Technische Universität Dresden, 01062 Dresden, Germany
e-mail: dirk.kuckling@chemie.tu-dresden.de

Abstract The synthesis of a series of photo cross-linkable terpolymers of poly(*N*-isopropylacrylamide) (PNIPAAm) with *N*,*N*-dimethylacrylamide (DMAAm) as a hydrophilic and 2-(Dimethylmaleimido)-*N*-ethylacrylamide (DMIAAm) as a chromophore is described for tailor-made swelling behavior of thin hydrogel layers. The polymers were spin coated on gold surface and swelling behavior thin hydrogel layers were investigated by the combination of Surface Plasmon Resonance and Optical Waveguide spectroscopy (SPR/OWS). It was observed that an increase in the hydrophilic monomer (DMAAm) content increases both transition temperature (T_c) and swelling of the hydrogel layers. In addition the collapsed hydrogel thickness was also dependent on the DMAAm mol-%. Adhesion promoter synthesized for the covalent attachment of hydrogel to substrate has been further investigated to optimize dynamics of adsorption and adhesion on the gold surface. UV illumination (> 310 nm) results in simultaneous crosslinking and chemical attachment of the hydrogel to the surface. Covalent attachment confines the swelling and deswelling in one dimension i.e. perpendicular to the substrate.

Keywords Hydrogel · Photo cross-linking · Poly(*N*-isopropylacrylamide) · Surface plasmon resonance spectroscopy

Introduction

Stimuli sensitive hydrogels are studied extensively because these gels demonstrate reversible volume phase transitions in response to external stimuli such as changes in temperature, pH, ionic strength or light [1, 2]. These properties make such hydrogels ideal for applications in drug delivery, in chemo-mechanical valves [3, 4], in separation systems [5] etc. [6–10]. One of the most intensively studied polymers in this field is Poly(*N*-isopropylacrylamide) (PNIPAAm), which exhibits a sharp phase transition in water upon heating above 32 °C [7]. It undergoes a temperature induced collapse from an extended coil to a globular structure, a transition revealed on the macroscopic scale by a sudden decrease in the degree of swelling in its gels. The volume phase transition has been described theoretically as resulting from a balance between free energy of mixing, rubber elasticity free energy and the osmotic contribution of the counterions in the case of ionizable comonomers [11–16]. The lower critical solution temperature (LCST) behavior of the PNIPAAm is strongly influenced by the nature of the comonomers. Hydrophobic comonomers decrease the transition temperature (T_c) and hydrophilic comonomers increase it. It has been generally observed that the LCST phenomenon disappears in the presence of a high content of hydrophobic or hydrophilic compounds.

The faster swelling/deswelling kinetics of thin hydrogel films is of interest especially from the application point of view. Recently there are some reports in the literature where photo cross-linking has been utilized for preparation of thin hydrogel films [17, 18]. In earlier reports the effect of the chromophore concentration and film thickness on the T_c of the thin films has been analyzed [18–20]. It has been observed that the physisorbed hydrogel layers eventually detach from the surface. To avoid such shortcomings, the thin hydrogel films should be covalently attached to the substrate. This may induce some changes in T_c and swelling ratios of the thin hydrogel films. Optimized polymers with high swelling and appropriate T_c ($\sim 35\,°C$) will then be used for future experiments involving cell attachment-detachment processes. Okano introduced PNIPAAm based hydrogels as a support for cell cultivation [21, 22]. The copolymerization of PNIPAAm with 2-carboxyisopropylacrylamide results in rapid cell detachment due to the increase in the hydrophilicity of such hydrogels [23].

Surface Plasmon Resonance Spectroscopy along with Optical Waveguide Spectroscopy (OWS) are suitable techniques for determination of film thickness, T_c, swelling ratio and anisotropy in swelling behavior of thin hydrogel layers. These techniques provide in dept information about the thin films [24]. The information obtained from SPR gives a direct measure of the local refractive index of the dielectric close to the surface. A subsequent Fresnel calculation gives either the thickness or the refractive index of the thin hydrogel film. For thicker films (dry thickness > 200 nm), the same system can also be used to couple into waveguide modes allowing the film thickness and the refractive index to be determined independently [25]. In addition to SPR/OWS in situ ellipsometric methods have been applied for characterization of hydrogel layers as well [26]. Moreover the thin hydrogel films prepared by techniques such as surface grafting [27, 28] and plasma polymerization [10] vary in T_c and swelling behavior even when the graft architecture and morphology of the film are well known [26]. Properties of stimuli sensitive layers like T_c and swelling ratio are different than bulk gels due to a constraint imposed on the film by the presence of a fixed substrate, which could be probed by SPR/OWS [19].

In this study we report the synthesis of a series of photo cross-linkable terpolymers of PNIPAAm with *N*,*N*-dimethylacrylamide (DMAAm) as a hydrophilic monomer and 2-(Dimethylmaleimido)-*N*-ethylacrylamide (DMIAAm) as a chromophore. We demonstrate the effect of incorporation of hydrophilic monomer on the T_c and swelling of hydrogel in the form of thin films. Adhesion promoter synthesized for the covalent attachment of hydrogel to substrate has been further investigated to optimize dynamics of adsorption and adhesion on gold surface by XPS, Contact angle and SPR. Hydrogels synthesized in this study have a dry thickness around 200 nm and its swelling behavior is investigated by the combination of SPR/OWS. This study provides a series of thin photo cross-linkable biocompatible films that might later be used for cell growth experiments.

Experimental Section

Materials. *N*-Isopropylacrylamide (NIPAAm, Aldrich) was purified by recrystallization from hexane. *N*,*N*-Dimethylacrylamide (DMAAm, Acros) was distilled prior to use. 2,2′-Azobis(isobutyronitrile) (AIBN) was recrystallized from methanol. 1,4-Dioxane, tetrahydrofuran (THF), and diethyl ether were distilled over potassium hydroxide. All other reagents were of analytical grade.

Synthesis. 2-(Dimethylmaleimido)-*N*-ethylacrylamide (DMIAAm) monomer was prepared according to the procedures mentioned in the literature [29]. *S*-Acetoxy (3-dimethylmaleimido)-propane thiol (ADP) (Scheme 1) as an adhesion promoter was synthesized according to the procedure mentioned in the literature [20].

SPR substrates (LaSFN9 glass) coated with 45 nm gold were immersed in 1 mM, 5 mM, 10 mM ethanolic solution of ADP for 24 h followed by thorough rinsing with ethanol and drying under nitrogen stream. The adsorption behavior was studied by grazing angle IR, contact angle, SPR (kinetic and angle scan) and X-ray photoelectron spectroscopy.

PNIPAAm terpolymers were obtained by free radical polymerization of NIPAAm, DMAAm, and DMIAAm (Scheme 2). Polymerization was initiated by AIBN using 1,4-dioxane as a solvent. DMAAm content was varied from 5 to 50 mol %. The total monomer concentration was 0.55 mol/L and the reaction was carried out at 70 °C for

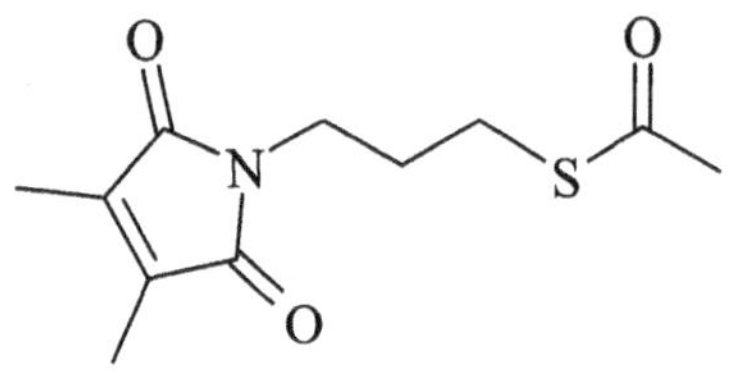

ADP

Scheme 1 Adhesion promoter for gold surface

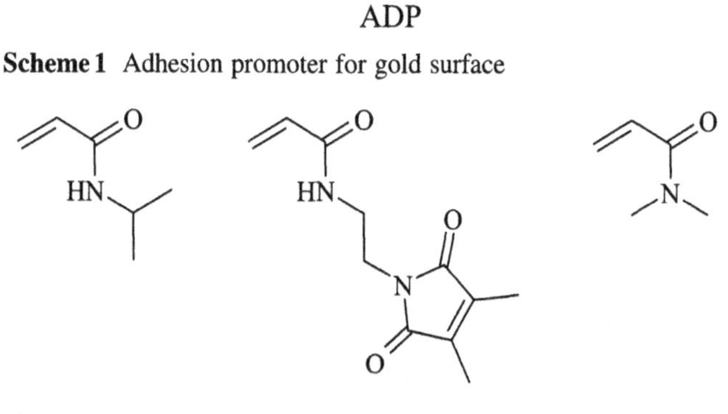

NIPAAm DMIAAm DMAAm

Scheme 2 Monomers used for polymer synthesis

Table 1 Influence of DMAAm content on T_c of NIPAAm photo cross-linkable polymers

Polymer	Composition (mol-%) NIPAAm/DMAAm/DMIAAm Feed	Polymer [a]	T_c [b] (°C)
NIPAAm5	90/05/05	91.9/5.4/2.7	28.2
NIPAAm10	85/10/05	84.2/12.9/2.9	30.6
NIPAAm20	75/20/05	75.6/21.8/2.6	33.7
NIPAAm30	65/30/05	65.1/32.7/2.3	39.2
NIPAAm40	55/40/05	56.9/40.6/2.6	45.3
NIPAAm50	45/50/05	44.7/52.6/2.7	53.4

[a] from ^{1}H NMR
[b] from DSC

7 h under nitrogen atmosphere. The polymers were precipitated in diethyl ether and purified by reprecipitation from THF into diethyl ether (1/3). Photo cross-linkable polymers of NIPAAm are referred as NIPAAm followed by the mol-% feed composition of DMAAm (for example NIPAAm10). Polymer composition and T_c values are mentioned in Table 1.

Characterization. The ^{1}H NMR spectra were recorded on a Bruker DRX 500 spectrometer (500 MHz). The solvent (acetone-d_6) was used as an internal reference. UV-VIS spectra were recorded on UV-VIS Perkin Elmer Lambda 19 spectrometer. For the estimation of photo cross-linker content at a wavelength of 305 nm, 1 mol % of polymer solution in water was used.

DSC measurements were carried out with a TA Instruments DSC 2920 to determine the T_c of the polymer solutions. The DSC thermograms of the polymer solutions (50 mg/l in deionised water) were recorded at a heating rate of 5 °C/min. The onset of the endothermic event was taken as T_c.

The molecular weight (M_w) and the molecular weight distribution (polydispersity M_w/M_n) of the terpolymers were determined by gel-permeation chromatography with a Waters instrument equipped with UV and RI detectors and using Waters' "Ultrastyragel" columns. The samples were measured at 30 °C in chloroform containing 0.1 vol % triethylamine as the mobile phase with a flow rate of 1 l/min.

Drop shape analysis system DSA 10 (KRÜSS) was used for contact angle measurements. X-ray Photoelectron Spectroscopy studies were performed on an ESCA 5700 from Physical Electronics. The S_{2p} spectra were recorded from a sampling area of ca. 1 mm^2 with a take-off angle of 45° and analyzer pass energy of 29.35 eV. Acquisition times were ca. 9 min with a base pressure less than 5×10^{-10} mbar. Spin coater model P 6700 series (Speedline Technologies) was used to prepare thin film (ca. 200 nm dry thickness) on SPR substrate with 250 rpm for 30 sec and 3000 rpm for 180 sec. Photo cross-linked hydrogel layers were prepared by spin coating 5 wt. % of polymer solution on modified SPR substrate. Thioxanthone (2 wt. % w.r.t. polymer) was used as a photo sensitizer. The films were dried under vacuum and cross-linked by UV irradiation using a 400 W UV lamp at a wavelength $\lambda > 310$ nm for at least 60 min.

Combination of Surface Plasmon Resonance (SPR) and Optical Waveguide Spectroscopy (OWS) was used for the simultaneous determination of refractive index and film thickness of the hydrogel layers in the Kretschmann configuration [24]. The resulting angle scans from the SPR instrument were fit to Fresnel calculations and different layers were represented using a simple box model. A detailed description of this process has been published previously [18].

Swelling experiments were carried out in phosphate buffered solution (PBS buffer) by connecting a peristaltic pump to the sample holder as flow through cell and temperature of hydrogel film was externally controlled by a temperature bath. The temperature inside the cell was measured using a thermocouple with an accuracy of 0.1 °C. The response time of the gels was on the order of seconds, which is much faster than the time scale of the temperature change in the flow cell. The equilibration time between each angular scan was verified in the SPR kinetic mode, monitoring the reflected intensity at a fixed angle as a function of time and ranged from 10 to 25 min.

Result and Discussion

Adhesion Promoter

S-Acetoxy-(3-dimethylmaleimido)-propane thiol (ADP) was synthesized for the covalent attachment of hydrogel to the substrate. The thioacetate linkage was used as a surface-active group and was responsible for the chemisorption on the gold surface. The dimethylmaleimido group of ADP held the hydrogel layer via [2+2]-cyclodimerization reaction with similar groups in polymer chains. The presence of adhesion promoter on the gold substrate was detected by SPR, XPS and contact angle measurements.

The kinetics of binding of ADP on the gold substrate was detected with SPR kinetic scans at a fixed angle of 56° in ethanol. The 1 mM adhesion promoter solution in ethanol was injected in the SPR cell. An increase in the reflected intensity was observed till 1 h indicating the adsorption of ADP. Changes in the refractive index of the gold surface due to the adsorption of ADP and increase in the total thickness (45 nm gold and monolayer thickness of ADP) were monitored by the plot of the reflected intensity versus angle scans (Fig. 1). Notable features observed from this angle scans were that the plasmon minima

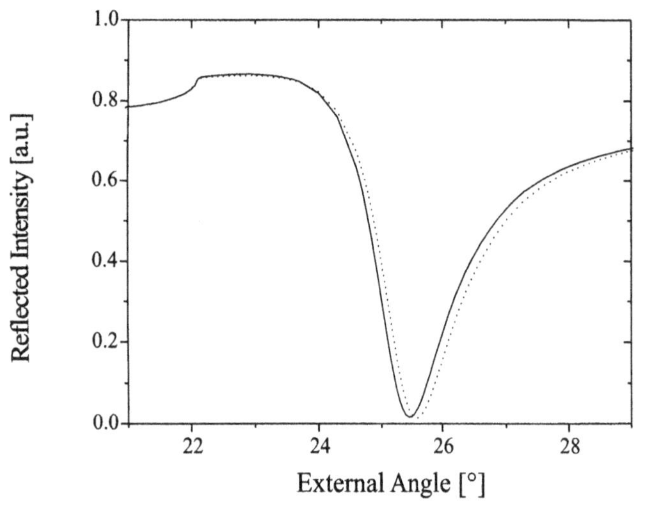

Fig. 1 SPR angle scans: Bare gold (*solid line*), 1 mM ADP adhesion promoter solution in ethanol adsorbed for 24 h on gold surface (*dotted line*)

for a bare gold surface which appeared at 25.4°, shifted to 25.6° in presence of ADP adsorbed on its surface. Apart from this observation, the refractive index of ADP was measured from a refractometer and its value was estimated to be 1.523. Taking this value into account the layer thickness of ADP can be estimated 1.43 nm.

Contact angle measurements were carried out to further investigate the adsorption of ADP on gold. The contact angle of gold surface increased from 57° to 64° upon adsorption of ADP. Increase in the contact angle indicates the presence of hydrophobic parts of ADP on the gold surface. It was observed that 1 mM solution of ADP was appropriate to obtain a monolayer of adhesion promoter.

To determine the electronic state of the sulphur atom in ADP on gold surface, the S_{2p} binding energies were measured. The binding mechanism of ADP to gold should involve the cleavage of the acetoxy group. XPS spectra showed the S_{2p} binding energy for ADP (after rinsing with ethanol) at 161.8 eV and this value matched with the value reported in the literature for binding of thiols to gold [30]. Thus, binding of ADP on gold results in the cleavage of the acetoxy moiety and formation of a surface bound thiolate species. In addition the XPS results indicated the absence of any surface adsorbed acetoxy moiety after ethanol washing. This confirms the attachment of ADP on gold surface through S – Au bond.

Photo Cross-linkable PNIPAAm Terpolymers

Synthesis and Characterization. The transition temperature (T_c) of NIPAAm copolymers can be modified by a hydrophilic or hydrophobic comonomer. This phenomenon has already been explained on the basis of intermolecular and intramolecular hydrogen bond formation between the solvent (water) and polymer chains [6, 7]. This study focuses on the synthesis and characterization of terpolymers of *N*-isopropylacrylamide (NIPAAm) with 2-(dimethylmaleimido)-*N*-ethylacrylamide (DMIAAm) and *N*,*N*-dimethylacrylamide (DMAAm) in different compositions. The polymers were synthesized by free radical polymerization using AIBN as initiator and 1,4-dioxane as solvent. The polymers were photo crosslinkable, biocompatible and exhibited a temperature sensitive behavior with wide range of phase transition temperature. DMIAAm in polymer chains resulted in photo cross-linking ([2+2] cyclodimerization) under UV irradiation. Incorporation of DMIAAm decreased the T_c of the NIPAAm by 6–8 °C, hence to counter this effect, hydrophilic DMAAm was introduced as a third component in various contents (5 to 50 mol %). The properties of resulting polymers are summarized in Table 1. Constant DMIAAm content of 5 mol % in feed mixture produced terpolymers with DMIAAm content in the range 2.3 to 2.9 mol %. The molecular weights (M_w) were estimated to be in the range of 37 000 with polydispersity indices (M_w/M_n) of 1.9.

Incorporation of DMAAm increases the T_c of NIPAAm copolymers [18]. It has previously been shown that the T_c and swelling of NIPAAm hydrogel layers is largely affected by varying the amount of chromophore [18, 31]. Terpolymers prepared from NIPAAm/DMAAm/DMIAAm were investigated by DSC, which is a convenient method to study the phase transition temperature of polymers. The phase transition temperature increases with the increase in the mol-% of hydrophilic DMAAm in NIPAAm terpolymers (Table 1).

Swelling Measurements. DSC provides information only related to the T_c and gives no information about the swelling behavior of thermoresponsive polymer. In addition, it has been reported that there is a difference in the T_c of NIPAAm copolymers measured by DSC and other techniques. In order to investigate the swelling behavior, NIPAAm terpolymers with different content of DMAAm and constant DMIAAm (5 mol %) were used for hydrogel formation. The dry film thickness of all hydrogel layers from 5 mol % polymer solutions was around 200 nm. Such surface attached films confined the swelling in one dimension. The swelling behavior of the cross-linked layer was determined by SPR and OWS [24].

The effect of temperature on SPR external angle of a photo cross-linked NIPAAm thin hydrogel layer in PBS buffer is shown in Fig. 2. The scan showed two important features: (a) SPR minima between 61° and 73° (b) waveguide mode between 46° and 48°. The SPR minima changed significantly around the transition temperature of NIPAAm photo cross-linked hydrogel thin layers. In addition the waveguide mode disappeared at higher temperatures. This is due to the fact that at higher temperature the polymer deswelled resulting in decrease in hydrogel thickness. The transition temperatures of PNIPAAm hydrogel

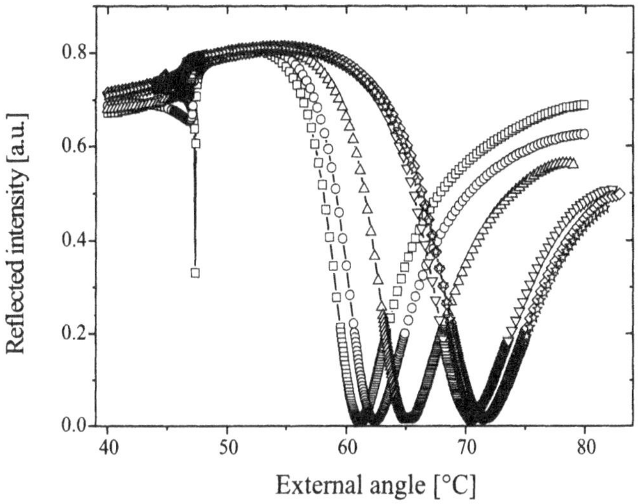

Fig. 2 SPR angle scan of a photo cross-linked NIPAAm hydrogel layer at various temperatures in PBS buffer (□ – 16.8 °C, ○ – 27.5 °C, △ – 32.6 °C, ▽ – 42.5 °C, ◇ – 51.8 °C, * – 61.5 °C)

layers can simply be determined by plotting the SPR minima with temperature. However, a better interpretation of the results was obtained by fitting the SPR angle scans to Fresnel calculations to determine refractive index (η) and film thickness (d) of the hydrogel layers.

The film thickness of hydrogel layers from 5 wt% solutions was in the order of 1 μm in the swollen state. Thickness (d) and refractive index (η) were determined without further assumptions, because these films could be fitted using a single layer model [19]. The refractive index (η) obtained from Fresnel calculation was converted to polymer fraction and volume degree of swelling ($1/\phi_p$).

The volume degree of swelling of the PNIPAAm photo cross-linked hydrogel layers plotted as a function of temperature in PBS buffer is shown in Fig. 3. It was observed that the volume degree of swelling decreased rapidly near T_c and remained almost constant at temperatures above and below T_c. Sigmoidal fits of the data in the Fig. 3 showed that T_c increases with the increase in the mol-% of DMAAm in NIPAAm hydrogel layers. This is due to the hydrophilic nature of DMAAm. The results are in agreement with the values obtained from DSC for these polymers. However, NIPAAm50 layers did not show a transition behavior in SPR experiments. The T_c values from the SPR measurements were found to be slightly higher than the DSC values (Fig. 4). The T_c of a similar system without photo cross-linker has been reported by Barker et al. [32]. Their results indicate that the T_c of a linear PNIPAAm copolymer is lower than the corresponding cross-linked copolymer of the same composition quoted by Shibayama et al. [33]. Clearly cross-linking of the polymer led to a more constrained structure increasing the critical temperature [19, 20].

As the mol-% of DMAAm increased in NIPAAm photo cross-linked polymer, the value of η (collapsed) and η (swollen) of hydrogel layer decreased. For NIPAAm5 hydrogel layer, η (collapsed) value of 1.451 above T_c and η (swollen) value of 1.378 below T_c was obtained. For NIPAAm40 hydrogel layer, η (collapsed) value of 1.414 above T_c and η (swollen) value of 1.354 below T_c was obtained. The refractive index (η) obtained from Fresnel calculation was converted to polymer fraction and volume degree of swelling ($1/\phi_p$). The resulting values for different hydrogel layers are summarized in Fig. 5. For NIPAAm40 hydrogel layer, $1/\phi_p$ of 8.2 was observed be-

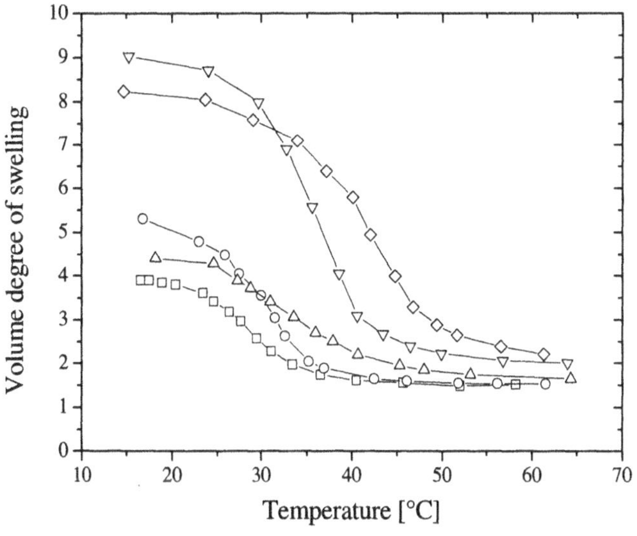

Fig. 3 Volume degree of swelling of the photo cross-linked PNIPAAm hydrogel layer with varying DMAAm content plotted as a function of temperature in PBS buffer (□ – NIPAAm5, ○ – NIPAAm10, △ – NIPAAm20, ▽ – NIPAAm30, ◇ – NIPAAm40)

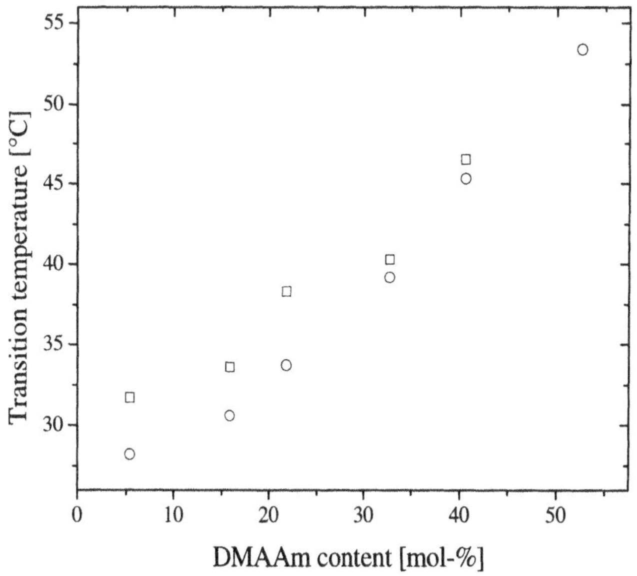

Fig. 4 Comparison of T_c of NIPAAm terpolymers in PBS buffer from (□) SPR and (○) DSC

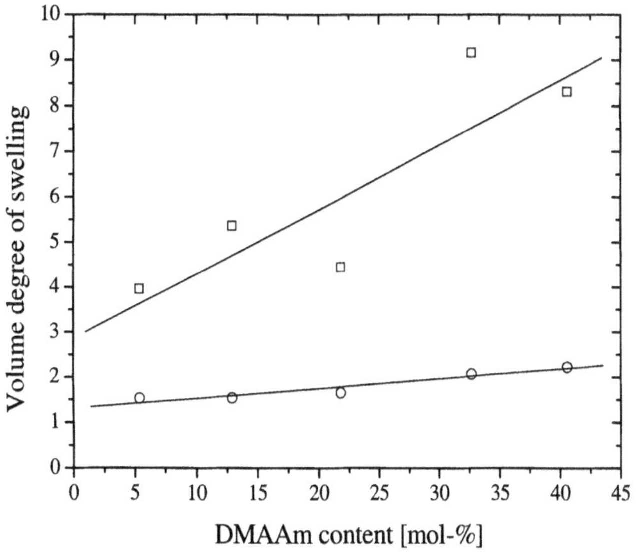

Fig. 5 Change in volume degree of swelling in PBS buffer at constant temperatures with varying DMAAm content (□ – 15 °C, ○ – 55 °C). Lines are drawn to guide the eyes

low T_c. Collapsed state of this hydrogel showed $1/\phi_P$ of 2.3. The hydrogel layer with 5.0 mol % DMAAm showed $1/\phi_P$ of 4.0 and its collapsed state showed $1/\phi_P$ of 1.6. These results clearly indicated that not only the T_c but also the volume degree of swelling in NIPAAm hydrogel layers was dependent on the mol-% of DMAAm. Higher the mol-% of DMAAm, larger was T_c and volume degree of swelling.

By raising the temperature above T_c, all hydrogel layers collapsed to different $1/\phi_P$ values. The collapsed state was also found to be dependent on the mol-% of DMAAm in hydrogel layers. Higher the DMAAm mol-% in hydrogel layers, larger is the $1/\phi_P$ in the collapsed state. It could be interpreted from these results that the DMAAm, which increases the swelling (below T_c) also results in a partially collapsed state at temperatures above T_c. This is due the hydrophilic nature of DMAAm, which retains water even after the PNIPAAm chains are collapsed.

SPR instrument was also used for the estimation of film thickness and swelling ratio (d/d_o) of the hydrogel layers. For NIPAAm hydrogel layers with a dry thickness of ~ 200 nm, the collapsed film thickness at temperatures above T_c was found to be dependent on the DMAAm mol-% in the polymer. For NIPAAm40 hydrogel layer, a d/d_o of 6.5 and 1.5 was calculated below T_c and above T_c, respectively. It was observed that the similar results were obtained from the swelling ratio and volume degree of swelling. The absence of waveguide modes at higher temperatures gave a slight error in the measurements of swelling ratios. T_c value was also determined by plotting d/d_o with temperature. NIPAAm hydrogel layer with higher DMAAm mol-% shows a higher transition temperature and higher swelling ratio.

The thin NIPAAm hydrogel layers were covalently attached to the surface, therefore the swelling was confined to one direction perpendicular to the substrate. The degree of anisotropy was calculated from ratio of $1/\phi_P$ and swelling ratio. This ratio was found to be between 1.2–1.3. Such a study describing the effect of hydrophilic comonomer on T_c of PNIPAAm provides an opportunity to formulate thermoresponsive hydrogel layers with desired transition temperature and swelling behavior for various applications.

Conclusion

PNIPAAm with different mol-% of DMAAm and a photo cross-linkable moiety were synthesized by free radical polymerization. These polymers were used to prepare thin hydrogel layers with wide range of phase transition temperature. Swelling behavior of the thin hydrogel layers was investigated by a combination of Surface Plasmon Resonance and Optical Waveguide Spectroscopy (SPR/OWS). Hydrogel layers were covalently attached to the gold surface by using ADP as an adhesion promoter. Adsorption of ADP on gold surface was confirmed by using XPS, contact angle measurements and SPR. Thickness and refractive index of the dry and swollen hydrogel layers was investigated. Volume degree of swelling in both swollen and collapsed hydrogel layer was dependent on the mol-% of DMAAm in PNIPAAm. These results provide vital information on the swelling behavior of surface attached hydrogel layer and showed the versatility of SPR instrument for studying thin hydrogel layer. In addition, such hydrogels layers are under investigation for cell attachment-detachment processes.

Acknowledgement The authors wish to acknowledge Deutsche Forschungsgemeinschaft (DFG) and European Graduate College (EGS 720) "Advanced Polymer Materials" for the financial support. The authors are also thankful to A. Rudolph for XPS measurements, M. Dziewiencki for contact angle measurements and K. Kretschmer for help with SPR experiments.

References

1. Tanaka T (1978) Phys Rev Lett 40:820–823
2. Tanaka T, Fillmore D, Sun ST, Nishio I, Swislow G, Shah A (1980) Phys Rev Lett 45:1636–1639
3. Osada Y, Gong JP (1998) Adv Mater 10:827–837
4. Arndt KF, Kuckling D, Richter A (2000) Polym Adv Technol 11:496–505
5. Galaev IY, Gupta MN, Mattiasson B (1996) Chemtech 26:19–25
6. Gehrke SH (1993) Adv Polym Sci 110:81–114

7. Schild HG (1992) Prog Polym Sci 17:163–249
8. Dong LC, Hoffman AS (1990) J Contr Release 13:21–31
9. Yakushiji T, Sakai K, Kikuchi A, Aoyagi T, Sakurai Y, Okano T (1998) Langmuir 14:4657–4662
10. Pan YV, Wesley RA, Luginbuhl R, Denton DD, Ratner BD (2001) Biomacromolecules 2:32–36
11. Shibayama M, Tanaka T (1993) Adv Polym Sci 109:1–62
12. Dusek K, Prins W (1969) Adv Polym Sci 6:1–102
13. Hooper HH, Baker JP, Blanch HW, Prausnitz JM (1990) Macromolecules 23:1096–1104
14. Ilavsky M (1982) Macromolecules 15:782–788
15. Flory PJ (1953) Principles of Polymer Chemistry. Cornell University Press, Ithaca, NY
16. Hirotsu S (1992) Macromolecules 25:4445–4447
17. Harmon ME, Jakob TAM, Knoll W, Frank CW (2002) Macromolecules 114:10471–10478
18. Kuckling D, Harmon ME, Frank CW (2002) Macromolecules 35:6377–6383
19. Harmon ME, Kuckling D, Frank CW (2003) Macromolecules 36:162–172
20. Harmon ME, Kuckling D, Pareek P, Frank CW (2003) Langmuir 19:10947–10956
21. Yamato M, Konno C, Kushida A, Hirose M, Utsumi M, Kikuchi A, Okano T (2000) Biomaterials 21:981–986
22. Okano T, Yamada N, Okuhara M, Sakai H, Sakurai Y (1995) Biomaterials 16:297–303
23. Ebara M, Yamato M, Hirose M, Aoyagi T, Kikuchi A, Sakai K, Okano T (2003) Biomacromolecules 4:344–349
24. Knoll W (1998) Ann Rev Phys Chem 49:569–638
25. Prucker O, Christian S, Beck H, Rühe J, Frank CW, Knoll W (1998) Macromol Chem Phys 199:1435–1444
26. Schmaljohann D, Nitschke M, Schulze R, Eing A, Werner C, Eichhorn KJ (2005) Langmuir 21:2317–2322
27. Okano T, Kikushi A, Sakurai Y, Takei Y, Ogata N (1995) J Contr Release 36:125–133
28. Liang L, Feng X, Liu J, Rieke PC (1999) J Appl Polym Sci 72:1–11
29. Ling L, Habicher WD, Kuckling D, Adler HJ (1999) Design Monom Polym 4:351–358
30. Skulason H, Frisbie D (2000) J Am Chem Soc 122:9750–9760
31. Kuckling D, Hoffmann J, Plötner M, Ferse D, Kretschmer K, Adler HJP, Arndt KF, Reichelt R (2003) Polymer 44:4455–5562
32. Barker IC, Cowie JMG, Huckerby TN, Shaw DA, Soutar I, Swanson L (2003) Macromolecules 36:7765–7770
33. Shibayama M, Mizutani S, Nomura S (1996) Macromolecules 29:2019–2024

Progr Colloid Polym Sci (2006) 132: 152–160
DOI 10.1007/2882_033

Published online: 8 March 2006

Gérald Guérin
Jose Raez
Xiao-Song Wang
Ian Manners
Mitchell A. Winnik

Polyferrocenylsilane Block Copolymers: Nanotubes and Nanowires through Self-Assembly

Gérald Guérin · Jose Raez ·
Xiao-Song Wang · Ian Manners ·
Mitchell A. Winnik (✉)
Department of Chemistry, University of Toronto, 80 St. George Street, Toronto, Ontario M5S 3H6, Canada
e-mail: mwinnik@chem.utoronto.ca

Jose Raez
National Institute for Nanotechnology (NINT), Edmonton, Alberta T6G 2V4, Canada

Abstract Polyferrocenyldimethylsilane (PFS) diblock copolymers with polyisoprene (PFS-PI) or with polydimethylsiloxane (PFS-PDMS) self-assemble in simple alkane solvents to form what appear by TEM to be dense flexible cylinders (nanowires) or "nanotube-like" structures. Typical widths are on the order of 20 to 30 nm, with variable lengths often greater than 10 μm. The structures that form, and the dimensions of the "tube-like" structures or wires, depend upon the composition of the polymers and the lengths of the blocks. Light scattering experiments show that the PFS-PDMS (block ratio 1:12) solutions aged at 25 °C contain long thin objects 950 nm in length. These structures probably rearrange on the TEM grid as the solvent evaporates to form elongated structures. Establishing a mechanistic connection between the objects present in solution and those seen in the TEM images, as well as understanding the factors that lead to these unusual self-assembled structures, remain a challenge.

Keywords Coil-crystalline block copolymers · Micelles in solution · Light scattering · Transmission electron microscopy · Cross-linkable micelles

Introduction

Block copolymers self-assemble to form nanoscale organized structures in a selective solvent. The most common structures are spheres, with the insoluble core surrounded by a solvent-swollen corona. In some instances, disk- or worm-like micelles form, and are of particular interest, since the control of their association can lead to a broad range of new applications [1, 2]. An important subset of block copolymer micelles are those which contain metal atoms, through covalent attachment or by complexation [3]. These structures are interesting because they take advantage of the intrinsic properties of their components, such as the mechanical properties of the polymer micelles and the optical and magnetic characteristics of the metal atoms. Moreover, the assembly permits the control of the uniformity in size and shape of the nanoparticles, and it stabilizes them.

We have a specific interest in the self-assembled structures formed by poly(ferrocenylsilane) block copolymers, such as poly(ferrocenyldimethylsilane-*b*-dimethylsiloxane) (PFS-PDMS) and (ferrocenyldimethylsilane-*b*-isoprene) (PFS-PI). The PFS block contains an iron atom in the main chain repeat unit. These polymers are particularly promising for novel applications, since they can be used as charge-transport materials and, by pyrolysis, as precursors to ferromagnetic ceramics [4–6]. Moreover, they can by synthesized with a very narrow molar mass distribution, with excellent control over chain length and composition [7]. An important feature of PFS is that the polymers bearing two methyl groups on the silane unit are crystalline, whereas polymers with two different substituents on each silane (methyl, ethyl; methyl, phenyl) are atactic and remain amorphous. This feature of the polymer composition has a strong influence on the type of self assembled structures that these polymers can form in solvents selective for the PI or PDMS block.

Both PFS-PI and PFS-PDMS are synthesized by two-step anionic polymerization. The synthetic approach for the preparation of PFS-PDMS is shown in Scheme 1 [8, 9]: *n*-butyllithium was used to initiate the polymerization of the strained silicon-bridged ferrocenophane in THF solution, while the second block was built by the subsequent addition of hexamethyltrisiloxane (D_3). The reaction was terminated with chlorotrimethylsilane. To obtain PFS-PI, the PI block was initiated with butyllithium, followed by the addition of the silicon-bridged ferrocenophane.

Morphology Observed by Transmission Electronic Microscopy (TEM)

One of the first samples we examined was PFS_{50}-$PDMS_{300}$ (the subscript refers to the degree of polymerization of each block), in which the PDMS block was 6 times longer than the PFS block. We discovered [10, 11] that this polymer self-assembled in hexane, a non-solvent for the PFS, to form elongated worm-like structures (Fig. 1a). The transmission electron microscopy (TEM) images in Fig. 1 are not stained. The Fe atoms in the PFS block provide sufficient contrast, and we infer that we see only the PFS domains of the self-assembled structures. Other samples, such as PFS-b-PI having the same block ratio (1 : 6) also form long fiber-like micelles (Fig. 1b) [11] and one PFS-b-PI sample in which the PI was the shorter block (block ratio 2 : 1) formed platelet structures [12].

Since one of the issues raised in this paper is whether the objects seen in the TEM images are a proper representation of the structures present in solution, we will describe briefly sample preparation strategies. For solutions of micelles in hexane, a very volatile solvent, samples for TEM studies could be obtained by aspirating a dilute solution directly onto a carbon-coated copper grid. Most of the solvent likely evaporated as the sample was deposited on the substrate. Alternatively, the TEM substrate could be dipped briefly into a dilute solution of the micelles and allowed to dry. This method also worked for less volatile solvents like decane. For decane, we could also place a small drop (a few μl) of solution on the grid and then touch the edge of the droplet with a Kimwipe to remove excess solvent. For several samples these methods were compared, and we observed the same morphology.

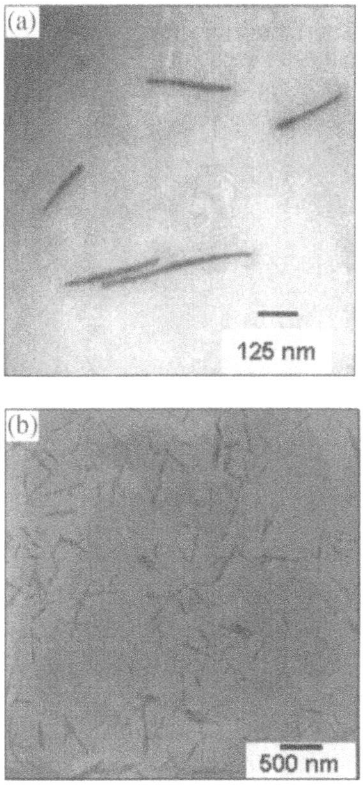

Fig. 1 Transmission electron micrograph of **a** PFS-*b*-PDMS and **b** PFS-*b*-PI micelles formed in hexane, prepared by direct dissolution. The sample was not stained

TEM images were obtained with a Hitachi model 600 electron microscope at 75 kV. Before every TEM session, the electron beam was aligned to minimize optical artifacts.

Influence of the Crystallinity

A common characteristic of these elongated structures is that the PFS block is crystalline. Crystallinity was detected by Wide Angle X-ray Scattering (WAXS) measurements on dry film samples prepared from micelle solutions. An example is shown in Fig. 2 for PI_{320}-PFS_{53}, where the WAXS pattern exhibits strong reflection peaks. The strongest reflection peak is related to a period of 6.42 Å and corresponds to the 6.3 Å spacing found in single crystals of the PFS pentamer, which is due to a Fe-Fe spacing in adjacent molecules [13].

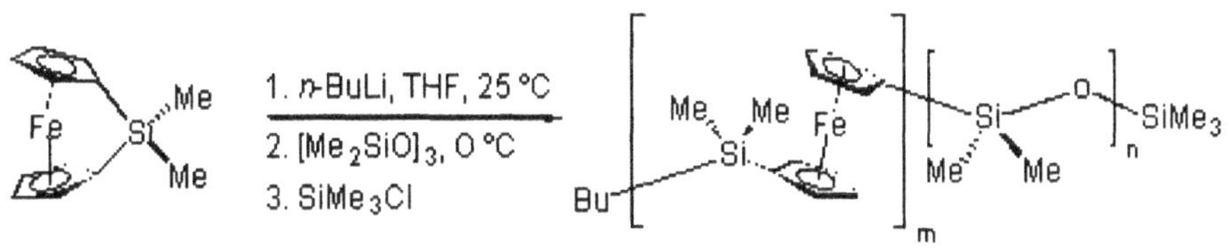

Scheme 1 PFS-*b*-PDMS

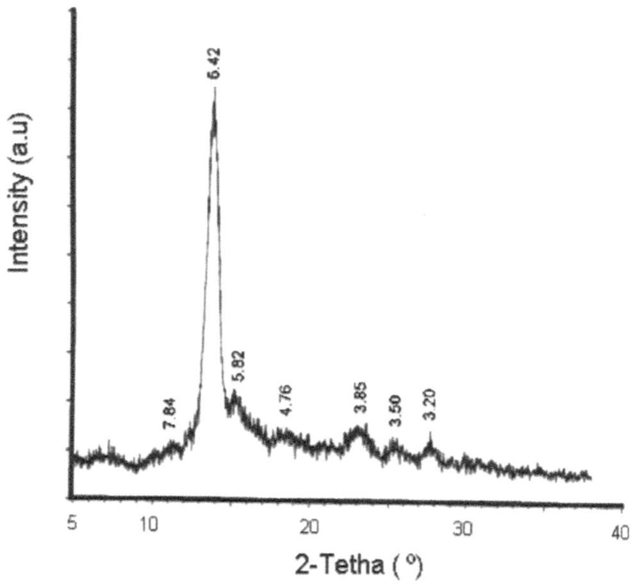

Fig. 2 WAXS pattern of wormlike micelles of PI_{320}-*b*-PFS_{53}. The related period are given above each apparent peak

We inferred that the unusual structures of PFS-b-PDMS and PFS-b-PI were due to the crystalline nature of the PFS block [11, 12]. In order to verify this conclusion, we slightly modified the poly(ferrocenyldimethylsilane) block by replacing one methyl group linked to the silicon by an ethyl group, leading to a non-crystallizable poly(ferrocenylethylmethylsilane) block (PFEMS) [11]. We found that PI and PDMS diblock copolymers with the non-crystallizable PFEMS block form only star-like micelles in n-alkane solvents. Moreover, further studies performed on PI copolymerized with polyferrocenylphosphine (PFP) which cannot crystallize [14] show only the formation of star-like micelles, in agreement with our previous observations.

Moreover, we observed by surface force microscopy (SFM) that, once deposed on a silicon substrate, the fiber-like micelles of PFS-PI and PFS-PDMS are much wider than high. This effect is expected because the SFM should detect both the PFS core and the surrounding PI or PDMS corona, and one expects the soft corona to spread along the substrate. More surprising is that the overall height detected by SFM is significantly smaller than the width of the PFS core seen in corresponding TEM images. For example, PFS_{53}-PI_{320} block copolymer micelles from hexane, in the SFM image are characterized by a height of 9 nm and a width of 60 nm [6] (Fig. 3a). What is most surprising is that in the TEM images of the micelles PFS core has a width of 20 nm. Thus it appears that the entire micelle including the PFS core is compressed on the surface of the substrate. In Fig. 3b we show that apparently continuous ceramic nanolines are produced upon exposure of these structures to an oxygen plasma. The ceramic lines are much thinner (ca. 2 nm high) than the micelles because of a loss of organic material during exposure to the plasma.

Influence of the Block Ratio

Another intriguing feature emerged when we examined PFS-PDMS diblock copolymers having a longer soluble block as well as an increased block ratio (1 : 12 and 1 : 18). Samples of the two polymers were prepared by the same protocol, i.e., subjected to the same thermal history. Dilute solutions in decane were heated for 30 min at 61 °C, cooled to room temperature over two hours and allowed to age for 24 hours. We have already shown, in Fig. 1a, that under these conditions, PFS_{50}-$PDMS_{300}$ formed dense rod-like structures in hexane. However, as

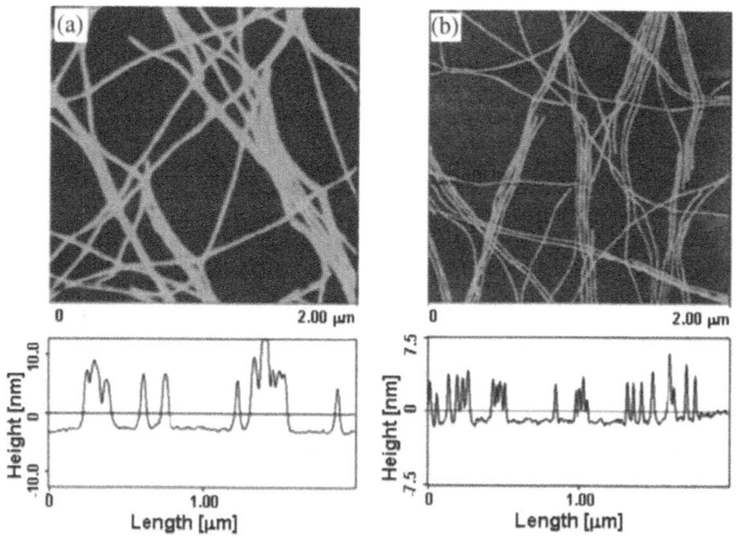

Fig. 3 Representative SFM height images of the cylindrical PFS_{50}-*b*-$PDMS_{300}$ **a** prior and **b** after O_2 plasma etching

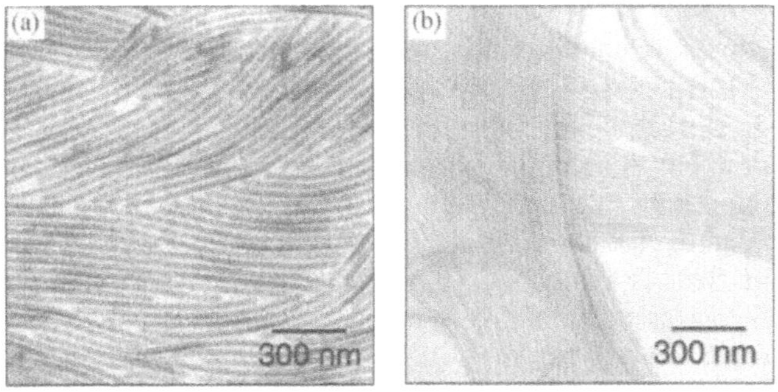

Fig. 4 TEM micrographs of **a** PFS_{40}-$PDMS_{480}$ (1 : 12) and **b** PFS_{54}-$PDMS_{945}$ (1 : 18) assemblies in *n*-decane after being slowly cool down from 61 °C to room temperature

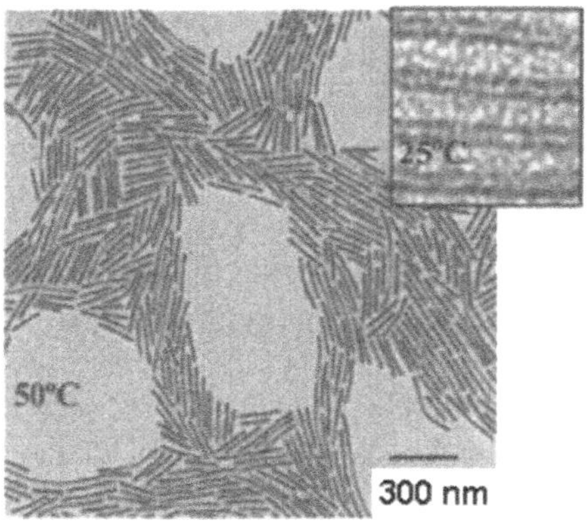

Fig. 6 TEM micrograph of PFS_{40}-*b*-$PDMS_{480}$ assemblies formed in *n*-decane at 50 °C. The inset shows "nanotube-like" structures at 25 °C

we can see in Figs. 4a and b, PFS_{40}-$PDMS_{480}$ (1 : 12) and PFS_{54}-$PDMS_{945}$ (1 : 18) demonstrated an even more complex behavior. From the weakly contrasted images of PFS_{40}-$PDMS_{480}$ and PFS_{54}-$PDMS_{945}$ samples, obtained by TEM, one sees what appears to be extremely long "nanotube-like" structures (Figs. 4a and b), suggesting a hollow cylindrical PFS core surrounded by the PDMS chains.

Influence of the Sample History

We also noticed that the formation of these structures is sensitive to the temperature history of the sample. For example, the sample shown in Fig. 4a, cooled from 61 °C to room temperature (ca. 23 °C) over 2 h and then aged, exhibited long "nanotube-like" structures [8, 15]. In contrast, a sample of the same solution of PFS_{40}-$PDMS_{480}$ in *n*-decane heated at 61 °C for 30 min and then rapidly quenched on the TEM grid, exhibited a mixture of star-like and short rod micelles (Fig. 5).

This high sensitivity to the temperature history encouraged us to study the influence of temperature on the morphology of self-assembled structure. We used the solution slowly cooled from 61 °C to 25 °C (see Fig. 4a) showing "nanotube-like" structures as our starting point, and then heated it to 50 °C. After one day of annealing at this temperature, we noticed that the structures had rearranged to form short dense rods (Fig. 6). The short dense rods seen in the image in Fig. 6, are clearly different from the long structures obtained at 25 °C shown in the inset of Fig. 6. Even more interesting, when the sample was slowly cooled again down to 25 °C, these short rods evolved back to the "nanotube-like" structures. From this reversibility, we infer that these two structures likely represent the equilibrium state of the system at these two different temperatures.

In order to obtain a better understanding of the nature of the structures present in solution at these two temperatures, we began a study of these solutions by light scattering. A classic problem in the field of self-assembly is to know whether the structures seen in TEM images are true representation of the structures present in solution, particularly for samples in slowly evaporating solvents like *n*-decane. The system may have time to rearrange as the solution becomes more concentrated while drying. The light scattering studies are not yet complete, and the following section summarizes results obtained to date.

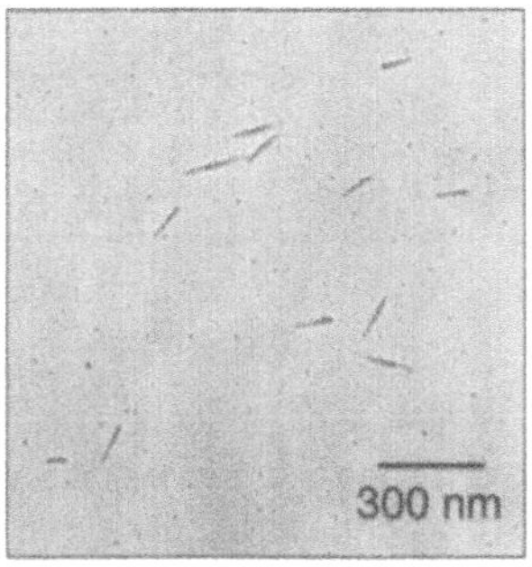

Fig. 5 TEM micrograph of PFS_{40}-$PDMS_{480}$ in *n*-decane heated at 61 °C for 30 min and then rapidly quenched to room temperature

Light Scattering Measurements

Most of our preliminary light scattering experiments were carried out on solutions of PFS_{40}-*b*-$PDMS_{480}$ in *n*-decane. The sample history being extremely important, we compared two different techniques to prepare the micelle solutions for light scattering studies. The first consisted of slowly injecting, through 0.2-μm pore filters, 0.1 mL aliquots of a PFS_{40}-*b*-$PDMS_{480}$ solution in tetrahydrofuran (THF, 10 mg/mL), which is a common good solvent for both blocks. No self assembly is expected, and thus no large self-assembled structures would be removed by the filter. After the complete evaporation of the THF from the cells, filtered *n*-decane was added to obtain a given concentration. After the cells were capped, the samples were placed in the pre-heated vat (60 °C) of the light scattering instrument to dissolve the polymer. The samples were then cooled to 25 °C.

The second technique consisted of heating a PFS_{40}–*b*-$PDMS_{480}$ sample in *n*-decane at 60 °C for 30 min in a closed vial. The solution was allowed to cool relatively quickly to room temperature and then injected into the Al-foil sealed cells through a 0.5-μm pore size filter. The solutions were allowed to equilibrate for 1 day at ambient temperature prior to carrying our light scattering experiments. In both cases, we obtained the same results, insuring us that the sample preparation does not interfere on the light scattering experiments. We also confirm that whatever structures that form when the sample was injected into the light scattering cell were small enough to pass through the filter.

Formation of Supramolecular Structures

Light scattering is a powerful tool to study the formation of supramolecular structures in solution, since, according to Eq. 1, the intensity scattered by a particle in a dilute solution is directly proportional to its mass:

$$i_{u,\theta} = I_0 \frac{4\pi^2 n_0^2}{r^2 \lambda_0^4 N_A} \left(\frac{dn}{dc}\right)^2 M c \tag{1}$$

Here $i_{u,\theta}$ is the intensity of the scattered light at an angle θ to the incident beam; I_0 is that of the polarized primary beam; r, the distance between the scattering molecule and the detector; λ_0, the wavelength of the laser beam in vacuum; n_0, the solvent refractive index; dn/dc, the specific refractive index increment of the solution, M, the molar mass of the scattering particle; and c, its concentration. For a dilute solution, at a given concentration, the scattered intensity can be rewritten as:

$$i_{u,\theta} = K' \left(\frac{dn}{dc}\right)^2 M \tag{2a}$$

where

$$K' = I_0 \frac{4\pi^2 n_0^2}{r^2 \lambda_0^4 N_A} c \tag{2b}$$

K' is a constant at constant solute concentration. When both single polymer chains and micelles are present in solution, populations A and B, respectively, Eq. 2a becomes:

$$i_{u,\theta} = K' \left(\left(\frac{dn}{dc}\right)_A^2 M_A x_A + \left(\frac{dn}{dc}\right)_B^2 M_B x_B \right) \tag{3}$$

where x_A and x_B are the mass fractions, respectively, of the single chains (population A) and the micelles (population B).

In order to verify that the intensity scattered by the single chains could be neglected in Eq. 3, we examined the behavior of the sample PFS_{54}-$PDMS_{945}$ with the largest content of soluble block. Here the block ratio PFS/PDMS is 1 : 18, and this among all of our samples should have the greatest tendency to exist with free chains present in solution. For this sample, we studied the evolution of the scattered light in two different solvents: hexane and benzene. Whereas hexane is a selective solvent for the PDMS block, benzene is a good solvent for both blocks. As we will see below, association takes place in hexane and evolves over time. No self-assembly to form micelles occurs in benzene [9]. Figure 7 shows the relative scattered intensity obtained as the function of time for two solutions of identical concentration. The scattered intensity for the hexane solution was normalized by dividing it by the scattered intensity obtained over the same time for a benzene solution of the polymer at the same concentration. This analysis presumes that the difference in dn/dc values for the single chains in benzene compared to hexane is much smaller than the increase in scattering contribution caused by the mass and mass-fraction contributions of the aggregated structures.

The large increase of the normalized scattered intensity observed for the hexane solution can be explained by the growth of the self-assembled structure as a function of time. At short times the intensity scattered in both solutions (in benzene and hexane) are nearly equal, indicating that, for this sample (PFS_{54}-$PDMS_{945}$), the micelle formation rate is slow. Moreover, from Eq. 3, it can be concluded that once the final structure is formed, there is a negligible contribution to the scattering intensity from single chains in solution. We can be confident that the scattering signal we measure is due to self-assembled structures.

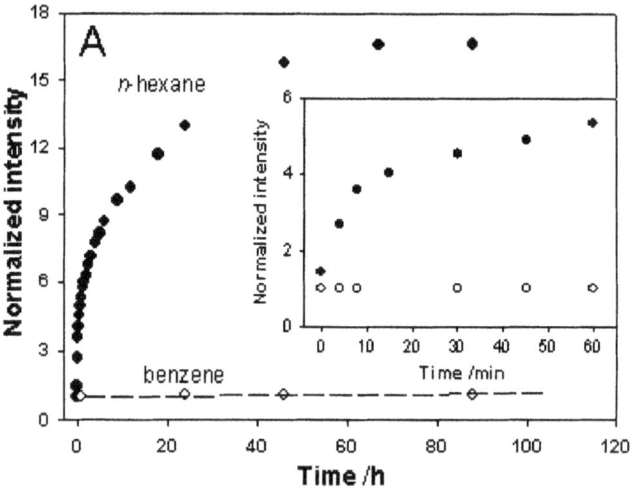

Fig. 7 Evolution of the light scattering intensity for PFS_{54}-*b*-$PDMS_{945}$ solutions at 1.0 mg/mL in *n*-hexane and in benzene (the *dashed curve* is used as a guide for the eyes). The measured intensity for each solution in hexane was divided by the measured intensity of the solution in benzene aged for an identical time. The *inset* shows the evolution of intensity over the first 60 min

Thermal Reversibility of the Morphology

In Figs. 5 and 6 we present evidence for a remarkable morphology transition with a change in temperature for PFS_{40}-$PDMS_{480}$ in decane. The "nanotube-like" structures that formed initially on cooling the solution to 25 °C (Fig. 5) evolved to form much shorter dense rod-like structures (Fig. 6) when the solution was heated to 50 °C. This process was reversible, and the rearrangement in both directions appeared to be slow. In this section, we describe how a sudden change in temperature affects the light scattering signal as a measure of the kinetics of the structure rearrangement. A solution of PFS_{40}-$PDMS_{480}$ in decane, allowed to equilibrate for one day at 25 °C, was suddenly immersed in the sample chamber vat of the light scattering instrument maintained at 50 °C. In Fig. 8 we plot the evolution of the normalized scattered light intensity (at an angle of 90°) as a function of time for this solution. At short times (less than 3 min) the normalized scattered intensity decreased sharply to a minimum value less than 10% of the initial value. At longer times, the intensity grew, reaching a plateau value at ca. 25% of the initial normalized intensity.

From the sudden decrease of the normalized intensity, we can infer that the long structures present at 25 °C rapidly dissociate or fragment at 50 °C, to form much smaller particles with substantially weaker scattering and then re-assemble as a function of time to form the rod-like structures shown in Fig. 6. A representative TEM image of the intermediate assembly is shown in the inset of Fig. 8. The sample was prepared by dipping a carbon-coated copper grid into the warm solution, after a short heating time (165 s). The image shows a mixture of very short cylindrical particles (rods and "nanotube-like" structures). While it is possible that some further structure evolution took place on the TEM grid as the decane evaporated, the presence of smaller objects in this image is in general agreement with the results obtained by the light scattering experiment.

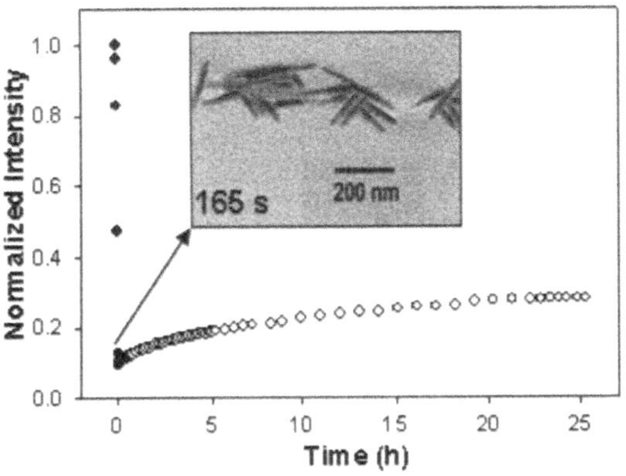

Fig. 8 Evolution of the light scattering (LS) intensity of PFS_{40}-*b*-$PDMS_{480}$ assemblies in *n*-decane (0.33 mg/mL) at 50 °C over a period of 24 h. The LS intensity was normalized to the LS intensity of the assemblies at 25 °C, after they reached equilibrium. Each *open symbol* represents a 30-second scan, while each *full symbol* represents a 10-second scan. The *inset* shows a TEM micrograph of the structure after an annealing time of 165 s at 50 °C

Static Light Scattering

In all experiments involving self assembly, it is important to establish whether the structures seen by microscopy correspond to those present in solution. Scattering experiments are powerful set of tools to examine the size, shape and dynamic properties of species in solution. We have begun to carry out static and dynamic light scattering (SLS and DLS, respectively) in order to learn about the size (mass, length), size distribution, stiffness, and diffusional properties of the self-assembled structures present in solution. We began our investigation with the study of dilute solutions of PFS_{40}-$PDMS_{480}$ in decane at 25 °C by SLS. From SLS, one obtains structural and dimensional information by measuring the angular dependence of the excess absolute time-average scattered intensity (the Rayleigh ratio, R_θ). For dilute solutions, R_θ is related to the sample concentration c, the second viral coefficient A_2, and the form factor $P(q)$ through the expression:

$$\left(\frac{Kc}{R_\theta}\right) = \left[\frac{1}{M_w P(q)} + 2A_2 c\right] \tag{4}$$

where $K = 4\pi^2 n_D^2 (dn/dc)^2/(N_A \lambda^4)$, and N_A, λ, dn/dc and n_D are Avogadro's number, the incident wavelength (632.8 nm), the refractive index increment, and the refractive index of the solvent, respectively. We measured a dn/dc value of 0.0465 mL/g ($\lambda_0 = 620$ nm) for the PFS_{40}-*b*-$PDMS_{480}$ micelles in *n*-decane at 25 °C.

SLS studies were performed at four different copolymer concentrations c (0.33, 0.57, 0.75 and 0.85 mg/mL). Because we learned from the kinetics studies that the sample equilibrium time was around 24 h, each solution was allowed to age for 24 h before data collection. Figure 9 shows the plot of Kc/R_θ as a function of q^2, at those four concentrations and the values extrapolated to zero concentration (open circles).

An interesting feature of Fig. 9 appears in the evolution of Kc/R_θ as a function of the polymer concentration. This value increases as the copolymer concentration decreases, leading to an apparent negative second virial coefficient, A_2. This observation is intriguing since micellar systems normally have positive values of A_2. Moreover, the downward curvature at increasing scattering angle indicates the presence of extended or rod-like objects, and the increasing downward curvature with increasing concentration may indicate a growth in size of the object present as the concentration is increased, in accordance with the apparent negative value of A_2. The relatively poor solvent quality of *n*-decane for PDMS may also contribute to the negative value of A_2.

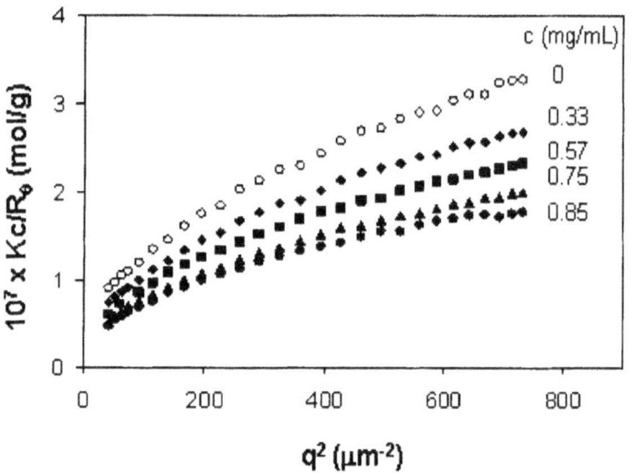

Fig. 9 Plots of Kc/R_θ as a function of q^2 of PFS_{40}-*b*-$PDMS_{480}$ assemblies in *n*-decane at 25 °C at four concentrations (0.33, 0.57, 0.75, 0.85 mg/mL) and extrapolated at $c = 0$ (*open symbol*) **N.B.** Each sample was dissolved in *n*-decane at 60 °C, cooled to room temperature, and then allowed to age 24 h at 25 °C

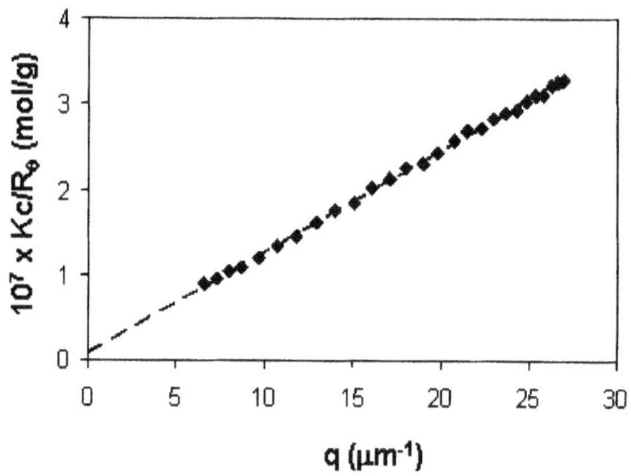

Fig. 10 Plot of $(Kc/R_\theta)_{c=0}$ as a function of q (*filled diamonds*) extrapolated to $q = 0$ (*dashed line*)

The determination of the molecular weight was complicated by the lack of data at low angles, and a simple Zimm plot analysis of the curve $(Kc/R_\theta)_{c=0}$ could not be performed in a straightforward way (as suggested by Eq. 4), since it would lead to an underestimated value of the micelle molecular weight. However, from the downward curvature of these different plots we were able to deduce that the scattering particles present in solution had a rod-like shape.

According to Holtzer [16], the Rayleigh ratio of rigid, rod-like scattering particles reaches a linear asymptotic behavior at large values of qL, given by:

$$\left(\frac{Kc}{R_\theta}\right)_{c=0} = \frac{1}{M_n}\left(\frac{2}{\pi^2} + \frac{1}{\pi}L_n q\right) \tag{5}$$

where M_n and L_n are the number average molecular weight and length of the rods, respectively. As one can see in Fig. 10, the plot of $(Kc/R_\theta)_{c=0}$ as a function of q can be easily fitted by a straight line, which is strong evidence of the rod-like nature of the PFS_{40}-*b*-$PDMS_{480}$ assembly in *n*-decane at 25 °C. From this representation, we calculated that $M_n \approx 2.6 \pm 0.3 \times 10^7$ g/mol and $L_n \approx 950$ nm. This result indicates that there is on average 570 block copolymer molecules per rod-like micelle, which corresponds to only 6 polymer molecules per 10 nm of length.

The lack of data at low angles not only make it impossible to obtain an unambiguous value for the weight average molecular weight, M_w, of the assembly from the Zimm plot, they also make it difficult to analyze the data to evaluate the polydispersity and the flexibility of the structures in solution. Nevertheless, when the experimental data at higher qL region are fitted to a model which assumes a Zimm-Schulz distribution of rod lengths (centered at $L = 950$ nm) [17], the best fit to the data implies a rather narrow distribution of rod lengths.

We want to point out that, although the length of the structures calculated from SLS is rather long, this length is still lower than the length observed on the TEM grids. This result seems to indicate that the structures present in dilute solution undergo end-to-end fusion when the solutions are dried on the TEM grid.

Dynamic Light Scattering

In a dynamic light scattering experiment, the measured intensity-intensity time-correlation function $g^{(2)}(t_c)$, where t_c is the delay time, is related to the normalized electric field correlation function $g^{(1)}(t_c)$, representative of the motion of the particles, by the Siegert relation [18]:

$$g^{(2)}(t_c) = 1 + \beta \left|g^{(1)}(t_c)\right|^2 \tag{6}$$

where β (≤ 1) is the spatial coherence coefficient for instrument beating efficiency. As t_c increases, $g^{(1)}(t_c)$ will decay. This decay carries information about the diffusion of the particles in solution, which is strongly influenced by the shape, size and size distribution of the scattering particles, and by their flexibility. As a consequence, DLS gives information complementary to that obtained by SLS about the assemblies present in solution.

The electric field correlation function can also be expressed in terms of a cumulant expansion:

$$\ln g^{(1)}(t_c) = -\Gamma t_c + \left(\frac{\mu_2}{2!}\right)t_c^2 - \left(\frac{\mu_3}{3!}\right)t_c^3 + \ldots \tag{7}$$

where Γ is the first cumulant, and μ_2 and μ_3 are the second and third cumulants, respectively. For samples characterized by a monomodal distribution of diffusing species, the ratio μ_2/Γ^2 is a measure of the width of the distribution A preliminary study of PFS_{40}-*b*-$PDMS_{480}$ in decane has been carried out for a solution at $c = 0.33$ mg/mL [17]. The distribution of rod lengths was evaluated at 90° using the cumulant expansion of $g^{(1)}(t_c)$, and we found that

$\mu_2/\Gamma^2 = 0.12$, confirming that the polydispersity of the rod lengths was rather narrow.

We can obtain further information about the sample from the angular dependence of the DLS signal, analyzed in conjunction with the rod length obtained from the SLS data. Since this length is known, we can calculate the translational and rotational diffusion coefficients of the assemblies in solution, by fitting the experimental DLS data with the Maeda and Fujime expression for the apparent diffusion coefficient [19]:

$$\Gamma/q^2 = D_t + (L^2/12)D_r f_1(k) - (D_{||} - D_{\perp})(1/3 - f_2(k)) \quad (8)$$

where $k = qL/2$, $D_{||}$ and $D_{\perp}$ are the diffusion coefficients for translation parallel and perpendicular to the rod's axis, D_r is the rotational diffusion coefficient and D_t is the overall translational diffusion coefficient of the rod. The functions $f_1(k)$ and $(1/3 - f_2(k))$ are weighting factors depending only on k. The term $f_1(k)$ accounts for the contribution of rotation to Γ/q^2. The term $(1/3 - f_2(k))$ accounts for the contribution of translational anisotropy. Moreover, the values of D_t, $D_{||}$, $D_{\perp}$ and D_r can be calculated via the Broersma equations [20–22], which give a relation between the diffusion coefficients and the hydrodynamic dimensions of a cylinder (the length and the hydrodynamic diameter). The dependence of the apparent diffusion coefficient, $D_{app} = \Gamma/q^2$, as a function of qL is shown in Fig. 11. From the fitting of the experimental data and the use of the Broersma equations, we deduce that the cross-sectional hydrodynamic diameter of the structure is ca. 20 nm for a concentration of 0.33 mg/mL in n-decane.

There are several ambiguities in the data analysis that can only be resolved with further experiments. Data at lower angles when added to those in Fig. 11 should provide a better fit to the hydrodynamic diameter of the rod-like micelles. In addition, DLS experiments on solutions of different concentrations are needed in order to be able to extrapolate the dependence of Γ/q^2 as a function of qL to zero concentration. In this way one can avoid any influence of possible concentration dependence of micelle size on the DLS results.

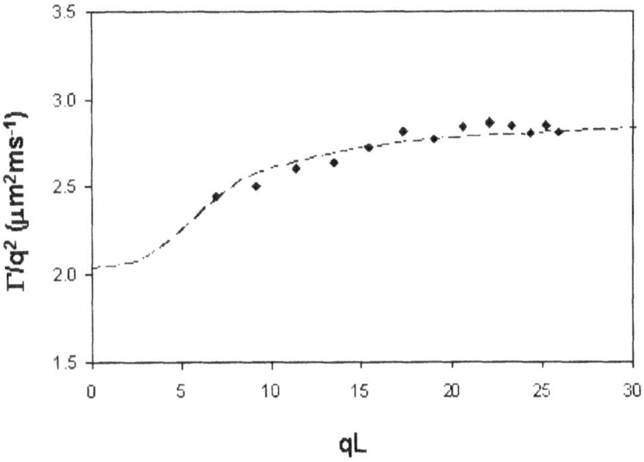

Fig. 11 Γ/q^2 versus qL, data (*full diamonds*) obtained at a polymer concentration of 0.33 mg/mL. The *dashed line* is a fit of Γ/q^2 using the Maeda equation (Eq. 8) with $L = 1050$ nm and $r = 9$ nm. We present the full curve to emphasize the change in curvature expected at low angles

Structure Conservation for TEM Experiments

There are many reasons to attempt the synthesis of micelles with a crosslinked corona [23, 24]. In the case of PFS block copolymer micelles, crosslinking the corona may enhance the thermal stability of the micellar structures on a substrate. PFS can form a magnetic ceramic upon pyrolysis, but in the absence of crosslinking, the yield is low. Corona crosslinking may allow the structures formed by the micelles to persist to high enough temperature that micelles such as those shown in Fig. 2 can be converted to magnetic ceramic nanolines upon pyrolysis. For PFS-PI, crosslinking was achieved by a platinum-catalyzed hydrosilylation reaction of the unsaturated sites in the PI component with tetramethylsiloxane, and our first examination of these materials indicated that they could be converted to continuous ceramic lines upon heating [4].

From the perspective of this paper, another reason to crosslink the micelles in dilute solution is that network formation in the corona should lock in place the structures present, and minimize the possibility of a structural transformation when the solution is transferred to a TEM grid and allowed to dry. A clever way to obtain this result for the PFS diblock copolymer system would be to modify the PDMS block, in the hope that the modified system also would self assemble into "nanotube-like" structures in n-alkane solvents. We syn-

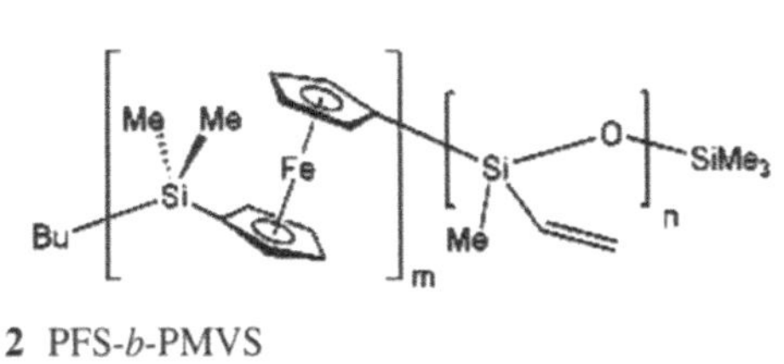

Scheme 2 PFS-*b*-PMVS

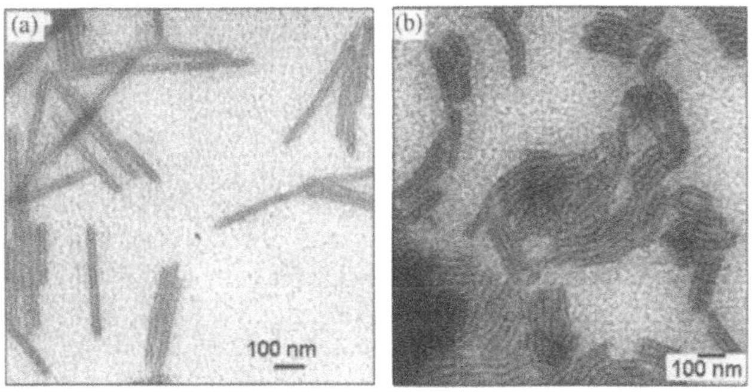

Fig. 12 TEM micrograph of PFS_{48}-$PMVS_{300}$ assemblies after shell-crosslinking. **a** from hexane and **b** fromTHF

thesized the polymer poly(ferrocenyldimethylsilane-*b*-methylvinylsiloxane) (PMVS, Scheme 2). This polymer meets our requirements, since it contains a reactive vinyl group in the repeat unit, and its structure is close to that of PDMS.

A TEM micrograph of a PFS_{48}-$PMVS_{300}$ is shown in Fig. 12a. To our good fortune, this polymer also forms "nanotube-like" structures in hexane. Crosslinking the corona can be achieved by Pt-catalyzed hydrosilylation with tetramethylsiloxane (to form PFS-xPMVS). In principle, the degree of crosslinking of the assembly can also be controlled by adjusting the amount of tetramethylsiloxane in the crosslinking reaction. As a test of the success of this reaction, we showed that the structure can be transferred in THF without dissociating, as detected by DLS and by TEM. Figure 12b shows that when a THF solution of the hydrosilylated micelles is dried on a TEM grid, "nanotube-like" structures form that are different in appearance from those obtained initially from hexane. THF is a good solvent for PFS. In THF, the PFS chains dissolve but remain attached to the xPMVS microgel network. Upon drying of the solution, the PFS chains phase separate from the xPMVS and form what appears to be a "connected-bead-like" structure in the interior of the xPMVS "tube." Films formed from THF solutions of PFS-xPMVS show no indication of any crystallinity of the PFS component, emphasizing that the PFS component of the structures formed in this way are rather different than those formed by self-assembly in alkane solvents.

The versatility of the crosslinked structures opens a new broad avenue for their application in several areas. As mentioned above, corona-crosslinked PFS assemblies demonstrate excellent shape retention upon pyrolysis and permit the formation of ceramic replicas. We have also recently shown that in a common good solvent, the PFS chains in the microgel interior of xPMVS can serve as a microreactor for the localized production of metal nanoparticles [25]. We are also about to begin light scattering studies of PMVS micelles, before and after corona crosslinking, and hope to use these experiments to learn more about the self-assembled structures formed in dilute solution.

Summary

The first poly(ferrocenylsilane)-based block copolymers were synthesized about 10 years ago. Experiments carried out in the intervening time have revealed the unusual and highly interesting structures formed by these polymers in different selective solvents that makes them still extremely attractive to study. Their ability to form long nanowires or "nanotube-like" structures has been confirmed both by TEM and light scattering experiments. Nevertheless, a lot of work needs to be done in order to fully understand this intriguing series of polymers. These studies promise new, unexpected, and exciting results.

Acknowledgement The authors thank NSERC Canada and the Province of Ontario under its EMK program for its support of this research.

References

1. Won Y-Y, Davis TH and Bates FS (1999) Science 283:960–963
2. Discher DE and Eisenberg A (2002) Science 297:967–973
3. Förster S and Antonietti M (1998) Adv Mater 10:195–216
4. Wang X-S, Arseneault A, Ozin GA, Winnik MA and Manners I (2003) J Am Chem Soc 125:12686–12687
5. Massey JA, Winnik MA, Manners I, Chan VZ-H, Ostermann JM, Enchelmaier R, Spatz JP and Möller M (2001) J Am Chem Soc 123:3147–3148
6. Cao L, Massey JA, Winnik MA, Manners I, Riethmüller S, Banhart F, Spatz JP and Möller M (2003) Adv Funct Mater 13:271–276
7. Ni Y, Rulkens R and Manners I (1996) J Am Chem Soc 118:4102–4114
8. Raez J, Manners I and Winnik MA (2002) J Am Chem Soc 124:10381–10395
9. Raez J, Manners I and Winnik MA (2002) Langmuir 18:7229–7239
10. Massey JA, Power KN, Manners I and Winnik MA (1998) J Am Chem Soc 120:9533–9540
11. Massey JA, Temple K, Cao L, Rharbi Y, Raez J, Manners I and Winnik MA (2000) J Am Chem Soc 122:11577–11584
12. Cao L, Manners I and Winnik MA (2002) Macromolecules 35:8258–8260
13. Rulkens R, Lough AJ, Manners I, Lovelace SR, Grant C and Geiger WE (1996) J Am Chem Soc 118:12683–12695
14. Cao L, Manners I and Winnik MA (2001) Macromolecules 34:3353–3360
15. Raez J, Barjovanu R, Massey JA, Winnik MA and Manners I (2000) Angew Chem Int Ed 39:3862–3865
16. Holtzer A (1955) J Polym Sci 17:432
17. Guérin G, Raez J, Manners I and Winnik MA (2005) Macromolecules 38:7819–7827
18. Siegert AJF (1943) Massachusetts Institute of Technology, Rad Lab Rep No 465
19. Maeda T and Fujime S (1984) Macromolecules 17:1157
20. Broersma S (1960) J Chem Phys 32:1626
21. Broersma S (1960) J Chem Phys 32:1632
22. Broersma S (1981) J Chem Phys 74:6989
23. Thurmond KB, Kowalewski T and Wooley KL (1996) J Am Chem Soc 118:7239–7240
24. Thurmond KB, Kowalewski T and Wooley KL (1997) J Am Chem Soc 119:6656–6665
25. Wang X-S, Wang H, Winnik MA and Manners I (2005) J Am Chem Soc 127:8924–8925

Progr Colloid Polym Sci (2006) 132: 161–167
DOI 10.1007/2882_042

Published online: 24 February 2006

Vera Bocharova
Ganna Gorodyska
Anton Kiriy
Manfred Stamm
Paul Simon
Ingolf Mönch
Dieter Elefant
Xudong Lou
Francois Stoffelbach
Christophe Detrembleur
Robert Jérôme

Synthesis and Deposition of Water-Dispersed Prussian Blue Nanocrystals on Polymers and CNTs

Vera Bocharova · Ganna Gorodyska ·
Anton Kiriy (✉) · Manfred Stamm
Leibniz Institute of Polymer Research
Dresden, Hohe Str. 6, 01069 Dresden,
Germany
e-mail: kiriy@ipfdd.de

Paul Simon
Max-Planck-Institut for Chemical Physics
of Solids, Nöthnitzer Str. 40,
01187 Dresden, Germany

Ingolf Mönch · Dieter Elefant
Leibniz Institute for Solid State and
Materials Research Dresden, IFW
Dresden, Helmholtzstraße 20,
01069 Dresden, Germany

Xudong Lou · Francois Stoffelbach ·
Christophe Detrembleur · Robert Jérôme
Center for Education and Research on
Macromolecules (CERM), University of
Liege, Sart-Tilman, B6, 4000 Liege,
Belgium

Abstract Nanoparticles can be used for decoration and functionalization of single polymer molecules that have been adsorbed to a solid substrate. Initial attempts to prepare Prussian Blue nanoclusters by a layer-by-layer deposition technique of hexacyanoferrate anions and ferric cations onto *isolated polycation chains* in water failed because of the desorption of the first layer upon deposition of the next one. A simple method for the preparation of charge-stabilized Prussian Blue nanoparticles of readily adjustable size is reported. Prussian Blue nanoparticles have been purified by addition of non-solvents and redispersed in water without aggregation. Thus formed Prussian Blue nanoparticles are crystalline and display a long-range ferromagnetic ordering at 5.1 K. Prussian Blue nanoparticles were selectively deposited along single polycation molecules to form a one-dimensional array or were attached to the surface of carbon nanotubes (CNTs) functionalized with poly2-vinylpyridine (P2VP). These nanoparticle-based nanostructures might be useful materials for manufacture of electrooptical devices, or mechanically robust ion-sieving membranes.

Keywords Prussian Blue · Nanocrystal · Carbon nanotubes · Magnetic properties

Introduction

Semiconductor and metallic nanoparticles have been extensively studied as active components in wide variety of basic research and technological applications due to their new or improved optical, electric, and magnetic properties compared to their bulk counterparts [1]. Therefore, exploitation of well-known materials in their new nanosized forms is strongly motivated area of research. Prussian Blue [2], an old pigment, is a coordination polymer formed by reaction of either hexacyanoferrate(II) anions with ferric (Fe(III)) cations, or hexacyanoferrate(III) anions with ferrous (Fe(II)) cations [3]. According to X-ray diffraction analysis, PB is a three-dimensional crystal of ferric and ferrous ions, which alternate at the sites of a cubic lattice [4]. The ferric ion is coordinated to the nitrogen atoms, and the ferrous ion to the carbon atoms, of the bridging cyanide ligands. The remaining charge is balanced either by potassium ions in the so-called "soluble" PB, or by ferric ions in the "insoluble" PB. The term "soluble", however, does not refer to the true solubility but only to the tendency of PB to form colloidal solutions [3].

A large family of cyano-bridged compounds with a cubic structure (PB analogues; PBs) is known for various interesting physical properties [3]. Neff et al. reported that the oxidation state of the iron centers could be controlled electrochemically, making possible dramatic color changes [4] that could be used in electrochromic devices (such as flat panel displays or switchable pictograms) [6,7]. PB shows a long-range ferromagnetic ordering at 5.6 K, whereas few PBs undergo magnetization at room temperature and even higher [8]. Hashimoto et al. showed that the magnetic properties of PBs could be modulated not only by the chemical composition but also by an optical or electrical stimulation (photomagnetism and electromagnetism, respectively) [9–11]. Finally, PBs exhibit remarkable ion-sieving properties as result of an open pore zeolite-like structure [12].

For the unique properties of PBs to be exploited, PBs must be deposited properly onto a solid support. It is highly desirable to prepare mechanically robust PBs films with controlled thickness, chemical composition and crystallinity, having ion-sieving membranes and electrochromic devices in mind [6], or to create regular patterns of PB-based single molecule magnets [13].

Classical methods of PB immobilization by casting [14], dip-coating [5] or electrochemical deposition do not allow film thickness, composition and/or mechanical properties to be controlled accurately. This problem has recently been overcome by the Langmuir-Blodget [15] and multiple sequential adsorption techniques [12,16] based on the stepwise adsorption of hexacyanoferrate anions and ferric (or ferrous) cations. Unfortunately, these methods are relatively expensive and time-consuming procedures. On the other hand, utilization of larger in size PB *nanoparticles* instead of small precursor *ions* would have a favorable effect on film assembly. However, only few synthetic methods of well-defined PB nanoparticles have been reported [10,17–19]. As a rule, these nanoparticles are protected by surfactants or embedded into matrixes, that is undesired for some applications. In our recent communication we reported on a straightforward approach to surfactant-free water-dispersible negatively charged PB nanoparticles simply by the addition of ferric chloride to an excess of potassium hexacyanoferrate(II) [17–19]. Thus formed mixture of PB nanoparticles with excess of unreacted hexacyanoferrate(II) salt were used to improve topography and phase contrasts of single adsorbed polycation chains in case of atomic force microscopy (AFM) imaging [20,21]. Nevertheless, several important issues such as stability, structural and magnetic properties of PB nanoparticles as well as their selective positioning still remain unresolved and are among the tasks of the present study.

Carbon nanotubes (CNTs) have shown exceptional stiffness, strength and remarkable thermal and electrical properties, which make them ideal candidates for the development of multifunctional material systems [22]. Nowadays, CNTs are dispersed within polymer in order to improve their mechanical and electrical properties [23]. Therefore, reinforcement of PB films by CNTs might be a strategy for manufacturing mechanically robust ion-sieving membranes, and for creating electro-conductive pathways in a poorly conductive PB matrix. Recently, we reported on the covalent modification of multiwalled CNTs by poly2-vinylpyridine (P2VP) [24]. The P2VP shell provides an improved dispersibility of CNT-*g*-P2VP in water and organic solvents and is a versatile scaffold for deposition of nanoclusters. Such modifications would be a method to modulate properties of CNTs nanotubes [25]. Furthermore, adsorption of CNT-*g*-P2VP onto oppositely charged surface results in transparent and highly conductive thin films, which can be used as high-surface area electrodes for fabrication of PB-based solid-state electrochromic devices with improved switching time [6,7].

This paper aims at reporting on the preparation, stability, structure, magnetic properties and positioning of PB nanoparticles.

Experimental Section

Materials

CNTs (MWNTs Thick purified from Nanocyl S.A., Belgium, 50 μm long, average inner diameter of 6 nm, outer diameter of 25 nm, and purity higher than 95 wt%) were modified by reaction of 0.3 wt% dispersion of CNTs in dry toluene (Aldrich) with a 100-fold wt excess of 2,2,6,6-tetramrthylpiperidinyl-1-oxyl (TEMPO) end-capped poly2-vinylpyridine (P2VP) [24,25] in toluene at 120 °C for 24 h [26]. Synthesis of poly(methacryloyloxyethyl dimethylbenzyl) ammonium chloride (PMB) [27] ($M_W = 6000$ kg/mol, polydispersity index of 1.6) was reported elsewhere.

Layer-by-layer Assembly on Isolated Polyelectrolyte Molecules

PMB molecules were deposited onto freshly cleaned Si-wafer in a stretched conformation by spin-coating of a 0.005 g/l solution in acidified water (pH 2, HCl) at 10 000 rpm. The substrate with the PMB molecules was then immersed in solutions of $K_4Fe(CN)_6$ (or $K_3Fe(CN)_6$) solutions (0.5–15 g/l) and KCl (0–50 g/l) at the same pH and for the same period of time as in the first part of the cycle. The substrate was analyzed by AFM after either half of cycle, a complete cycle or several cycles.

PB Clusters

Dispersion of PB clusters was prepared by mixing vigorously a solution of $K_4Fe(CN)_6 \cdot 3H_20$ (1.18 mmol/l, Aldrich) in acidified water (HCl, pH 2.0) and an equal volume of a solution of $FeCl_3$ (Aldrich) at the same pH. Concentration of the $FeCl_3$ solution was either 0.148 mmol/l for the preparation of PB1 clusters (smaller size), or 0.296 mmol/l for intermediate PB2 clusters, or 0.444 mmol/l for larger PB3 clusters. A small amount of tetrahydrofuran (Aldrich) was added to the freshly prepared dispersions of PB in order to separate the PB clusters from the unreacted $K_4Fe(CN)_6$ and KCl. PB clusters were collected by filtration, washed with a water-tetrahydrofuran solution (2 : 1), and redispersed in acidified water (HCl, pH 2).

Deposition of the PB Nanoclusters

Si-wafers (Wacker-Chemitronics) were first cleaned with dichloromethane (Aldrich) in an ultrasonic bath for 5 min (3 times), followed by cleaning with a solution prepared by mixing of 4 volume parts of deionized water, 1 volume part of 25% NH_4OH (Aldrich), and 1 volume part of 30% H_2O_2 (Aldrich) at 60 °C for one hour. This NH_4OH : H_2O_2 solution must be handled cautiously because of violent reaction with organic compounds. Samples were finally exposed to 50% sulfuric acid for 15 min and then rinsed repeatedly with water purified through a Millipore (18 MΩxcm) filter. Clean Si-wafers onto which either PMB molecules or CNT-*g*-P2VP nanotubes were deposited (conc. of the CNT-g-P2VP dispersion was 0.1 g/l), were dipped in the freshly prepared dispersion of PB2 (PB3) clusters for 3 min at 25 °C, followed by washing with water and drying under an argon flow. Samples for TEM investigations were prepared by depositing a drop of the PMB or CNT-*g*-P2VP solution onto a standard TEM grid (Plano, Wetzlar, Germany) covered with a holey carbon film. After drying, the PB2 dispersion in acidified water (0.5 g/l) was added on the spot and removed with a filter paper one minute later. Finally, the grid was washed by adding a drop of water and removing it with a filter paper.

AFM Measurements

A multimode AFM instrument and a NanoScope IV-D3100 (Digital Instruments, Santa Barbara) were operated in the tapping mode. Silicon tips with a radius of 10–20 nm, a spring constant of 30 N/m and a resonance frequency of 250–300 kHz were used.

TEM Measurements

TEM images were recorded with a Philips CM 200 FEG at 200 kV. They were processed by the Digital Micrograph program (Gatan, USA).

Magnetic Properties

Dry powder of the Prussian blue nanoparticles (PB3, ~ 15 nm) was compacted by epoxy glue. The magnetization measurements were performed in a SQUID magnetometer (MPSM – Quantum Design).

Ellipsometry

The thickness of the polymer films was measured by an SE400 ellipsometer (SENTECH Instruments GmbH, Germany) with a 632.8 nm laser at a 70° incident angle.

Results and Discussion

Layer-by-layer Deposition of the Prussian Blue Precursors on Isolated Polyelectrolyte Molecules

Prussian Blue was tentatively deposited onto isolated polycations by the well-known layer-by-layer (or multiple sequential adsorption) technique. The stepwise adsorption of hexacyanoferrate anions and ferric (or ferrous) cations was actually effective in fabricating PB films [12]. In this work, the first adsorption step of hexacyanoferrate anions along positively charged PMB chains were successful as confirmed by the increase of the chain thickness by approximately 0.7 nm (Fig. 1a).

However, whenever the hexacyanoferrate pre-adsorbed PMB chains on the Si-substrate were dipped into the ferric chloride solution, the hexacyanoferrate anions were completely removed leaving bare PMB molecules undetectable at the rough Si-wafer surface. The complete removal of the previously deposited hexacyanoferrate layer was systematically observed whatever the salt, e.g., $FeCl_2$, $NiCl_2$, $CoCl_2$, $CuCl_2$, $PdCl_2$, and $AuCl_3$, in a broad range of pH and ionic strength. When the deposition cycle of hexacyanoferrate and $FeCl_3$ onto pre-adsorbed PMB molecules was repeated several times, local growth of relatively big clusters was occasionally observed (Fig. 1b). The failure of the layer-by-layer deposition of PB clusters can be explained as follows. The hexacyanoferrate anions interact expectedly with the positively charged units of the

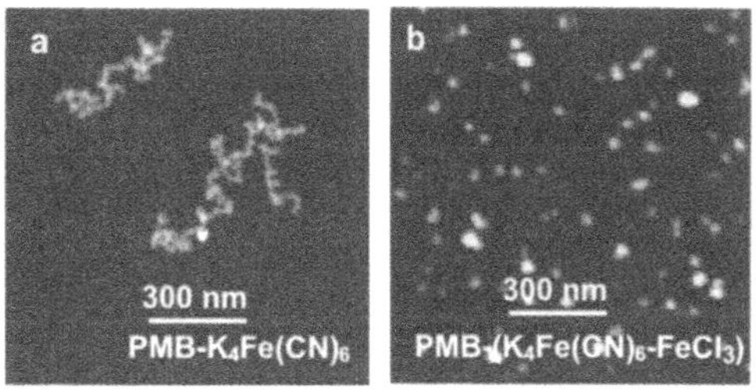

Fig. 1 **a** AFM image of PMB molecules adsorbed onto Si-wafer after dipping into a 5 g/l solution of $K_4Fe(CN)_6$ in acidic water and washing with water (the height of the worm-like structure is about 0.6 nm). Further dipping into a $FeCl_3$ solution removes the hexacyanoferrate anions and the PMB molecules are unobserved (image not shown). **b** AFM image of the sample shown in **a** after the fifth cycle of the sequential dipping into $K_4Fe(CN)_6$ and $FeCl_3$ solutions ($FeCl_3$ being the outermost deposit layer): PB clusters (10 nm height) are randomly located and the PMB molecules are unobserved

PMB molecules. In the next step, an excess of ferric cations interact with the pre-adsorbed hexacyanoferrate particles and overcharge them. The accordingly formed PB nanoparticles are positively charged by a shell of ferric cations, which facilitate their detachment from the similarly charged PMB chains. The observations in this work are consistent with Tieke et al. who reported on the non-linear increase of the PB film thickness with the number of dipping cycles [12]. This irregular growth was pronounced up to the 6-th cycle and whenever the dipping time was relatively long. Thus, although the layer-by-layer method is useful to prepare PB films onto *polyelectrolyte multilayers*, it is ineffective in case of deposition onto *isolated polycationic molecules*.

Synthesis, Stability of Prussian Blue Nanoclusters and Deposition onto Polycations

In order to prepare clear dispersions of PB nanoparticles diluted acidic solutions (pH = 2) of $FeCl_3$ and $K_4Fe(CN)_6$ (used in excess) were mixed [20]. Upon addition of tetrahydrofuran until a H_2O-tetrahydrofuran volume ratio of 2, PB nanoparticles precipitated quantitatively and selectively. They were filtrated and washed with a water-tetrahydrofuran solution (2 : 1). Remarkably enough, they are easily redispersed in water with a constant size. A drop of a PB dispersion in water (0.5 mg/l) was deposited on spin-stretched PMB chains onto a pre-adsorbed Si-wafer. After washing and drying, the sample was analyzed by AFM in the tapping mode. Figure 2 shows typical AFM images of the beads-on-string morphology at the surface of the PMB chains. PB nanoparticles appear to be selectively and regularly attached along the polycation chains, which indicates that this method should be successful anytime a substrate is covered by a tiny amount of polycations.

Range of PB dispersions were prepared by changing the mixing ratios and the stirring time. The PB nanoparticles were analyzed by AFM and the images were statistically treated. In most cases, a narrow size distribution is observed (PDI = 1.1–1.2), which depends on the $K_4Fe(CN)_6/FeCl_3$ mixing ratio. The average diameter of the PB nanoparticles is approximately 3.7 nm when the $K_4Fe(CN)_6/FeCl_3$ molar ratio is 8/1, $\sim$ 4.8 nm for the molar ratio of 4/1, and 10–15 nm for the molar ratio of 3/1. As a rule, the size of the PB nanoparticles remains quasi constant with time. Obviously, the observed water-solubility of the clusters and stability of the dispersions are due to excess of hexacyanoferrate-anions attached to the surface of PB nanoparticles and caused their negative charge.

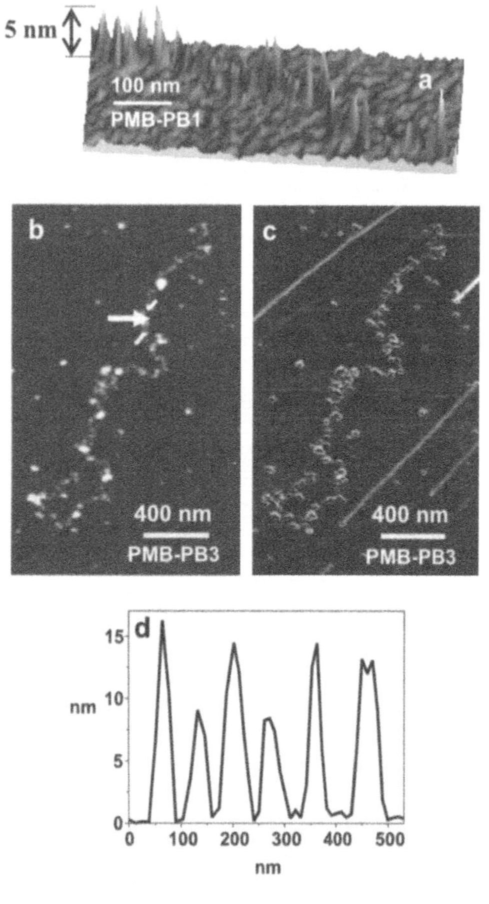

Fig. 2 Representative three-dimensional (**a**) and two-dimensional (**b**,**c**) topography **a**,**b** and phase **c** AFM images of PB clusters adsorbed along PMB chains deposited onto a Si-wafer: **a** PB1 ($\sim$ 4.8 nm); **b**–**d** PB3 (10–15 nm). **c** Cross-section is taken along the line in the image **b**, see the *white arrow*

Magnetic Properties

The magnetic properties of the dry powder of the PB hexacyanoferrate (PB3, $\sim$ 15 nm) were investigated. Figure 3 shows the field-cooled magnetization curve at external magnetic field of 10 G.

The critical temperature (T_c) where Prussian Blue nanoparticles turn to a ferromagnetic compound was found to be $\sim$ 5.1 K that is slightly less then the value for the bulk PB (5.5 K). This result is qualitatively consistent with

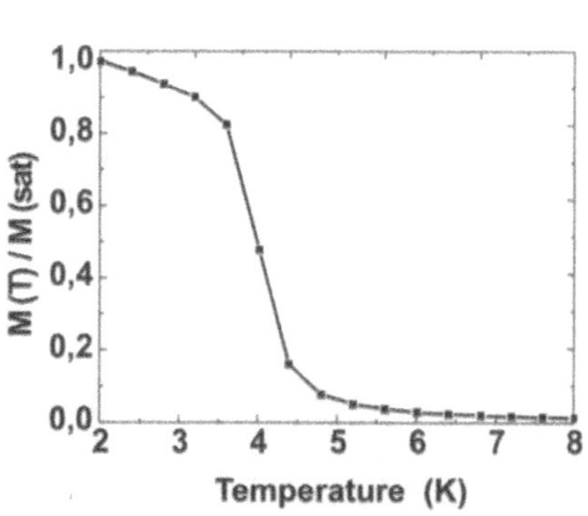

Fig. 3 Field-cooled magnetization versus temperature curve for the PB3 clusters at an external magnetic field of 10 G

earlier reported observations of Zhou et al. [19] and Uemura et al. [18] for PB nanoclusters and PB nanowires embedded into the polymer or alumina matrixes, respectively. It is believed that the decrease of Curie temperature in PB nanoparticles comparably to the PB bulk comes from the diminution of the average number of nearest magnetic interaction neighbors [19]. In our case the decrease of the T_c is less pronounced, because the data represents the magnetic properties of *aggregated* PB3 nanoclusters only separated by rather thin hexacyanoferrate shell. Investigation of the magnetic properties of well-separated PB nanoparticles (embedded into the polyelectrolyte matrix), as well as a dependence of magnetic properties on the size of PB nanoparticles will be reported elsewhere.

Deposition of Prussian Blue Nanoclusters on Carbon Nanotubes

P2VP macroradicals were added to multi-walled carbon nanotubes (CNTs), which was proved to be an effective "grafting to" method [26]. In contrast to other methods of CNTs modification, the macroradical addition does not require any hard oxidative pre-treatment of the CNTs, which preserves their original size. Deposition of an amorphous shell at the surface of CNTs was confirmed by TEM observations. Figure 4a shows modified nanotubes with closed ends and a relatively smooth surface (see the insets of Fig. 4a). High-resolution images (Fig. 4b) reveal the multiwall crystal lattice of the carbon nanotubes (3.43 Å) that have an average inner diameter of 6 nm, outer diameter of 22 nm and a polymer shell of about 2 nm. On the assumption that density of the graphite and P2VP is 2.2 g/cm^3 and 1 g/cm^3, respectively, the 2 nm thick polymer shell amounts to approximately 13.6 wt % that nicely fits the TGA data (12 wt %).

CNT-*g*-P2VP forms a stable dispersion in acidified water (HCl, pH = 2) as result of the hydratation and expansion of the protonated P2VP chains. CNT-*g*-P2VP was decorated by water-dispersible Prussian Blue nanoparticles after previous deposition onto solid supports (Si-wafers or TEM-grids). Figure 5 shows an AFM image of CNT-*g*-P2VP before and after decoration by PB2. As expected, the roughness of the nanotube surface is significantly increased. Low-magnification TEM images (Fig. 6a) confirm that the carbon nanotubes are covered by particles of approximately 10 nm.

This size is larger than the size of the pure PB2 estimated from the AFM observations (i.e., the height of the particles because the width is systematically overestimated by the tip-broadening effect). This difference is consistent with the embedding of the PB2 in the P2VP layer. The core-shell morphology of these particles was confirmed by high resolution TEM (HR TEM). As seen from the Fig. 6,

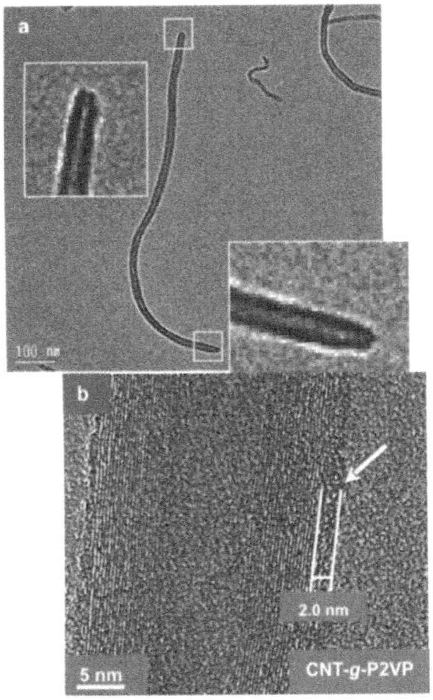

Fig. 4 **a** TEM images of CNT-*g*-P2VP; high-magnification images (*insets*) show the closed ends of the CNT-*g*-P2VP. **b** The HR TEM image shows the multi-wall crystal lattice of the carbon nanotube (3.43 Å); the *arrow* indicates a 2 nm thick amorphous layer, presumably of P2VP

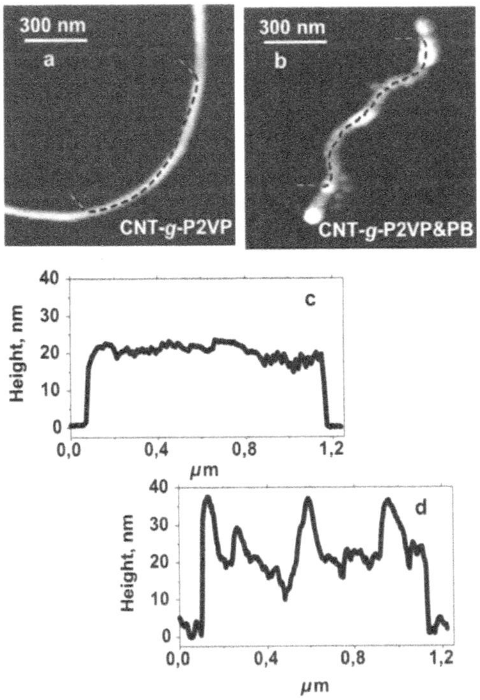

Fig. 5 Representative AFM image (**a**) and cross-section (**c**) of CNT-*g*-P2VP; AFM image (**b**) and cross-section (**d**) of the CNT-*g*-P2VP after PB deposition. The cross-sections **c**,**d** are repeated along the tubes shown in images **a** and **b**

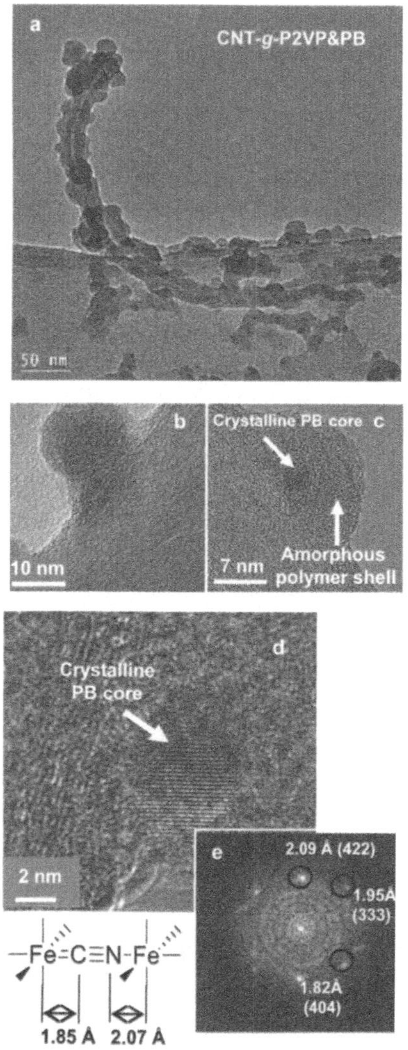

◀ **Fig. 6** **a** TEM images of the CNT-*g*-P2VP after deposition of PB2 clusters. **b,c** HR TEM images of the core-shell structure of the particles consisting of a dense crystalline core (diameter of 3–5 nm) surrounded by a few nanometer-thick amorphous shell. HR TEM image (**d**) and selected area diffraction pattern (**e**) confirm the crystalline structure of the PB clusters (lattice distances of 2.09 Å (Fe – N), 1.96 Å and 1.83 Å (Fe – C) that correspond to the (422), (333) and (404) reflections, respectively)

Prussian Blue nanoparticles are in the close contact with the nanotube surface that would be helpful for efficient charge injection from CNTs-based electrodes into the PB layer in electrochromic devices. A dense crystalline core with 3–5 nm diameter is surrounded by a few nanometer-thick amorphous shell (Figs. 6b–d). The lattice of the PB nanodots is resolved with spacings of 2.09 Å, 1.95 Å and 1.82 Å that correspond to (422), (333) and (404) reflections (Figs. 6d–e) [28]. The first reflection can be assigned to the Fe – N interatomic distance, and the last one to the Fe – C distance.

In conclusions, water-dispersible, charge-stabilized, surfactant-free Prussian Blue nanocrystals were prepared by mixing solutions of ferric chloride and excess of potassium ferrocyanide. The average size was readily controlled by the molar ratio of the two reagents. Thus formed PB nanoparticles are crystalline and display long-range ferromagnetic ordering at 5.1 K that is slightly lower than the magnetization temperature of the bulk PB. PB nanoparticles can be selectively attached to single polycation molecules, or to the surface of CNTs pregrafted with P2VP. These Prussian Blue based nanostructures might be useful materials for manufacture of electrooptical devices or mechanically robust ion-sieving membranes.

Acknowledgement The authors are grateful for financial support from the "Deutsche Forschungsgemeinschaft" (DFG) and the "Fonds National de la Recherche Scientifique (FNRS)", Belgium, within the ESF EUROCORES/SONS program (02-PE-SONS-092-NEDSPE) and SFB 287 "Reactive Polymers" (B1 project). P.S. thanks Prof. H. Lichte and Dr. M. Lehmann for the opportunity to carry out TEM observations at the Triebenberg High-Resolution Laboratory in Dresden.

References

1. Klabunde KJ (2001) Nanoscale materials in Chemistry. John Wiley & Sons
2. Miscellanea Berolinensia ad incrementum scientiarum (1710) Berlin, p 377
3. Sharpe AG (1976) The chemistry of Cyano Complexes of the Transition Metals. Academic Press, New York
4. Keggin JF, Miles FD (1936) Nature (London) 137:577
5. Neff VDJ (1978) Electrochem Soc 125:886
6. DeLongchamp DM, Hammond PT (2004) Adv Funct Mater 3:224
7. DeLongchamp DM, Hammond PT (2004) Chem Mater 16:4799
8. Garde R, Villain F, Verdaguer M (2002) J Am Chem Soc 124:10531
9. Sato O, Iyoda T, Fujishima A, Hashimoto K (1996) Science 271:49
10. Moore GJ, Lochner EJ, Ramsey C, Dalal NS, Stiegman AE (2003) Angew Chem 115:2847; Sato O, Kawakami T, Kimura M, Hishiya S, Kubo S, Einaga Y (2004) J Am Chem Soc 126:13176
11. Ohkoshi S, Fujishima A, Hashimoto K (1998) J Am Chem Soc 120:5349
12. Pyrash M, Toutianoush A, Jin W, Schnepf J, Tieke B (2003) Chem Mater 15:245
13. Sessoli R, Gatteschi, Caneschi A, Novak MA (1993) Nature 365:141
14. Tosshima N, Liu K, Kaneko M (1990) Chem Lett:485
15. Ravaine S, Lafuente C, Mingotaud C (1998) Langmuir 14:6347
16. Millward RC, Madden CE, Sutherland I, Mortimer RJ, Fletcher S, Marken F (2001) Chem Commun:1994

17. Vaucher S, Li M, Mann S (2000) Angew Chem 112:1863; Catala L, Gacoin JPB, Rivere E, Paulsen C, Lhotel E, Mallah T (2003) Adv Mater 15:826
18. Uemura T, Kitagawa S (2003) J Am Chem Soc 125:7814
19. Zhou P, Xue H, Luo H, Chen X (2002) Nano Lett 2:845
20. Kiriy A, Gorodyska G, Minko S, Tsitsilianis C, Jaeger W, Stamm M (2003) J Am Chem Soc 125:11202; Kiriy A, Gorodyska G, Kiriy N, Separovich R, Lupytsky R, Minko S, Stamm M (2005) Macromolecules 38:501
21. Nearly at the same time, Hammond et al. have used a similar method to prepare PB particles, and they have published a comprehensive account of the use of PB-based films in electrochromic devices [6, 7]
22. Baughmann RH, Zakhidov AA, de Heer WA (2002) Science 297:787
23. Dresselhaus MS, Dresselhaus G, Avouris P (2001) Carbon Nanotubes: synthesis, structure, properties and applications. Springer, New York
24. Lou X, Detrembleur C, Sciannamea V, Pagnoulle C, Jérôme R (2004) Polymer 45:6097; Lou X, Detrembleur C, Pagnoulle C, Jérôme R, Bocharova V, Kiriy A, Stamm M (2004) Adv Mater 16:2123
25. Kong J, Franklin NR, Zhou CW, Chapline MG, Peng S, Cho KJ, Dai HJ (2000) Science 287:622
26. Chalari I, Pispas S, Hadjichristidis N (2001) J Polym Sci A: Polym Chem 39:2889
27. Kiriy A, Gorodyska G, Minko S, Jaeger W, Štěpánek P, Stamm M (2002) J Am Chem Soc 124:13454
28. Weiser HB, Milligan WO, Bates JB (1942) J Phys Chem 46:99; Buser HJ, Schwarzenbach D, Petter W, Ludi A (1977) Inorg Chem 16:2704; Robin MB (1962) Inorg Chem 1:337

Progr Colloid Polym Sci (2006) 132: 168–169

AUTHOR/TITLE INDEX

Progr Colloid Polym Sci (2006) 132: 170–171

KEY WORD INDEX